Artificial Intelligence
Principles and Applications

Artificial Intelligence
Principles and Applications

Second Edition

VINOD CHANDRA S.S.
Department of Computer Science
University of Kerala
Thiruvananthapuram

ANAND HAREENDRAN S.
Department of Computer Science and Engineering
Muthoot Institute of Technology and Science
Ernakulam

PHI Learning Private Limited
Delhi-110092
2026

In fond memory of ***Shri Asoke K. Ghosh*** *(October 1942 – February 2024), Founder Chairman and Managing Director of PHI Learning, whose vision endlessly inspires.*

The Legacy Continues....

Published by Pushpita Ghosh, PHI Learning Private Limited, Rimjhim House, 111, Patparganj Industrial Estate, Delhi-110092 and Printed by Gopsons Printers Pvt. Ltd., A-14, Sector-60, Noida, Gautambudh Nagar, Noida, Uttar Pradesh-201301.

₹850.00

ARTIFICIAL INTELLIGENCE: PRINCIPLES AND APPLICATIONS, Second Edition
Vinod Chandra S.S. and Anand Hareendran S.

ISBN-978-93-89347-83-8 (Print Book)
ISBN-978-93-89347-84-5 (e-Book)

Contents

Preface

Artificial Intelligence (AI) is the most rapidly growing field, which provides more and more insights into how to make the human life easy. Defining AI is a herculean task; for us, AI represents a systematic method of creating intelligence in the machine through appropriate activities and dedicated functions with strong mathematical background. It remains as one of the most vibrant and challenging areas in computer science. It is mainly used in problem-solving, application development and automation process. This book (earlier titled as *Artificial Intelligence and Machine Learning*), now in its second edition, has 17 chapters, which covers almost all areas of artificial intelligence with examples, case studies and exercises. The highlight of the book is its application oriented case studies for project development. The basic motto while designing the book is to make the book as simple as it could be, so that even a beginner is able to understand every concept.

The first chapter is an introduction to artificial intelligence, its timeline and applications. Following this, Chapter 2 covers various intelligent agents. Type, structure and various types of agents like autonomous, nature inspired and planning agents are explained in detail. Chapter 3 covers the basic problem-solving aspects. Various heuristic search techniques like graph, greedy, branch and bound are explained under Chapter 4. Knowledge representation and reasoning which act as the platform for understanding various artificial intelligence problem are explained in detail in Chapters 5 and 6. Types of reasoning, case based reasoning, model based reasoning, etc. are explained with example in Chapter 7. Reasoning which is the primary goal setting in every artificial intelligence strategy needs to be analysed in depth. This chapter helps the reader to understand each and every technique in detail.

Learning is the process by which the environment acquires new knowledge or improves itself with its experience. Learning, its various aspects, and brief of machine learning are mentioned in Chapter 8. The various types of learning, namely supervised and unsupervised are explained in detail in Chapter 9. Various case studies are also included which throw light on research problem for curious readers. Chapter 10, deals with reinforced learning methods. Markov decision problem, Q learning, TD learning, etc. are also mentioned in this chapter. One major highlight of the book is Chapter 11—nature inspired learning. Nature inspired learning and the strategies developed from it for application deployment will be a new revolution. This chapter is also included with various case studies.

Game playing is an important topic in artificial intelligence. Alpha beta pruning, iterative deepening, etc. are included in Chapter 12. Major applications of the current era are somehow connected with the natural language processing. Whether it be a text to speech conversion or a real-time image recognition or it be a universal translator—the application of natural language processing is applying everywhere. A detailed explanation on the techniques and principles of natural language processing is provided in Chapter 13. Each and every book of artificial intelligence explains about planning. In this book, a research/application oriented planning in the computing world is explained in-depth. Each planning methodology is explained with case studies for better understanding. The last portion of the book is dedicated to artificial intelligence applications. Chapter 15 deals with expert systems, characteristics and system development. Application and case studies of expert system are also included in this chapter. Chapter 16 explains the major application of artificial intelligence. Mathematics of robots, line follower robot and its principles are explained in detail in this chapter. The last chapter is a fantasy walk by the authors. The next generation applications and the various possible technological changes are mentioned in Chapter 17. The video lectures for all the chapters are available at https://www.phindia.com/Artificial_Intelligence_Chandra_Hareendran

Various example/research oriented problems are discussed throughout the book, which are useful for undergraduate and postgraduate students. Our primary aim was to cover various artificial intelligence strategies and principles in a single book. The mentioned futuristic aspects will also help the students in coming up with dedicated and interesting applications in the future.

The broad coverage of the book supports various undergraduate, specialised master's courses on advanced topics. The syllabi of more than 100 universities were reviewed while selecting the chapter titles and text contents. The book covers case studies for most of the artificial intelligence strategies and techniques. It is highly desirable to have a basic introductory course in discrete mathematics, predicate calculus and knowledge engineering before going through the book. Case studies and application are explained with an assumption that the readers will have an understanding of basic data structures including trees, graphs and search procedures using stacks and queues. We have used Pascal like pseudo codes describing the algorithms. These notations use the control structures of Pascal along with English descriptions of the tests and operations.

This book will be helpful for artificial intelligence/machine learning students. There are many textbooks that provide introduction to artificial intelligence but the following features distinguish this text from others:

1. This text is organised in such a way that the entire artificial intelligence timeline is covered. Beginning with the inception, the book goes on describing the applications of next generation of AI. The examples and case studies explained in the book are those problems that everyone faces day in and day out. This will help the readers for better understanding.
2. Throughout the book, we have presented an application-focussed and hands-on approach to learning the sections. Each method provides exercises at the end of each chapter.

3. In all major chapters we have included case studies for each and every technique that is being illustrated.
4. Most of the text is supported with highlighted illustrations. We have followed a student-researcher mind approach to solve problems.

A book that summarises the various AI principles and concepts with a focus to the future is what this book is all about.

Vinod Chandra S.S.
Anand Hareendran S.

Acknowledgements

We express our sincere gratitude to our guru Achuthsankar S. Nair, who ignited many a spark for this textbook. He has been a significant influence in our research work, spending precious time with us on long discussions, technology dissertation, and tips for improving our writing skills. We also thank all members of the machine intelligence research group (www.mirworks.in) for giving valuable suggestions and support. We are deeply indebted to Directors of Research and Development Centers and Government Organizations at Thiruvananthapuram, who permitted us to do consultancy studies. Our research articles helped us build confidence in writing books, and we are thankful to all the reviewers and research journal publishers. We have visited many Universities and Colleges for discussions regarding the syllabi and question patterns. We are grateful to all those Universities and College professors who have contributed much to our efforts. We extend our heartfelt thanks to our dear students, who helped shape the sections based on their requirements. We are also deeply thankful to our family and friends who gave us immense support while molding this work. Last but not least, we thank our publishers PHI Learning, for the open-minded support they offered during editorial and production stages.

Finally, we would like to thank all the readers of our previous books, without whose tremendous support, this one wouldn't have happened.

Vinod Chandra S.S.
Anand Hareendran S.

CHAPTER 1 Introduction

Intelligence is the ability to learn and cope or the ability to contemplate, think, and reason. Human intelligence is the most complex entity forever, when we consider intelligence. The intelligent behaviour of human brain has resulted in today's world. The concept of Chinese room argument shows intelligent behaviour.

Consider the situation, where an English speaker who does not know Chinese is locked up in a room. The room is full of boxes of Chinese symbols (which can be considered as a database) and a book of instruction for manipulating the symbol (which can be treated as the program). The people outside the room send Chinese symbol which is unknown to the person in the room. These are the questions (input), and by following the instruction he gives reply in English (output). Thus, the program enabled the person to pass a test for understanding Chinese but the reality is that he does not understand even a word in Chinese. The people located outside the room may think that there is an intelligent mapping in the room for Chinese to English.

It was in 1980; John Searle proposed this Chinese room argument. It was totally against the possibility of artificial intelligence. The argument was based on the thought experiment in which the person sitting inside the room who knows only English follows the English instructions and decodes the Chinese character. But for people outside the room it appears that the person inside the room knows Chinese. This argument was put forward just to conclude the fact that, a suitably programmed computer will appear to converse in a natural language but it will not be capable to understand the language. Thus, Searle was trying to baseline the fact that computers merely use syntactic symbols to manipulate strings without understanding the semantics.

From this context, we may have a feeling that the Chinese room is equipped with intelligence. Searle's argument shows a straight challenge to proponents of artificial intelligence. A broad implication has been placed for Searle's argument for practical and computational theories of sense and of mind. As an outcome, there have been many crucial replies to the argument.

Definition: Artificial intelligence is a branch of computer science that studies how to program computers to exhibit apparently intelligent behaviour.

Artificial Intelligence (AI) is the science and engineering of making intelligent machines, and especially intelligent computer programs. Since ancient times, human beings have been fascinated with the idea of creating intelligent machines. With the arrival of much efficient computers, combined with over 50 years' research in artificial intelligence programming techniques, this dream of ours is fast becoming a reality. Researchers are modelling systems that can imitate human thought, understand speech, beat a chess player, and can do countless other feats never possible before. Many areas of science contribute to the subject of artificial intelligence such as a combination of computer science, physiology, and philosophy, etc.

Two scientists are remembered for the term 'artificial intelligence'. Alan Turing [Figure 1.1(a)] who published a paper 'The Turing test for Machine intelligence' in 1950. This may be the first approach in the field of artificial intelligence. Turing was highly influential in the development of computer science and providing a formalisation of the concept of algorithms and computation with the Turing machine, playing a significant role in the creation of the modern computer. John McCarthy [Figure 1.1(b)], who coined the term in 1956, defines it as 'The science and engineering of making intelligent machines'? According to McCarthy:

(a) Alan Turing (b) John McCarthy

Figure 1.1

It (AI) is the science and engineering of making intelligent machines, especially intelligent computer programs. It is related to the similar task of using computers to understand human intelligence, but artificial intelligence does not have to confine itself to methods that are biologically observable.

This definition includes the study of human intelligence through the assembly of artificial intelligence systems. The word 'artificial' is in the definition because artificial intelligence is about studying and constructing man made intelligence. This can help in our understanding of human intelligence. Artificial intelligence is not necessarily about replicating human intelligence.

1.1 Definitions of Artificial Intelligence

In order to understand artificial intelligence, we should consider other definitions given by scientists as follows:

- The science of making machines do things that would require intelligence if done by men (Minsky, 1968).

- Artificial intelligence is the study of how to make computers do things which, at a moment, people do better (Rich, 1983).
- The study of mental faculties through the use of computational models (Charniak and Mc Dermott, 1995).
- The study of the computations that make it possible to perceive, reason, and act (Winston, 1992).
- The art of creating machines that perform functions that require intelligence when performed by people (Kurzweil, 1990).
- A field of study that seeks to explain and emulate intelligent behaviour in terms of computational processes (Schalkoff, 1990).
- The branch of Computer science that is concerned with the automation of intelligent behaviour (Luger, 1993).

The underlying assumption in the study of artificial intelligence is the intelligence of mind (for search purpose, for distinguishing, etc.) that can be incorporated by programming under the assumption of physical symbol hypothesis. The physical symbol system consists of a set of entities called symbols which are physical patterns that can occur as components of another entity called symbol structure (expression).

To distinguish objects there is one test called 'Turing test', proposed by Alan Turing in 1950. Turing test (Figure 1.2) was designed to prove a satisfactory operational definition of intelligence. If we keep a computer in one room and a person in another room, by asking questions to both, one can easily find out whether the computer has been updated to the level of human intelligence or not. Turing test intentionally avoided direct physical communication between the human and the computer. It is not necessary for a personal physical simulation for intelligence measurement. Turing test was focussed to area weak equivalence and strong equivalence. Weak equivalence is a system for human and computer that are equivalent in results (output), but they do not arrive at those results in the same way. Strong equivalence for human and computer uses the same internal processes to produce results.

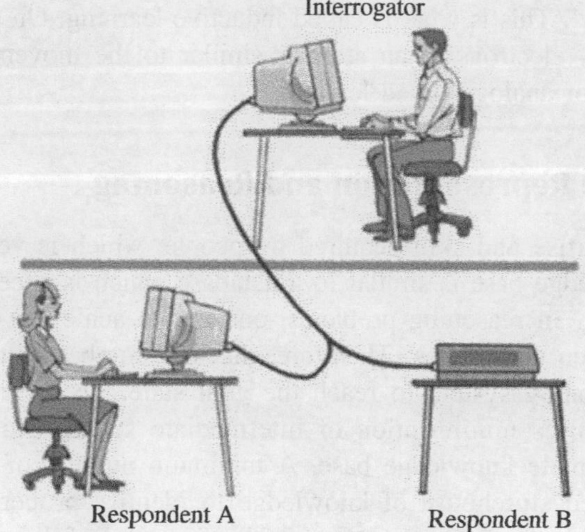

Figure 1.2 Turing test.

1.2 Topics of Artificial Intelligence

The subject of artificial intelligence was originated from game playing and theorem proving programs. From a number of parent disciplines, it gradually enriched with theories and solutions. Being a new discipline of science, the importance of topics covered under the subject changes very much with time. Artificial intelligence covers areas in industrial applications including process plants. Many subsystems are included in this area. Major topics, which we find important and valuable to understand the subject, are outlined as follows:

1.2.1 Learning Systems

Learning systems need special mention among the subject of artificial intelligence. According to Herbert Simon, *learning is any process by which a system improves performance with experience*. We can illustrate the concept of learning by taking the reference of a natural problem. Consider the situation where child starts to learn pronunciation from his mother. When mother tries to teach the child, the pronunciation of 'A', the hearing system of the child receives it and the voice system tries to imitate it. There will be a huge difference in the pronunciation of mother and child. This is going to be referred as error signal from now onwards. This error signal is received by the child's learning auditory nerves system and an actuation signal is generated through the motor nerve for correcting the pronunciation. The process is continued until the error rate becomes negligibly low. Finally, the 'A' will be stored and we can say that the child has learned. The learning process discussed here can be treated as an adaptive learning process where the parameters are continuously adjusted to get results close enough to the training dataset. There are many types of learning methods, which can adapt knowledge in different way. For supervised learning problems, Artificial Neural Networks (ANN) are gaining importance. ANN represents the electrical analogue system of the biological nervous systems. Inductive learning and analogy based learning are also other common learning methods. For example, 'Pigeon flies', 'Parrot flies', 'Eagle flies' and the system learns 'birds fly'. This is what is called inductive learning. On the other hand, learning the fact that motion of electrons in an atom is similar to the movement of planets in solar system is an example of analogy based learning.

1.2.2 Knowledge Representation and Reasoning

Knowledge is an expertise and skill acquired by people, which is very hard to represent in computers. The knowledge base is similar to a database which is necessary for expert system building and reasoning. In reasoning problems, one has to achieve a pre-defined target state from one or more given start states. Therefore, the less number of transitions gives a high efficiency in the reasoning system to reach the goal state. To increase the efficiency of a reasoning system requires minimisation of intermediate states, which ultimately calls for an organised and complete knowledge base. A minimum number of search is needed for a complete and organised storehouse of knowledge to identify proper knowledge at a given problem state. This yields the right next state on a leading edge of the problem-solving process.

Therefore, organisation of knowledge is of supreme importance in knowledge engineering. A variety of knowledge representation techniques are used in artificial intelligence such as, production rules, predicate logic, frames, filler and slots, semantic nets, etc. For a particular type of representational schema, the selection of knowledge depends on both choice of users and nature of applications. The testing of satisfiability of a goal from a given set of data and knowledge is the main concern of reasoning problem.

1.2.3 Planning

Planning is a significant area in artificial intelligence. Though the problems of reasoning and planning seemingly share many similar issues, they have a basic difference in the definitions. Then again, problem planning involves the determination of methodology from which a successful result could be achieved from the known initial states. Extensive applications in robotics and navigational problems are key components of automated planning. Planning focus in the area of decomposition of an original problem into appropriate subunits and use interaction on these subunits during the problem-solving process. The major components of a planning system include heuristic functions, set of rules in the problem state and a deduction process. The deduction process will find a solution to a problem when it has found a sequence of operators that travel an initial state to the goal state.

1.2.4 Knowledge Acquisition

Acquisition or the 'elicitation' of knowledge is uniformly hard for human being as it is for computer machine. Knowledge acquisition includes making of new pieces of knowledge from a given knowledge network, setting active structures for existing knowledge, learning knowledge from the environment and fine tuning of knowledge. Using machine learning approaches, automated acquisition of knowledge is possible and is a dynamic area of current research in artificial intelligence. A set of knowledge string mechanisms is required to store the acquired knowledge. Frames, conceptual dependency, scripts, semantic nets, etc. are such data structures for knowledge store. The captured knowledge is stored into these frameworks accordingly. Each of these data structures is useful to the corresponding knowledge representation scheme.

1.2.5 Intelligent Search

Generally search problems encountered in computer science are of a deterministic nature. This means, order of visiting elements of the search space is known. For example, in our basic graph traversal algorithms such as Depth First Search (DFS) and Breath First Search (BFS), one knows the node sequence based on its visiting in a tree. Search problems in artificial intelligence are non-deterministic and the order of visiting the elements in search space is completely dependent on given datasets. There are a variety of search techniques used for mining databases depending on its usage. A*, Best-first search, greedy search, BFS, DFS, etc. are some examples.

1.2.6 Logic Programming

From the past century, logic symbols are grown up with respect to their use. Mathematicians are used to designing various tools for representing logical statements by symbolic operators. Proportional logic is such an attempt which deals with a set of binary statements (propositions) associated with Boolean operators to handle complex situations of the real-world logic of propositions regularly enriched. It is named as predicate logic or first-order logic. A logic based programming using predicate logic is a classical variety example for computer language. PROLOG (PROgramming in LOGic), is a computer language that supports logic programs. It was widely used for artificial intelligence programming. Logic programming is one of the prime areas of research in artificial intelligence that covers knowledge representation structures. The eventual aim of this research is to expand the PROLOG compiler to handle spatial temporal models and maintain a parallel programming environment. LISP (LISt Processor) is another artificial intelligence programming language that can handle strings. These languages are old languages in programming sense because present language shells can perform much better than above stated languages. At present, R programming, Python and Java are offering various supporting packages to the machine learners. Also, there are many machine learning benchmarking tools available as open source for the programmers.

1.2.7 Soft Computing

Soft computing is a popular problem-solving technique that makes an effort to mimic the intelligence found in nature. Statistical learning, fuzzy logic methods, genetic or evolutionary algorithms, fractals, probabilistic reasoning and chaos theories are the most important constituents of soft computing. Most of the soft computing techniques are inspired from nature.

The human brain can process millions of data, learn from experience, generalise from learning rules, recognise patterns and make decisions. It is in effect a good engineering model that performs these tasks as well as it can use heuristic approximations, low precisions or less generality, depending on the problem to be solved. We want to transfer some of these abilities into our designed model, algorithms, machines and artificial intelligence systems. That is, to solve a given task, based on previous experience, with reasonable accuracy at reasonable cost and time.

According to Zadeh (1994) the guiding principle of soft computing is

Exploit tolerance for imprecision, uncertainty and partial truth to achieve tractability, robustness and low solution cost.

In general, soft computing is a collection of computing techniques and tools, shared by closely related disciplines that comprise artificial neural nets, fuzzy logic, genetic algorithms, inductive logic programming, belief calculus and so on. These tools are used separately as well as jointly depending on the type of domain of applications.

1.2.8 Management of Imprecision and Uncertainty

Data and knowledge bases in many typical artificial intelligence problems, such as reasoning and planning, are often contaminated with various forms of incompleteness. The incompleteness in data and knowledge bases are always a problem for reasoning and planning in artificial intelligence problems. This incompleteness of data can be called as imprecision. Imprecision generally occurs in database for lack of appropriate data, and poor authenticity level of sources. It is an uncertainty when the incompleteness of knowledge is referred that originates in the knowledge base due to lack of certainty in the pieces of knowledge. A complex problem in reasoning is the presence of imprecision of data and uncertainty of knowledge.

There are numerous techniques and tools developed for reasoning under incomplete data and knowledge. Some of the techniques employed are fuzzy, stochastic and belief network models. In every problem, the system always undergoes change of states. The probability of sum of the transition from one state to another will always be strictly unity in case of stochastic systems. In fuzzy reasoning, the sum of the membership value of transition from a given state to the next state may be greater than or equal to one. Fuzzy reasoning has a useful role in uncertainty problem-solving. The belief network model updates the stochastic fuzzy beliefs that assigns certain facts embedded in the network. This process may stop until a condition of equilibrium is reached. When equilibrium is reached there is no more change in the network. Fuzzy Petri net is a fuzzy tool and technique that can apply into a specialised belief network. This is used for handling both imprecision of data and indecision of knowledge by a unified approach.

1.3 Timelines of Artificial Intelligence

The major research activities carried out during 1950's include game playing and theorem proving. The concept of state space approach is a major tool for solving problems in artificial intelligence. This approach was developed during this period and is used currently too. The period of classical artificial intelligence research marked its beginning in 1950 by the publication of Shannon's paper on chess (1950) and ended with the publication by Feigenbaum and Feldman. It was during this period when Alan Turing proposed 'Turing's test', that was a major method for testing machine's intelligence.

The passionate period

The period from the mid 1960s to the mid 1970s is termed 'the romantic period'. Romantic period of artificial intelligence is known to be 'machine understanding'. During this period, the major activity was to make the machine 'understand', by which they usually meant the understanding of natural languages. In 1972, Winograd's developed SHRDLU system, which is a program capable of understanding a non-trivial subset of English by representing and reasoning about a restricted domain that needs special mention. It explains reasoning by considering a world consisting of toy blocks. During this period, Quillian developed 'semantic nets', a knowledge representation scheme using special structures. Mervin Minsky, in 1968 used semantic nets, which made a great contribution from the point of view of information processing. Minsky's knowledge representation formalisms using frames, was another contribution during this period.

The modern period

The modern period of artificial intelligence started from the latter half of the 1970s to present day. More complex practical problems were solved and even tried to be solved in this period. A popular expert system named as MYCIN of Stanford University could diagnose and prescribe medicines for infectious bacteriological diseases.

Newtonian machines is an expert system that deals with real-life problems. To solve real-world problems, researchers are also engaged in theoretical research on artificial intelligence including heuristic search, uncertainty modelling, machine learning and various reasoning. As a summary, artificial intelligence period includes research on both various theories and practical aspects.

These periods roughly summarise the history of artificial intelligence, hence we are able to understand artificial intelligence history by periods, and its evolution. The summary of incidents and milestones in artificial intelligence is as follows:

1917: Karel Capek suggested a new verbal term 'robot' (in Czech 'robot' means 'worker').

1928: John von Neumann's suggested Minimax theorem, which is used in some game playing programs.

1943: Neural network architectures for intelligence were proposed by McCulloch and Pitt.

1950: Isaac Asimov proposed 'I, Robot'. In the same year, Shannon proposed chess program. Alan Turing proposed Turing test in his paper 'Computing Machinery and Intelligence'.

1954: Isaac Asimov proposed 'The Caves of Steel' This was the first Robot science fiction.

1955: Newell, Shaw, and Simon develop 'IPL-11', first artificial intelligence language.

1956: Newell, Shaw, and Simon created 'The Logic Theorist', a program that solves mathematical problems. In this year, demonstration of the first running artificial intelligence program, the Logic Theorist (LT) was written by Allen Newell, J.C. Shaw and Herbert Simon at Carnegie Mellon University (CMU). Also, first commercial robot, from the Unimation company founded by George Devol and Joseph Engelberger, was based on Devol's patents.

1957: Newell, Shaw and Simon proposed the General Problem Solver (GPS) meaning end analysis. Norm Chomsky wrote 'Syntactic Structures' for language processing.

1952–62: First checkers game program was written by Arthur Samuel at IBM. This was a major achievement for the sufficient skill that challenged a world champion.

1958: John McCarthy at MIT, invented the LISP language. Herb Gelernter and A proved a theorem in geometry developed by Nathan Rochester at IBM that exploits a semantic model of the domain in the form of diagrams of 'typical' cases.

Early 60's: Margaret Masterman at Cambridge University, designed semantic nets for machine translation.

1961: First symbolic integration program (SAINT) was written by James Slagle at MIT, which can solve calculus problems.

1962: Unimation, first industrial robot company was founded.

1963: Thomas Evans at MIT developed ANALOGY, and computers can solve the same analogy problems as are given in IQ tests. During the same year, Ivan Sutherland at MIT introduced the idea of interactive graphics into computing. At the same time, first collection of articles about artificial intelligence (Computers and Thought) was proposed by E.A. Feigenbaum and Julian Feldman.

1964: Danny Bobrow at MIT showed that computers can solve algebra word problems correctly by understanding natural language. On the same year, logical representation of knowledge for question answering systems demonstrated by Bert Raphael at MIT.

1965: Resolution method is mathematical procedure that was invented by J. Alan Robinson that allowed programs to work efficiently with formal logic as a representation language. At the same time, ELISA was developed by Joseph Weizenbaum at MIT, an interactive program that carries on a dialogue in English on any topic.

1966: Ross Quillian demonstrated a new knowledge representation scheme, i.e. semantic nets. First machine intelligence workshop and influential annual series were organised.

1967: An expert system named Dendral was demonstrated at Stanford University to interpret mass spectra on organic chemical compounds. Dentral was considered one of the first successful knowledge-based programs for scientific reasoning. In this year, Joel Moses demonstrated the power of symbolic reasoning for integration problems in the *Macsyma* program at MIT. This was a knowledge-based program in mathematics. Its use was successfully demonstrated. A scientist at MIT, Richard Greenblatt built a knowledge-based chess-playing program, i.e. MacHack which was good enough in tournament play.

Late 60s: Doug Engelbart at Stanford Research Institute (SRI) invented the mouse.

1968: Marvin Minsky and Seymour Papert published perceptrons in the ANN. This demonstrated restrictions of simple neural nets.

1969: SRI introduced a new robot. This robot was demonstrated by combining locomotion, perception and problem-solving. Roger Schank at Stanford University defined conceptual dependency model for natural language understanding. Another milestone in the artificial intelligence was International Joint Conference on Artificial Intelligence (IJCAI). This conference is a continuous and successful event presently.

1970: SCHOLAR is interactive program for computer-aided instruction developed by Jaime Carbonell. This program fulfilled on semantic nets as the representation of knowledge. Augmented Transition Networks (ATN's) proposed by Bill Woods described a representation schema for natural language understanding. ARCH is an example in the world of children's blocks written by Patrick Winston at MIT.

Early 70's: An influential Natural Language Processing, proposed by Jane Robinson and Don Walker at SRI.

1971: Ability of computers to understand English sentences was demonstrated by Terry Winograd at MIT. A robot arm named SHRDLU carried out instructions typed in English, then moved the arm based on that instruction.

1972: PROLOG was developed by Alain Colmerauer, another popular artificial intelligence language. PROLOG has its roots in first order logic and is declarative. Most of the expressions are represented in terms of relations as facts and rules. A computation is carried out by running a query over these relations.

1973: Freddy, a robot developed by the Assembly Robotics group at Edinburgh University. This robot was capable of using vision to locate and assemble models. This robot was one of the first robots that could assemble wooden models using vision to identify and locate the parts. A robot with six electro-mechanically driven axes developed by KUKA Robot Group was the first industrial robot.

1974: A rule based system for knowledge representation and inference in the domain of medical diagnosis was developed by Ted Shortliffe. This expert system was known as MYCIN from the Stanford University artificial intelligence laboratory. This may be the first expert system with professional components. At the same time, first planning program was proposed by Earl Sacerdoti. It was popularly with techniques of hierarchical planning.

1975: Marvin Minsky published a new knowledge representation structure named 'Frames'. Another learning program called Dentral produced new results in chemistry, as the part of expert system shell component. A Programmable Universal Manipulation Arm (PUMA) was developed by Victor Scheinman.

Mid 70's: A new approach that discourses modelling was established by Barbara Grosz at SRI. This model used Natural Language Processing (NLP) environment. Xerox cooperation developed the Smalltalk language. This used the usage of icon-oriented interfaces and power of object-oriented programming. MIT published "primal sketch" for visual perception is another step in artificial intelligence research.

1976: Doug Lenat's Artificial Modelling (AM) program (Stanford University) demonstrated the discovery model which is loosely-guided search for interesting conjectures. Randall Davis at Stanford University demonstrated the power of meta-level reasoning.

Late 70's: A group of engineers at Stanford University demonstrated the power of the ARPAnet for scientific collaboration.

1978: Tom Mitchell at Stanford University, developed the version spaces for describing the search space of a concept formation program. One of the cornerstones of artificial intelligence known as 'satisfying' was proposed by Herb Simon. He got the Nobel Prize in Economics for the theory of bounded rationality. The MOLGEN program was demonstrated at the Stanford University as an object-oriented representation of knowledge. This program can be used to plan gene-cloning experiments.

1979: EMYCIN program was developed by Bill VanMelle at Stanford University. Through his program, he demonstrated the generality of MYCIN's representation of knowledge and style of reasoning. This was the general model for many commercial expert system 'shells'. INTERNIST, a knowledge-based medical diagnosis program was developed by Jack Myers and Harry People at University of Pittsburgh. This knowledge building was based on clinical knowledge. CHI system for automatic programming was developed at Stanford University. The same university developed a computer-controlled, autonomous vehicle. A non-monotonic logic's formal aspect of truth maintenance was developed by John McCarthy at Stanford University. Raj Reddy founded Robotics Institute at Carnegie Mellon University.

1980's: LISP Machines are designed and marketed. First expert system shells for commercial applications was developed in the same year.

1980: A new blackboard model was published. This was a framework for the HEARSAY-II speech understanding system.

1981: The connection machine was designed by Danny Hillis, a massively parallel architecture that brings new power to artificial intelligence, and for general computation.

1983: An artificial intelligence research project, SOAR was developed by John Laird and Paul Rosenbloom. SOAR showed intelligent behaviour used as a general cognitive architecture for developing systems. Another work was James Allen's Interval Calculus, the first widely used formalisation of temporal events.

Mid 80's: Neural networks became widely used with the back-propagation algorithm.

1985: An autonomous drawing program was developed by Harold Cohen.

1987: Marvin Minsky published 'The Society of Mind'. In this paper, he described that the human mind is a collection of cooperating agents.

1989: ALVINN (An Autonomous Land Vehicle in a Neural Network) was developed by Dean Pomerleau at CMU. ALVIN was a computer controlled system that drove a car coast-to-coast. ALVINN was designed as a single hidden layer peceptron system using back-propagation algorithm for learning purpose.

1990's: Major advances in all areas of artificial intelligence were noted in this decade. There were significant demonstrations in machine learning, computer games, case-based reasoning, uncertain reasoning, intelligent tutoring, multi-agent planning, scheduling, data mining, natural language understanding and translation, computer vision, virtual reality, and other topics. Collaborative research showed a significant development in the development of humanoid robot.

Early 90's: Gerry Tesauro developed a backgammon program that used reinforcement learning. This game was powerful at the level of a championship game playing with the help of world class players.

1997: The Deep Blue chess program beated Garry Kasparov (world chess champion). Another event was the first official Robo-Cup soccer. This match featured with 40 teams of interacting robots and over 5000 spectators.

Late 90's: Artificial intelligence created a role in web-crawlers. Artificial intelligence based information extraction programs popularly-used in the world-wide-web. Another milestone was demonstration of an 'intelligent room' and emotional agents at MIT's.

1999: Remote agent was developed as an artificial intelligence system. This system had a significant role in many escalations.

2000: Commercially developed and marketed were the 'Smart toys' or interactive robot pets. Another robot showed with a face that expressed emotions developed by Cynthia Breazeal at MIT.

2002: Researchers at the MIT developed 'iRobot'. They also introduced 'Roomba', a vacuum cleaning robot.

2004: The Defence Advanced Research Projects Agency (DARPA) sponsored a competition for autonomous (driverless) vehicles.

2005: An artificially intelligent humanoid robot designed by Honda's ASIMO robot was able to walk faster than a human. ASIMO was created at Honda's Research and Development Wako Fundamental Technical Research Centre in Japan. The technology based on tracking web activity or media usage brought artificial intelligence to marketing. Another development was 'Blue Brain'. It was born, a project to simulate the brain with molecular detail by reverse-engineering the mammalian brain down to the molecular level.

2006: Artificial intelligence was applied to the development of a cognitive radio engine. Artificial intelligence was used to understand biological intelligence.

2008: Researchers at the University of Alberta solved checkers game.

2009: Largest and strongest industrial robot with six axes was developed by FANUC Robotics America. The most popular application of artificial intelligence was the driverless car. It was during this year, Google built its first autonomous car.

2010: Gaming was a powerful stream where we could atomise the environment and provide the player a feel of it. Microsoft launched kinext for Xbox 360. It was able to track the human movement and was able to play as a second player. It basically made use of a 3D camera and Infra Red (IR) communication. This also enabled to play wireless.

2011: Cognitive computing started to become the talk of the town. IBM Watson computer defeated television game show Jeopardy. Natural language processing applications became commonly available for users. Apple's Siri used in smartphone apps used NLP to answer questions make recommendation, and perform action. After its introduction, Google came up with Google Now (2012) and Microsoft's introduced Cortana (2014) for NLP.

2013: For image recognition applications, Carnegie Mellon University released NEIL (Never Ending Image Learner) to constantly compare and analyse relationship between different images.

2016: When artificial intelligence became prominent, one area where we could make sure machines overtake humans is on gaming platform. The alpha version of Google DeepMind's AlphaGo defeated 3 times European Go champion, thus proving machine's learning curve to have significantly increased.

2017: Artificial intelligence, as it was exposed heavily, there were chances to use them unethically, thus Asilomar conference on Beneficial artificial intelligence was held for assuring the ethical usage of artificial intelligence. Tensor flow used for implementing Deep learning was overtook by Google DeepMind's AlphaGo Zero. During 2017, a propositional logic boolean satisfiability problem (SAT) solver proved a long-standing mathematical conjecture on Pythagorean triples over the set of integers.

2018: How well can a machine outscore humans? Alibaba language processing artificial intelligence topped humans at Standford University in reading and comprehension test. This was seen as one of the major leading steps in the NLP application. Google duplex was an application of TALK, where users could book appointments over telephone. Similar application which has got all credits is Los Angeles Times which judges the artificial intelligence's voice to be a nearly flawless imitation of human sounding speech.

2019: Digital transmission for intelligent digital mesh was connected between humans, robots, devices, content, and services in 5G technology. Business platforms were embedded with the help of artificial intelligence for smart business operation.

Today: ?

1.4 Branches of Artificial Intelligence

Some common branches of artificial intelligence are the following. The main branches are useful in research and development for problem-solving and theorem solving.

Symbolic artificial intelligence

It is a collection of all methods in artificial intelligence research based on high-level representations in problem domain or a search domain. During 1950s through 1980s, symbolic artificial intelligence played dominance in the research. There are many projects explored during these years in the artificial intelligence research. An expert system is an example generated from a symbolic artificial intelligence. Most modern research is focused to specific subproblems, while the symbolic artificial intelligence focussed to produce general human like intelligence in machines. After 1980s, the scientists believed that the symbolic systems cannot be able to imitate all human processes like robotics, perception learning, etc.

Statistical artificial intelligence

Inspiration of mathematics is a diversification in artificial intelligence applications. The statistical approaches included in the learning process give a progressive approach in artificial intelligence research field. In real-world applications, mathematical models designed as a computational model perform much better.

Computational intelligence

Computational intelligence can solve most of the real-world problems rather than a traditional mathematical approach. The principles of soft computing techniques are built based on the requirements from real-world problems. After symbolic period, researchers are set much interested in computation intelligence based problem-solving. Artificial neural networks, support vector machines, fuzzy logic, heuristic search, etc. are some selected methods built on computational intelligence obtained from real-world cases. The Markov model is developed as a statistical model in 1900s, but computational intelligence modified this statistical technique into hidden Markov model, highly used as a machine learning approach.

According to John McCarthy, the following are selected branches in artificial intelligence:

1. **Logical artificial intelligence:** What a program knows about the world are represented by sentences of some mathematical logical language. To achieve its goals, the program decides what to do by inferring that certain actions are appropriate.
2. **Search:** Artificial intelligence programs often examine large numbers of possibilities. For example, these examine the moves in a chess game or inferences by a theorem proving program. Discoveries that can answer the question as to how to do this even more efficiently, are being made continuously. Various search techniques are highly useful in the development of intelligence application, which cut the time delays at the time of searching.
3. **Pattern recognition:** When a program makes clarification of some variety, it is often programmed to compare what it sees with a pattern. For example, to find a face, a vision program may try to match a pattern of eyes and a nose in a scene. More complex patterns like a natural language text, a chess position, or the history of some event are also studied. More complex methods are required to find specific patterns. Various pattern recognition algorithms are useful for pattern detection.
4. **Representation:** In some way, facts about the world are represented. Usually languages of mathematical logic are used to represent facts. Most of the facts are represented in symbolic representation. These representations can be used in the artificial intelligence algorithms for making a solution. Facts are highly connected with rules when they are in action with real problems.
5. **Inference:** It is deriving logical conclusions from a true premise known or assumed. In some cases, mathematical logical deduction is adequate, but new methods of non-monotonic inference have been added to the logic set. One of the simplest kinds of non-monotonic reasoning is default reasoning. This makes a conclusion by default, but the conclusion can be withdrawn if there is evidence to the contrary. For example, we heard a word 'bird', we inferred that it can fly and has wings. Is it true for 'penguin'?, so we reverse conclude in the case of a penguin. This conclusion may have lead to constitute the non-monotonic character of the reasoning. Set of conclusions that can be drawn from a set of premises give monotonic increasing function of the premises.
6. **Common sense knowledge and reasoning:** From 1950s onward, this is one of the areas of artificial intelligence that is farthest from human-level in spite of the fact that it has been an active research area. It is necessary for new ideas while there has been considerable progress in developing systems of non-monotonic reasoning and theories of action. An example of such system is the CYC system that contains a large but spotty collection of common sense facts.

7. **Learning from experience:** Artificial intelligence programs can learn from experience. There are many learning approaches to artificial intelligence based on specialised application. The logic facts are expressed through learning rules or learning laws. Programs can only learn what facts or behaviours their formalisms can represent, and unfortunately almost all learning systems are based on very limited abilities to represent information.
8. **Planning:** Planning in artificial intelligence deals with general facts of the world and its effects of actions. For a particular solution and statement of a goal achievement, planning in artificial intelligence played its important role. For a sequence of actions, the strategy of planning is common for real-world example.
9. **Epistemology:** Epistemology deals with the philosophy of knowledge. In artificial intelligence programs, it involves the study of kinds of knowledge that are required for solving problems in the world.
10. **Ontology:** Ontology is the study of kinds of things that exist. In artificial intelligence, the programs and sentences deal with various kinds of objects, and we study what these kinds are and what their basic properties are. Emphasis on ontology-related programs was initiated in 1990s.
11. **Heuristics:** To identify or discover the idea embedded in program, we can use certain functions of Mathematics. Heuristic is a term used in artificial intelligence. It has many meanings. For example, in shortest path identification problems, heuristic is a function $h(n)$ defined to measure as an estimate of the cost of cheapest path from a particular node to the goal node.
12. **Genetic programming:** Genetic programming is a soft computing technique for getting programs to solve a task by mating random and selecting fittest in millions of generations. Genetic Programming is evolved from a machine learning technique like genetic algorithms used to optimise a population of computer programs.

1.5 Applications of Artificial Intelligence

Every branch of science, engineering and technology shares the tools and techniques available in the domain of artificial intelligence. However, for the sake of convenience of readers, we can see here a few typical applications, where artificial intelligence plays a significant and decisive role in engineering automation.

1.5.1 Game Playing

In game theory, a formal model of interactive situation is defined. There may be two or several players in the system out of which one player is the decision problem. The players define their preferences, information and strategic actions when they meet to game. It always influences the outcome of the game.

Game playing is one of the leading domains where artificial intelligence has been applied with great success. Games can generate extremely large search space. A large and complex powerful technique determines the alternatives to explore in the problem space. These techniques are based on heuristics (proceeding to a solution by trial and error) and constitute a major area of artificial intelligence research. The best example is a computer chess program, which could defeat even the world chess champion. A computer game can be formally defined as a kind of search problem with initial state, set of operators, terminal state and a utility function.

1.5.2 Expert Systems

An expert system is a software that manipulates encoded knowledge to solve problems in a specialised domain that normally requires human expertise. An expert system is an artificial intelligence program in which the system's knowledge is obtained from an expert source such that the machine can offer intelligent advice or take an intelligent decision in solving a problem. Here artificial intelligence performs a decision-making task much faster than the humans can. For example, MYCIN is one of the first expert systems, which could diagnose infections and suggest treatments, sometimes better than the medical practitioners. Developing an expert system means extracting, from human expert knowledge that can be considered experts, rules, strategies and procedures to solve a pre-determined group of problems, aggregating this knowledge into a computer software system. This process is known as knowledge engineering and the person responsible for it (the knowledge engineer) must understand this.

The main components of an expert system shell are knowledge base, inference engine and user interfaces. There are rule based, knowledge based and hybrid expert systems. One of the major bottlenecks in the development of expert system is the acquisition of expert's knowledge about the problem. Typically expert's knowledge is elicited through the questionnaires, interviews protocol analysis or combination of some of these. These direct knowledge acquisition methods can be very time and effort consuming. Modern expert systems use self-learning methods to dynamically learn new rules that can be validated by an expert and can be added to the knowledge base. There are many applications incorporated with expert systems.

1.5.3 Natural Language Processing

Natural Language Processing (NLP) is a technique that builds ability of machines to read and understand the languages that a humans speak. NLP is a fusion field of computer science and linguistics. This field is disturbed with the interactions between computers and natural languages. Powerful natural language processing system is able to acquire new knowledge by reading the text that understand situations. This area is a hotcake for researchers. Major area of NLP includes information retrieval, information extraction, speech processing, stemming, text simplification, text-to-speech, text-proofing, natural language search, query expansion and truecasing.

The important tasks of NLP are automatic summarisation, co-reference resolution, discourse analysis, machine translation, morphological segmentation, named entity recognition, natural language generation, natural language understanding, Optical Character Recognition

(OCR), Parts-Of-Speech (POS) tagging, parsing, question answering, relationship extraction, sentence breaking, sentiment analysis, speech recognition, speech segmentation, topic segmentation, word segmentation and word sense disambiguation. Major application domains of NLP include text processing (For example, word processing, e-mail, spelling and grammar checkers), interfaces to databases (For example, query languages, information retrieval, data mining, text summarisation), expert systems (For example, explanations, disease diagnosis) and linguistics (For example, machine translation, content analysis, writer's assistants, language generation).

Machine learning algorithms are used to apply in NLP tasks. A large number of features that are generated from the input data are used by all of these algorithms. Parts-of-speech is a typical feature that might be the identity of the word being processed. The POS tag is usable to identity the words immediately to the left and right of the word, and whether the word being considered or its immediate neighbours are content words or function words. All these algorithms are different but the same nature of rules are generated.

1.5.4 Image Understanding

Many of the artificial intelligence programs are engineering to solve some useful problems without regard to their similarities with human cerebral architecture. A two dimensional array of pixels containing grey levels can be used to receive digital image recognised by a video camera. To interpret a scene, the image is passed through three basic processes like low, medium and high-level vision. The low-level visions pre-process the image by filtering from noise. The system in medium-level deals with the improvement of details and segmentation. The segmentation of image is partitioning the image into objects of interest. The high-level vision system includes yet another three steps, first is recognition of the objects from segmented image then labelling the image and finally interpretation of the scene. It is in the high-level vision system that we make use of the artificial intelligence tools and techniques. Pattern classification helps in the recognition of objects from its image. It is presently realised by the supervised learning algorithms. On the other hand, interpretation process requires knowledge-based computation.

1.5.5 Robotics

Artificial intelligence is applied in robotics, in order to see, hear, and to react to other sensory stimuli. Robotic is employed in tasks requiring a high degree of precision or that are risky for humans. Another important factor is the size of the space within which the robot makes possible sequences of moves. Even an extremely simple robot is capable of generating a vast number of potential move sequences.

The major area which is going to flourish in the 21st century is robotics. With the latest techniques of deep neural networks and convolution networks, various prime tasks are been mechanised. So, it is not that far away, we should see a world of related humans with less emotions. The robots will definitely take up the man powered tasks to a great extent.

1.5.6 Machine Learning

Machine learning is a branch of artificial intelligence that deals with the design and development of algorithms. So, machine learning plays as a scientific discipline which is concerned with the design and development of algorithms that allow computers to evolve behaviours based on empirical data, such as from sensor data or databases. The major focus of the study is to make the system automatically learn from the available data, environment and to train itself for making intelligent decisions. It thus helps in creating a system which is otherwise difficult to manually construct with lot of ease and reduced effort. The learning is one of the most important components of intelligent behaviour. A learner takes advantage of training examples (data) to capture characteristics of interest of their unknown underlying probability distribution. After the training, learner should check itself by testing. The training and testing data can be seen as examples that illustrate associations between observed variables.

A key focus of machine learning research is to automatically learn to recognise complex patterns, and based on data it makes intelligent decisions. The difficulty lies in the fact that a set of all possible behaviours is too large to be covered by the set of observed examples. Decision tree learning, association rule learning, support vector machines, hidden Markov models, neural networks, inductive logic programming, evolutionary learning algorithms and fuzzy logic are some examples of machine learning algorithms. Only selected methods are discussed here and the following chapters give a detailed description of various machine learning techniques.

Neural networks

Neural Networks or Artificial Neural Networks (ANN) are modelled from the biological inspiration of the human brain. ANN is composed of number of artificial neurons that implies that the learning information is stored through the node weights. But we cannot say that all learned knowledge is perfect. The ability to generalise problems is the advantage of ANN. First, we model the network then lean it and finally test it. The performance of network depends on the training on which how well it can generalise. This depends on how well the network was designed and trained. Another criterion of ANN is dataset selection for its training. Researchers who are concerned in computerised prediction/classification problems are aware of ANN.

A neural network is an interconnected group of nodes, similar to the vast network of neurons in the human brain. Computers are programmed to simulate intelligence by copying the types of physical connections that occur in human brains. Neural computing systems mimic the brain through a network of highly interconnected processing units. These in turn afford learning capabilities of the computers and enable them to recognise, and to understand very complex or mysterious patterns.

Inductive logic programming (ILP)

It is a technique of defining the problem by checking the results of a function. The process is successful only if there are sufficient examples and how complex the functions are. Even if recursively defined, inductive logic programming algorithms can learn simple logic programs.

Applying to the process with real-life data rather than computer generated functions, more composite programs will prove challenging to learn. Inductive logic programming is a subfield of machine learning which makes use of logic programming for rule creation. Thus, it makes the systems capable of predicting results by learning the background knowledge and by learning various examples.

Inductive logic programming, in simple words, can be mentioned as the process of reasoning from specific observations to create a general hypothesis. There is a logic program 'B' which represents the background knowledge, then a set of positive (E+ve) and negative (E-ve) examples. From these, a hypothesis 'H' or a rule is generated. We can explain ILP in simple way. There are mainly two strategies for learning a rule, Top-down approach and Bottom-up approach. Top-down approach starts with the most general rule or empty rule, and similar constrains are added until all positive examples are covered. FOIL is a system that works on Top-down strategy. The latter approach starts with the most specific rule and constrains are removed in order to add more positive results. GOLEM is an ILP system which works on the Bottom-up approach. Inductive logic programming is mainly used for prediction and classification problems.

Decision tree learning

A decision tree is a structure that allows learning of opinions in binary and is more related to objects based on their attributes. From a series of examples, a classification model can be built from the continuous learning by an algorithm. This classification model is the decision tree that can handle example correctly or nothing is done. The results are displayed with the structure of tree, if it is correct.

Bayesian networks

Bayesian networks are tree structure that built on decision based learning opinions. There are options like good or bad related to objects, based on their attributes like length, colour, age, etc. From a series of examples, the learning algorithm builds a decision tree which can classify new examples. No modification is carried out, if the new examples are handled correctly. Otherwise, the tree structure is modified until the correct results are displayed. Major challenge of the Bayesian network is handling errors in values (noise), getting the algorithm to perform well on very large sets of data and determining the optimal fit of the tree for training and testing data.

Association learning

Association mining is finding frequent patterns, associations, correlations or causal structures among sets of items or objects in transaction databases and other information repositories. When given a set of transactions, this will find rules that will predict the occurrence of an item based on the occurrences of other items in the transaction. It is necessary to satisfy a user-specified minimum support and a user-specified minimum confidence at the same time for association rules. Association rule generation is usually split up into two steps: (i) Minimum support is applied to find all frequent item sets in a database. (ii) These frequent item sets and the minimum confidence constraint are used to form rules.

Association rule learning is a machine learning technique which discovers relations between variables. In this method, rules are formed for problem scenario. Association rules examine the causal correlations between non-fatal and fatal events. For example, an association rule is in the form of $e_1, e_2, ..., e_k \rightarrow f, conf$, where e_i is a non-fatal event, f is a fatal event and *conf* is the corresponding confidence value.

Hidden Markov Model (HMM)

HMM is a statistical model used in pattern recognition, speech and language processing. In HMM, the system being modelled consists of a set of hidden states, visible states and undergoes a Markov process. Hidden states are non-observable states. In a normal Markov process, the states are directly visible to the observer. In HMM, on consumption of each input symbol, a state may transit to another state or remain in the same state and emit a visible symbol with a probability. HMM is operated in Transition probability matrix, Emission probability matrix and an Initial probability vector. There are algorithms that are used for training and evaluation of the HMM.

1.5.7 Pattern Recognition

It is "the act of taking in raw data and taking an action based on the category of pattern". Most research in pattern recognition is about methods for supervised learning and unsupervised learning.

Major aim of the pattern recognition is classification of data (patterns) that is based either on a priori knowledge or on statistical information extracted from the patterns. Classifications of patterns are usually groups of measurements or observations, defining points in a suitable multidimensional space. This is in distinction to pattern matching, where the pattern is rigidly specified. Some of the application areas of pattern recognition are Optical character recognition, Handwriting recognition, Speech recognition and Face recognition.

1.5.8 Virtual Reality

Virtual reality is an attempt for creating natural interaction. For this purpose, variety of specialised input and output devices are necessary. Normally, a tracker is used as the input device. It is capable of reporting its location space and orientation. The tracking device can either be optical, magnetic or acoustic. Tracker also comes along with the traditional input devices like joystick or mouse. The attempt to create a truly natural input device was fulfilled by the development of data gloves. Data glove had numerous sensors that is capable of reading the angle of each of the finger joints. Thus, by wearing the gloves, user can interact with the virtual world through hand gestures. For creating the visual world experience stereoscopic displays are used. Two views of the simulated world are generated; one for each eye and stereoscopic device is used to show the correct view to each eye. Figure 1.3 shows a virtual reality environment.

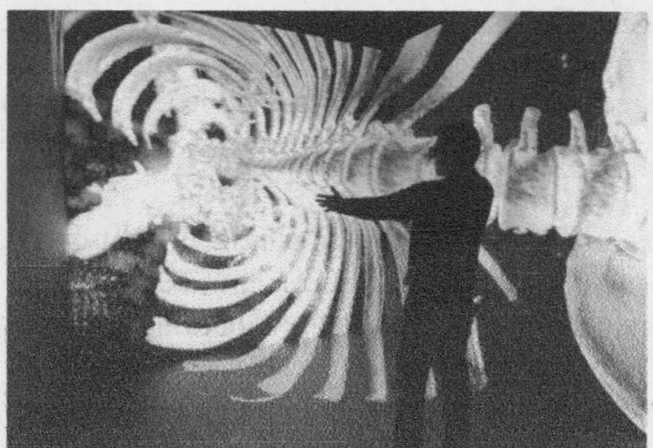

Figure 1.3 Virtual reality environment.

In a variety of ways we can apply the virtual reality environment. In research, virtual environments are used to visually explore whatever physical world phenomenon is seen. It can also be used for training personnel for work in dangerous environments like nuclear power stations or, with expensive equipment best done through simulation. For example, air plane pilots, training in flight simulators. Virtual reality can enable medical personnel to practice new surgical procedures on simulated individuals. In the field of entertainment also virtual reality is used, it offers highly engaging way to experience imaginary worlds and to play games. When designing prototypes for new products, virtual reality provides a way for experimentation with these designs.

Hardware

A variety of specialised input and output devices is required to achieve the virtual reality field. One of the most important input devices used in a virtual environment is a tracker. This device is capable of reporting its location in space and its orientation. The mode of operation of the tracking devices can be optical, magnetic, or acoustic. The tracker can be combined with input device, such as a mouse or a joystick to perform in its virtually real environment. Another virtually real environment is the data glove. It is natural input device outfitted with sensors that can read the angle of each of the finger joints in the hand. A user wears such a glove to interact with the virtual world through hand gestures. This movement includes pointing or making a fist to the application. Using stereoscopic displays, real-world visual experience is approximated. For each eye of the viewer, simulated world is generated. A stereoscopic display device is used to show the correct view to each eye. There are a variety of such simulated input media created for the user to justify the virtually real environment.

1.5.9 Computer Vision

In the era when machines can see, it is able to extract information from its vision. Computer vision plays here to recognise the object as an image and identify the tasks. It is considered as a scientific discipline concerned with the theory behind artificial systems that extract information from images. The image data can be in the form of picture, video sequences, multi-dimensional data from a medical scanner or views from multiple cameras.

Computer vision seeks to apply its theories and models to the construction of computer vision systems because it is a technological discipline. Examples of applications of computer vision include systems for:

1. Controlling processes, for example, an industrial robot or an autonomous vehicle.
2. Detecting events, for example, for visual surveillance or people counting.
3. Organising information, for example, for indexing databases of images and image sequences.
4. Modelling objects or environments, for example, industrial inspection, medical image analysis or topographical modelling.
5. Interaction, for example, as the input to a device for computer-human interaction.

1.5.10 Nature Inspired Computing

A new computing technique is required for the researchers for solving artificial intelligence problems that have not yet been covered. Nature inspired computing is one of such computing techniques that absorb ideas by observing how nature behaves in various situations to solve complex problems. Research on this computing technique has opened new branches such as neural networks, artificial immune systems, biological computation, evolutionary computation, swarm intelligence, and so on.

The emergence of nature inspired computing was made possible by the advancement of computer science and the remarkable growth of computing techniques. In almost all fields of Biology, Physics, Engineering, Economy, and Management, nature inspired computing techniques are applied. The models are swarms, colonies, or other natural similes, software agents that are suitable for modelling particularly complex and dynamic systems. Algorithms that find optimal paths to food sources have already been developed. These developments are mainly inspired by ant colonies that exhibit swarm intelligence. This is called Ant Colony Optimisation (ACO), highly used in optimisation problems. Similarly Particle Swarm Optimisation (PSO) is used in variety of real world problems. Molecular electronics attempts to develop solid-state components similar to molecular structures. A biological molecule such as DNA has inspired researchers to work on DNA-based computing.

A typical nature inspired computing system is based on self-organisation and complex systems obtained from nature. It is derived as a computing system operated by population of autonomous entities surrounded by environment. Nature inspired computing are techniques that can be applied to wide range of problems like classification, prediction or optimisation. The nature inspired computing techniques are adapted from the unseen data that is useful for learning and can handle incomplete data. The computation techniques are decentralised in control of computational activities.

One of the subsets of nature inspired computing is biologically inspired computing. There are three important differences between traditional computing systems and biological information processing systems. For higher level operations, the components of biological systems respond slowly but implement. Major ability of the biological system is to assemble and greatly enable own higher interconnection densities. Biological systems implementation is not a planned one. The biologically inspired computing is a field of study that loosely knits together subfields related to the topics of connectivity, social behaviour and emergence. The biologically inspired computing's related to the field of artificial intelligence. It can also link to machine learning as well as fields of Biology, Computer science and Mathematics. To summarise, nature inspired computing system is the use of computers to model nature and simultaneously the study of nature to improve the usage of computers.

Major difference between bio-inspired computing and artificial intelligence is in how it takes more evolutionary approach to machine learning. In traditional artificial intelligence, programmer is a creator and makes something and imbues it with its intelligence. Bio-inspired techniques involve a method of specifying the set of simple rules. These rules and a method of iteratively applying those rules are adhered to by simple organisms. Some forms of complex behavior arise after several generations of rule applications that usually make some cases. Complexities of such works are very high.

1.5.11 Intelligent Control

In a process control system, the controller is designed from the known models of a process and the required control objective. The existing techniques for controller design are not valid when the dynamics of plant is not completely known. In such situations rule-based control is mostly used and for controller realisation an expert engineer sets the production rules. The predecessor part of the rules in a rule-based system is then searched against the dynamic response of the process plant parameters. The selected rules whose antecedent part matches with the plant response are selected and fired. There are situations where more than one rule is friable. In such cases, a set of strategies made by the controller is used to resolve it. The other extreme situation may also occur, when no rule matches with the plant response. Such situations are handled by fuzzy logic which is capable of matching with the antecedent parts of rules with the plant response. In various industrial plants, fuzzy control has been successfully implemented. For example, for the power control in a nuclear reactor, various controllers are implemented. While, design a controller, issues in the process control of a particular plant is estimated. This will attempt to follow the actual plant response, when both the plant and the estimator are jointly excited by a common input signal. The new tools for plant estimations which are recently developed are fuzzy and artificial neural network-based learning.

1.6 Artificial Intelligence Tree

We can just summarise that the artificial intelligence is looked at as a tree (Figure 1.4) with

Fruits: Applications.

Branches: Expert systems, Natural language processing, Speech understanding, Robotics and sensory systems, Computer vision, Neural computing, Fuzzy logic, Genetic algorithm and so on.

Roots: Psychology, Philosophy, Electrical engineering, Management science, Computer science, Linguistics, Mathematics and Statistics.

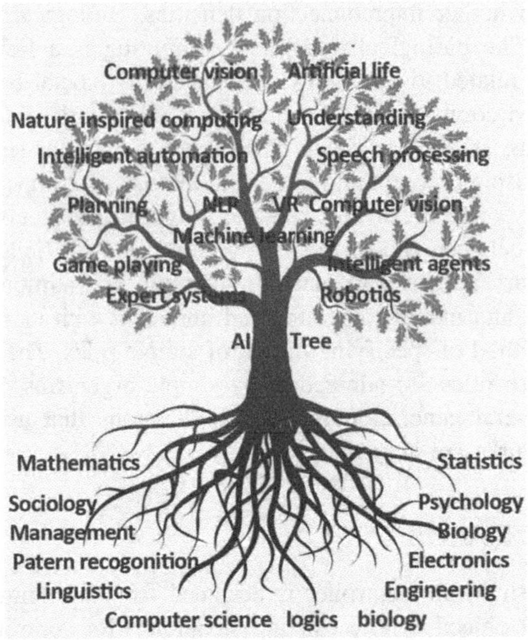

Figure 1.4 Artificial intelligence tree.

Weak and strong artificial intelligence

Machines can start their own thinking process using complex algorithms and software. This is one of the tasks in strong artificial intelligence. An independent decision can be conclude from the accepted input is considered in strong artificial intelligence. Today's computer games are designed as a strong artificial intelligence software support. For example, poker artificial intelligence which can teach itself to adapt to and outsmart the skills of human opponents.

In weak artificial intelligence system, machines cannot do anything on their own and the human interference is heavily involved therein. That is, pre-programmed software is involved in the applications while they are designed. Weak artificial intelligence is also known as narrow artificial intelligence capable to perform a single task and it is designed to pull data from specific dataset. A pre-determined, pre-defined range, even if it appears to be much more sophisticated, involves operations in narrow artificial intelligence. iPhone's Siri and Amazon's Alexa are examples of weak artificial intelligence software.

1.7 Exercises

1. Define artificial intelligence.
2. What is the difference between intelligence and artificial intelligence?
3. Explain the architecture of artificial intelligence system.
4. Define intelligence. What is the intelligent behaviour of a machine?

5. Describe the Turing test for artificial intelligence and justify its validity from a modern standpoint.
6. Write a note on the birth and development of present day artificial intelligence research.
7. Write a note on important events in the history of artificial intelligence research.
8. What are the characteristics of intelligence?
9. What is artificial intelligence technique?
10. What are the knowledge characteristics?
11. How can you analyse a problem?
12. Explain the concept of artificial intelligence problem characteristics in detail.
13. List out the various machine learning techniques.
14. There are various machine learning techniques. How will you select an appropriate one for your problem.
15. What is intelligent control? What is its role in artificial intelligence?
16. Why is planning important in artificial intelligence?
17. What is IQ? Do computer programs have IQs?
18. How far is artificial intelligence from reaching human-level intelligence? When will it happen?
19. What about other comparisons between human and computer intelligence?
20. What is 'artificial intelligence and artificial technique'? Briefly explain how artificial intelligence technique can be represented. List out some of the task domains of artificial intelligence.
21. What is the difference between strong artificial intelligence and weak artificial intelligence?
22. What are the most common AI terms you should know?
23. How will artificial intelligence change the future?
24. Which are the most powerful artificial intelligence companies?
25. What are some common benefits of artificial intelligence technology?

CHAPTER 2

Intelligent Agents

Intelligent agent is a program or a software which is capable of making decisions based on the inputs which it receives or through some experiences. Information is normally gathered in an autonomous manner in a regular schedule. Intelligent agents become more prominent when they start adapting to the environment in real-time with less error rate. They can be compared to an autonomous computer program or a bot which responds to user instructions.

Machine learning process produces numerous intelligent models either in the form of software or adaptive programs. These models are very useful for various applications. For example, a spam filter in the e-mail server will act as an intelligent software component. We call this system as an intelligent agent. Basically, these agents are software components incorporated with an application.

In another way, anything which can perceive its environment through sensors and react to the environmental stimuli through actuators, can be termed as an agent (Figure 2.1). A human agent is eyes, ears, and any other organs. For sensors; hands, legs, mouth, and other body parts for actuators. A robotic agent is any one of cameras and infra-red range finders for sensors and various motors for actuators. Intelligent agents are described as an abstract functional system similar to a computer program. It was the scientist trios, Smith, Cypher, and Spohrer in the year 1994 who described an agent as a persistent software entity dedicated to a specific purpose.

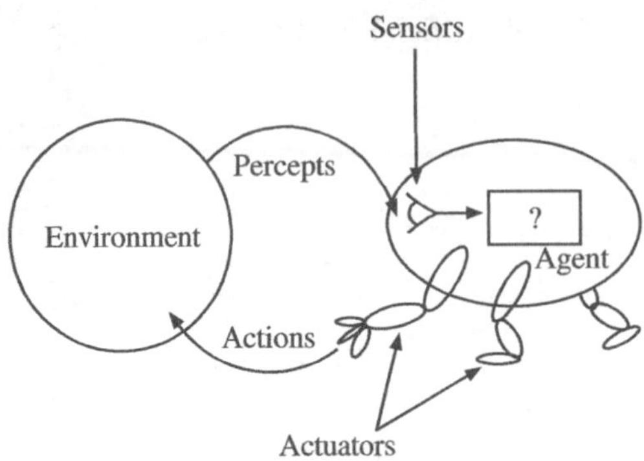

Figure 2.1 An agent.

Intelligent agents can be called as software entities which process some operations on behalf of an end user or another program with some degree of independence. While doing so, agents employ some knowledge or represent the user's goals or desires. One of the major differences between a program and an agent is in its size, purpose, persistence and autonomy. A program size is smaller than that of an agent. Programs are multifunctional whereas agent is used for particular purpose. The duration of agents life does not fully depend on a user launch and quit. Autonomy of an agent does not need the user's input to function. Agents are divided into five classes based on their degree of perceived intelligence and capability (Figure 2.2).

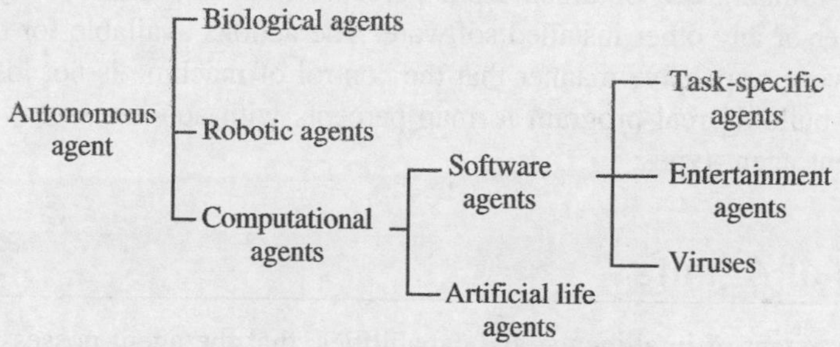

Figure 2.2 Classification of agents.

A biological agent performs biological activities in human cell. For example, an enzyme or a virus will make a biological function in the human cell. Robotic agents automatically perform their operations like a crawler in the web architecture.

Computational agents are classified into software and artificial agents. Virtual reality is an application of artificial life agents. Agent softwares assisting specific tasks are called 'task specific agents'. An environment agent is a program that replicates in the network like a spam.

2.1 Structure of Agents

From the different types of agents that have been explained, we can conclude how agents responds to the input that is received. But how exactly the process works? It is the job of artificial intelligence to program the agent in such a way that it delivers the actions mapped for each perceived input. Artificial intelligence induced program running on some hardware/machine which is termed as the architecture of agent. Architecture can be some embedded system, or a computer, sensor camera or any software which triggers the agent behaviour. Thus, we can say that an intelligent agent comprises of the architecture and in-built artificial intelligence program.

Before designing an agent, we have to understand the type, action, goal and environment in which it works. Only if these parameters are explained, we can have a well-defined agent responding accurately. Consider the creation of a medical diagnosis system. Major objective is to find the symptoms from patients and to provide answers to patient's queries. This is achieved by various actions like test, questions, and treatments. The environment is patient or hospital and goal can be seen as healthy people with reduced treatment costs. Similarly we can account for various agent types like power plant controller, satellite image analysis system, interactive English tutor etc. Each and every system needs to have all the parameters well explained.

Let us discuss a sample example of having a Uber taxi agent. We need to first identify the input, action, goal and environment. The inputs come from GPS, camera, speedometer, topology, etc. Based on inputs, the agent needs to specify actions like break, steer, accelerate or stop. Major goal of the taxi is to maximise profit through a safe and comfortable trip with least distance. The environment can be the customer, road, other fellow travellers on the road.

Basic informations like coordinates of the car on road, other vehicles, speed by which the vehicles are cruising are obtained from percepts from the car. It may be through the cameras, odometer or any other installed software. The actions available for the driver need to be mapped in a very responsive manner that the control of machine is not lost anytime during the journey. For building real program to map percepts with actions we have to consider the following different agent types:

2.2 Types of Agents

According to the extent of intelligence and capabilities, that the agent possesses, it is classified as the following:

Simple reflex agents

Decisions are taken based on the current perception and not the history of perceptions. This is basically termed as a condition–action rule (*if* < *condition* > *then* < *action* >). There is a lookup table of actions from which actions are taken for every possible state of environment. If the environment has *n* variables, each with *t* possible states, then the table size is t_n. This will work only for a small number of possible states in the environment and works if environment is fully observable. The simple reflex agent is shown in Figure 2.3.

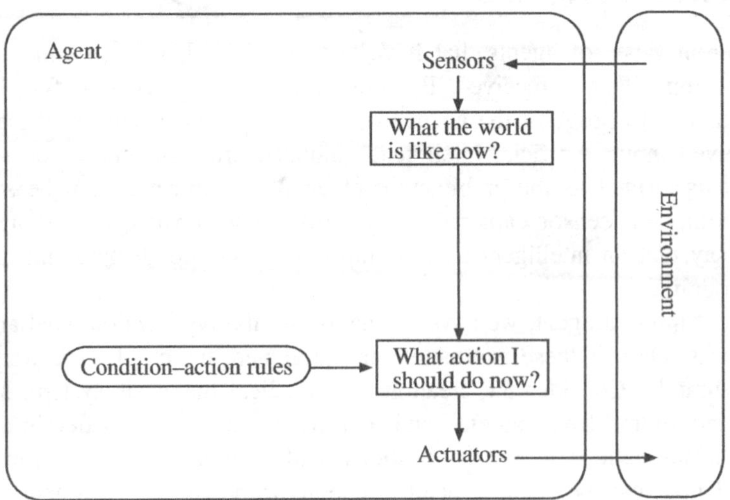

Figure 2.3 Simple reflex agents.

Simple reflex agents have admirable property of being simple, but they turn out to be of very limited intelligence. The agent in Figure 2.3 will work only if the correct decision can be made on the basis of only the current perception. That is, only if the environment is fully observable. Even a little bit of unobservable environment can cause serious trouble. For simple reflex agents operating in partially observable environments, infinite loops cannot be avoided. Escape from infinite loops is possible if the agent can randomise its actions.

Model-based reflex agents

If the world is not fully observable, the agent must remember observations in parts of environment which it cannot currently see. This needs to have an internal representation of the world or internal state. Since this representation is a model of the world, we call it as model-based agent. The model-based agent is shown in Figure 2.4.

Most effective way to handle partial observability is to keep track of the part of world it cannot see now. That is, the agent should maintain some sort of internal state that depends on the percept history and thereby reflects at least some of the unobserved aspects of the current state. We need some information about how the world evolves independent of the agent and next we need some information about how the agent's own actions affect the world. Knowledge about 'how the world works' is called a model of the world. An agent that uses such a model is called a model-based agent.

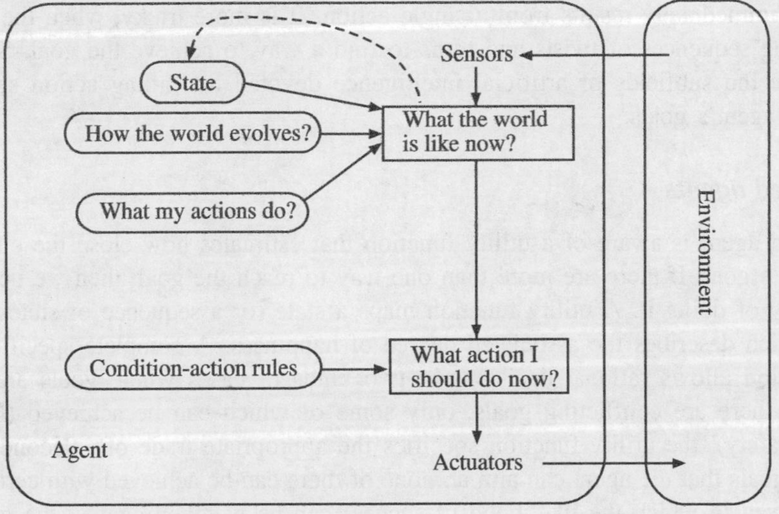

Figure 2.4 Model-based reflex agents.

Goal-based agents

The agent has a purpose and action to be taken that depends on the current state and on what it tries to accomplish (the goal). In some cases, the goal is easy to achieve. In others, it involves planning, shifting through a search space for possible solutions, developing a strategy. The goal-based agent is shown in Figure 2.5.

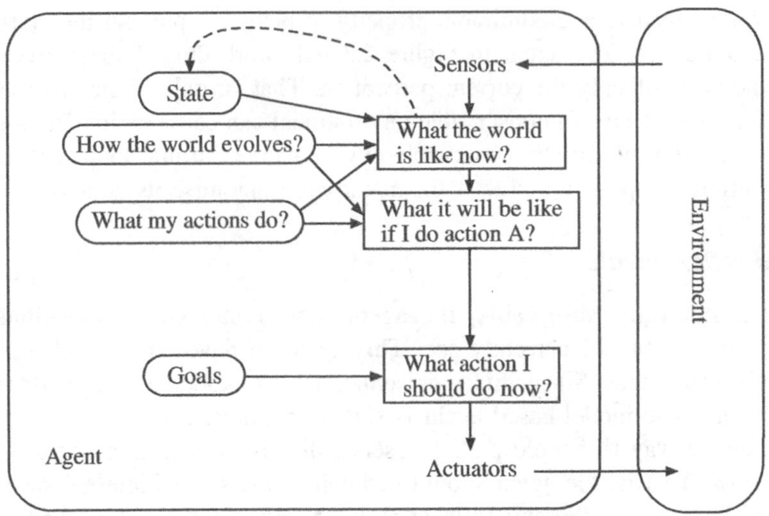

Figure 2.5 Goal-based agents.

The agent sometimes needs some sort of goal information that describes situations that are desirable. The agent program can combine this goal with information about the results of possible actions. Sometimes goal-based action selection is straightforward, when goal satisfaction immediately results from a single action. It is more tricky, when the agent has to consider long sequences of twists and turns to find a way to achieve the goal. Searching and planning are the subfields of artificial intelligence devoted to finding action sequences that achieve the agent's goals.

Utility-based agents

Utility-based agent is aware of a utility function that estimates how close the current state is to the agent's goal. If there are more than one way to reach the goal, then we need to choose the best way of doing it. A utility function maps a state (or a sequence of states) with a real number, which describes the associated degree of happiness. A complete specification of the utility function allows rational decisions in two kinds of cases where goals are inadequate. First, when there are conflicting goals, only some of which can be achieved (For example, speed and safety), the utility function specifies the appropriate trade off. Second, when there are several goals that the agent can aim at, none of them can be achieved with certainty. Utility provides a way in which the likelihood of success can be weighed against the importance of goals. The utility-based agent is shown in Figure 2.6.

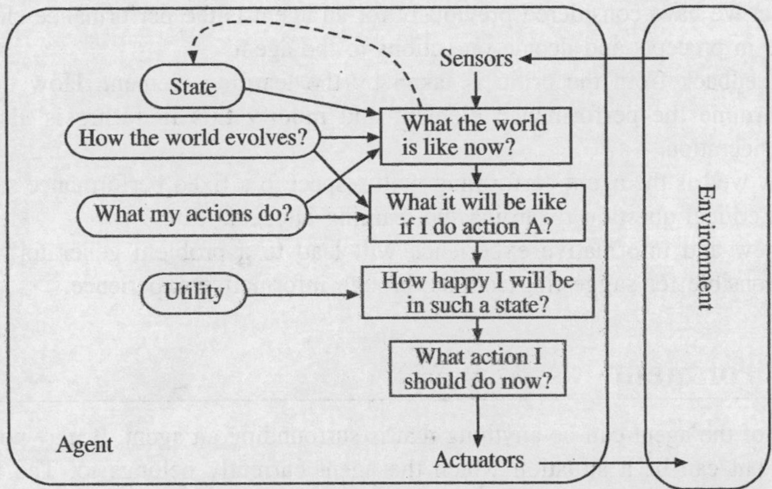

Figure 2.6 Utility-based agents.

Learning agents

These agents are capable of acquiring new competence through observations and actions. The basic components are learning element (modifies the performance element), performance element (selects actions), feedback element (critic) and exploration element (problem generator). Learning agents operate initially in an unknown environment and become more competent than their initial knowledge. A typical learning agent is shown in Figure 2.7. The agent does not require to be hand programmed. Major parts of a learning agent are:

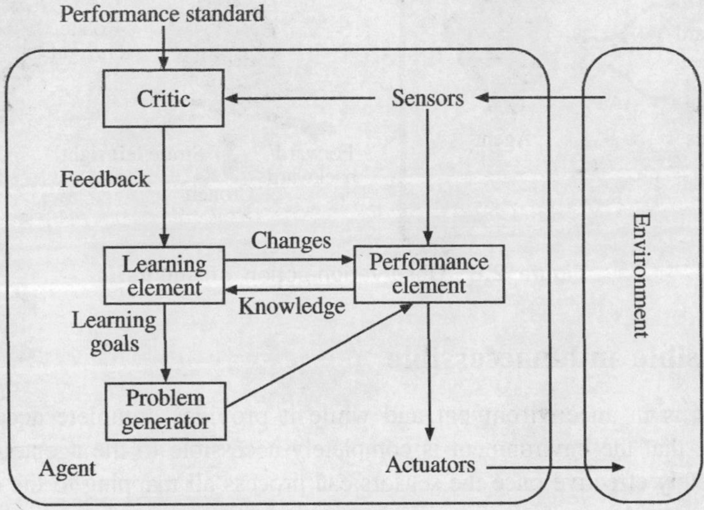

Figure 2.7 Learning agents.

(a) What we have considered previously for an agent is the performance element. It will take in precepts and decide on actions to the agent.
(b) A feedback from the critic is taken by the learning element. How the agent will determine the performance element and modify this in future is also taken into consideration.
(c) How well is the agent performing with respect to a fixed performance standard? This is a critical question regarding the learning element.
(d) A new and informative experience will lead to a problem generator. This will be responsible for suggesting actions through informative experience.

2.3 Environment

Environment of the agent can be anything that is surrounding an agent. It may not be any part of the agent but can be a situation which the agent currently belongs to. The link between agent and environment can be stated in a single statement, "actions are done by the agent on environment which in-turn provides percepts to the agent". Environment comes with various features. As per Russell and Norvig, there are eight different flavoured environments. A generalised observation–action image is shown in Figure 2.8. This shows how agents observe or perceive the inputs and how the inputs are mapped with the corresponding actions.

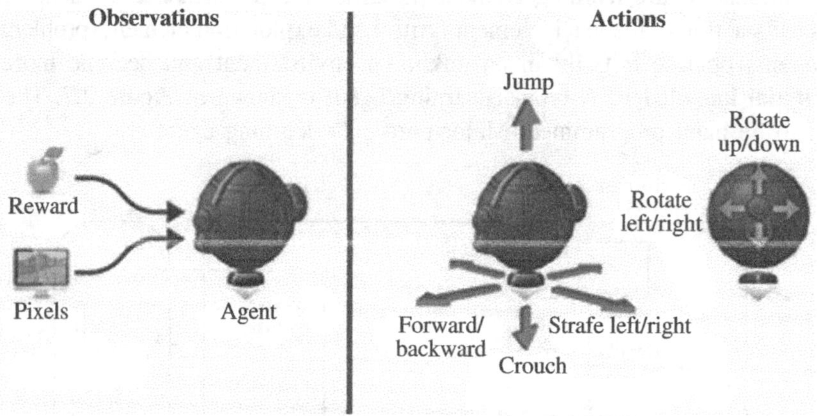

Figure 2.8 Observation-action of agents.

2.3.1 Accessible and Inaccessible

When the agent is in an environment and while it provides complete access to its states then we can say that the environment is completely accessible to the agent. An environment becomes completely effective once the sensors can process all mapping to the choice of action of the agent. Agents also prefer an accessible environment because it may need to keep track of the world in another memory if it is not completely accessible. An empty room whose state can be defined by its temperature is an example of an accessible environment. Inaccessible environment is generally vague. It only states the place of action to the mapped objectives.

Information about an event happening with respect to an agent can be termed as an example to inaccessible environment. Figure 2.9 shows a general schematic of how agent and environment interact. While working on inaccessible environments the agents should also be prepared for possible failures.

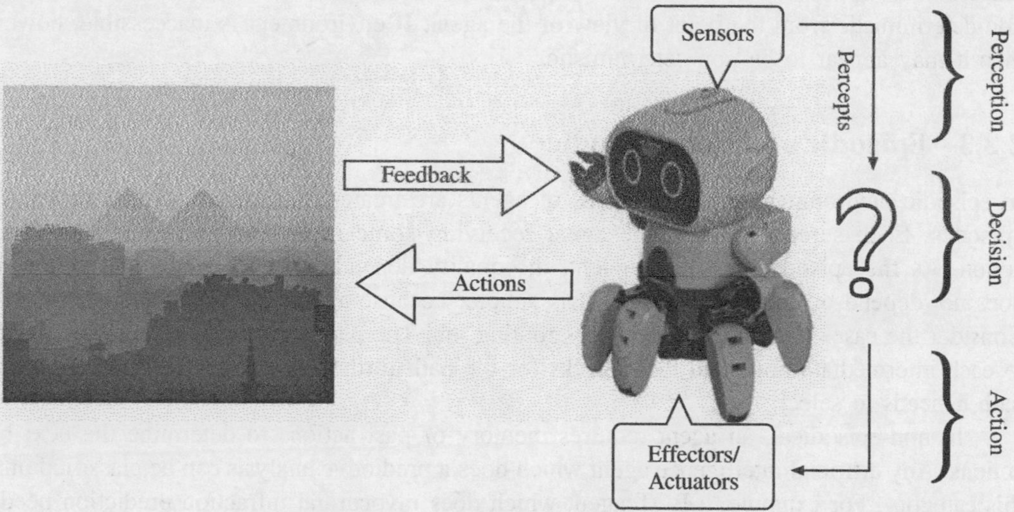

Figure 2.9 Agent and environment.

There is yet another smaller variation termed as partial accessible environment. Military planning can be termed as partial since the details are not fully available but the agent knows what is happening. The major details which are showcased may also not be always true. This makes the agent confused in taking decisions. An action taken in response to a feedback in an accessible environment may not work successfully in a partial or inaccessible environment.

2.3.2 Deterministic and Stochastic

If the next state of environment is completely determined by the current state and the actions selected by the agents, then we say that the environment is deterministic. In a deterministic, fully observable environment, agent does not need to worry about uncertainty. In a real-world situation, no environment of an artificial intelligence is deterministic, because the artificial intelligence itself has a built-in method for selecting states that has to account for some randomness. Given the current state of the environment and an action of artificial intelligence agent, the artificial intelligence can know with certainty the next state of the environment.

f(state i, action i) = state f

So, the resulting state does not change given the same state and action at all times. This is applicable for both single and multi-agent environments. For example, in the case of tic-tac-toe having this environment state and two artificial intelligence agents (black and white) and it is currently the white's turn. Choosing the rightmost bottom square always guarantees a win for the white artificial intelligence agent. There are no instances where that action in this situation does not guarantee a win for said artificial intelligence agent.

If environment is inaccessible, however then it may appear to be non-deterministic. This is particularly true if the environment is complex, making it hard to keep track of all the inaccessible aspects. Thus, it is often better to think of an environment as deterministic or non-deterministic from the point of view of the agent. If environment is inaccessible, however then it may appear to be non-deterministic.

2.3.3 Episodic and Non-episodic

In episodic environment the experiences of agents are treated as separate screens or separate episodes. Each screen contains the agent receiving some input and performing a mapped action. As the episodes are independent, the quality depends on each individual action and does not depend on the previous one. It is simpler because agents are given short-term goals. Consider the case of an agent which does routing job. The path selections are episodic. It goes to each intermediate node and then checks for the path further. Agent can fully determine the path it needs to select.

In non-episodical, an agent requires memory of past actions to determine the next best actions. Any artificial intelligence agent which does a predictive analysis can be classified under this category. For example, a BOT agent which does myocardial infraction prediction need to keep track of all the prevailing medical conditions and suggest the health related solutions. Such an agent cannot work independently with the environment. It is highly dependent and continuous. Your.MD is a free platform that offers actionable health information based on highly accurate sources and lets the user make the best choices for his health. It is basically a symptom checker powered by artificial intelligence. Figure 2.10 depicts Your.MD BOT working in a non-episodic environment.

Figure 2.10 Your.MD BOT.

2.3.4 Static and Dynamic

Dynamic environments are most challenging for the agents. While deliberating an action to perceived signal the environment changes. In dynamic environment, agents should track the past action history to an earlier version of the new environment. The actions which turn positive and negative are stored for further performance improvement. Static environment is easier to handle. It can just look at the world around at that instance and make a decision on its action. This hardly changes and goal or objective to be accurate is more high. Driving a vehicle is an example of a dynamic environment whereas solving a *Sudoku* puzzles is an example of a static environment.

2.3.5 Discrete and Continuous

When the agent is in a limited number of clearly defined action box, then we can say that it is a discrete one. Game playing can be seen as a discrete environment. A chess match which happens across 64 different squares with a limited number of action is a perfect example for discrete environment. Continuous environment can be tracking the satellites which move in a pre-defined continuous locus with a constant speed. There is also some unpredictability in paths, however still it has a continuous defined environment. There are situations where this classification is seen very closely to deterministic and stochastic, just because of the limitation of search space. Still there is a clear distinction between the both.

2.3.6 Single Agent vs Multi-agent

If only one agent is involved in an environment, and operating by itself then such an environment is called single agent environment. However, if multiple agents are operating in an environment, then such an environment is called a multi-agent environment. The agent design problems in multi-agent environment are different from single agent environment.

For example, consider an agent which is designed specifically for tracking movement in dark. The sensors are so placed that it gets activated when the environment turns dark. The agent constantly deploys signals to check the conditions and once it finds appropriate environment, the motion detection cameras are automatically triggered to find the movements. In case of a multi-agent environment, there are multiple instances which need to be triggered for an agent. It can be some background knowledge about the states or can be the relative information from another one. Activities deployed on neural nets can be seen as a major example for this type of environment.

2.3.7 Known vs Unknown

Known and unknown are not actually a feature of an environment, but it is an agent's state of knowledge to perform an action. In a known environment, the results for all actions are known to the agent. While in unknown environment, agent needs to learn how it works in order to perform an action. It is quite possible that a known environment to be partially observable and an unknown environment to be fully observable. It is just like a system which does trial and error to perform tasks.

For example, in solitaire card games, we know the rules but we are still unable to see the cards that have not yet been turned over. Conversely, an unknown environment can be fully observable. In a new video game, the screen may show the entire game state but we still do not know what the buttons do until we try them.

2.4 Autonomous Agents

Agents can be of any specification. In the era of machine learning, all research points to the development of automated process. Thus, importance of autonomous agents is also increasing. While designing the autonomous agents it is necessary to be dependent on various input sensors and the mapping to its actions. The systems should be so placed that each and every action will point to the final goal or objective.

An autonomous agent is "a system situated within and a part of an environment that senses that environment and acts on it, over time, in pursuit of its own agenda and so as to effect what it senses in the future".

This definition captures all of the basic features of intelligent agents except their sociability. It provides a good approximation of the basic features of the large variety of intelligent agents now under development. Let us take a close look at this definition.

Intelligent agents sense their environments as shown in Figure 2.11. We have seen this feature of agents in the first definition. But here the sensory data or percepts include not only data about other objects, but also about the impact of agent itself on the state of affairs in the environment. The sensors could be organic, like eyes, ears and their neural processors, or artificial, like video and audio processors integrated with a digital computer. The environment could be a very limited domain, like a blocks world, or very sophisticated one, like the stock market or a set of asteroids. The sensors must be appropriate to the sort of objects with which the agent is designed to interact. Whatever the sensors, history of all percepts sensed by an agent is critical to its future interaction with the environment.

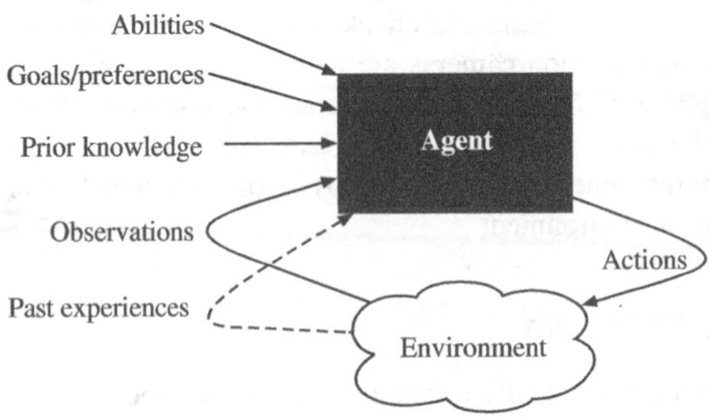

Figure 2.11 Agent and attributes.

Set of all percepts that an agent has to start with plus those gained by interaction with the environment (experience) is the percept sequence of the agent. An intelligent agent consults its percept sequence and the current state of affairs (which may be considered part of the total percept sequence) in light of its goals before taking action. This means an intelligent computer agent (like human agent) consults past experience and current situation before deciding what course of action will further its goals.

In order to save computational power, it is possible to narrow down the search for relevant percepts by using shortcuts. These shortcuts, or heuristics, group the percepts into classes of events so that the agent need not consult all classes of events, but only those that might assist in attaining its goals given the current state of affairs. If I am looking for two seats in a crowded movie theatre, I do not check every seat one at a time to see if it is empty. I use a simple heuristic: look for gaps between people's heads and check to see if those gaps indicate empty seats. This shortcut will save my time so that I do not miss the entire movie.

Now, we have some understanding of a percept sequence, we may ask: In what way does the agent's own actions form a part of this sequence. The second definition states that the agent is considered a part of the environment. This means the agent inhabits a world or domain. If I am waiting in one line for a slice of pizza, my own behaviour impacts on the overall length of the line. It may happen that because I have joined the line, another cashier is called to work in second line. Thus, my own impact on the world has changed the state of affairs to require a different sort of behaviour from all agents behind me as well as myself. Another example is familiar to anyone who enjoys jumping in water. If I make a big splash in the pool, I then get to experience the series of ripples generated by my antecedent action. Another example comes from the medical field. If a drug dispensing agent has already administered half its prescribed dose to a patient, the amount already given alters the amount to be given in the future (unless there is a computational malfunction, in which case the tragedy occurs). So, when an agent acts on the environment, it senses the impact of its own acts, along with other events that fall within its domain.

The intelligent agent also reacts to its environment. Just as the generic agent has both sensor and effector, so too does the intelligent agent. But here things get a bit more interesting. A thermostat can react to its environment by turning the heater off when it senses that the air temperature reaches the threshold of 72 degrees Fahrenheit. However, thermostat is not an intelligent agent because a mechanical thermostat does not have a program. It is a dynamic system that is immediately related to the temperature of environment. By complicating the relation between sensing the environment and effecting the environment, we can build up our concept of agency to the level of intelligence and autonomy. To see this clearly, we will distinguish between a reflex agent, a goal-based agent, and a utility-based agent.

2.5 Nature Inspired Agents

In case of nature inspired agents, the system always tries to learn from the environment. High response systems are normally built using the nature inspired agents. It starts from a simple optimisation problem to radar tracking systems. Agents and multi-agent systems have become increasingly relevant for developing distributed and dynamic intelligent environments. The ability of software agents to act somewhat autonomously links them with living animals and humans, so they seem appropriate for further discussion under nature inspired agents.

The scope of projects that can be deployed in this field is immense. It can start from a simple routing problem to activities in day to day scenario. Nature inspired agents are now a days most common in medical fields. It is basically used as a medical assistance to both patients and doctors. Let us have a brief discussion on a smell detection agent. How do dogs trace path? It is through the smell. The smell of hormones are very specific and it guides them, here the smell acts as the agent. The agent provides input to perform specific action. The goal can be described as the end point where the dog needs to reach. Hence, a total agent cycle is completed.

Figure 2.12 represents a generalised architecture of any nature inspired agent. The major power of a computing methodology around is the ability to learn from the environment for betterment. For each and every problem there are *n* number of solutions, which the system performs each with some sort of intuition. Agent through this process tries to achieve the goal set by the environment. For reasoning mechanism, the system will undergo a 4R cycle — revise, retain, retrieve and reuse. Each of these phases will have a definite algorithm and plan of action to achieve the final goal.

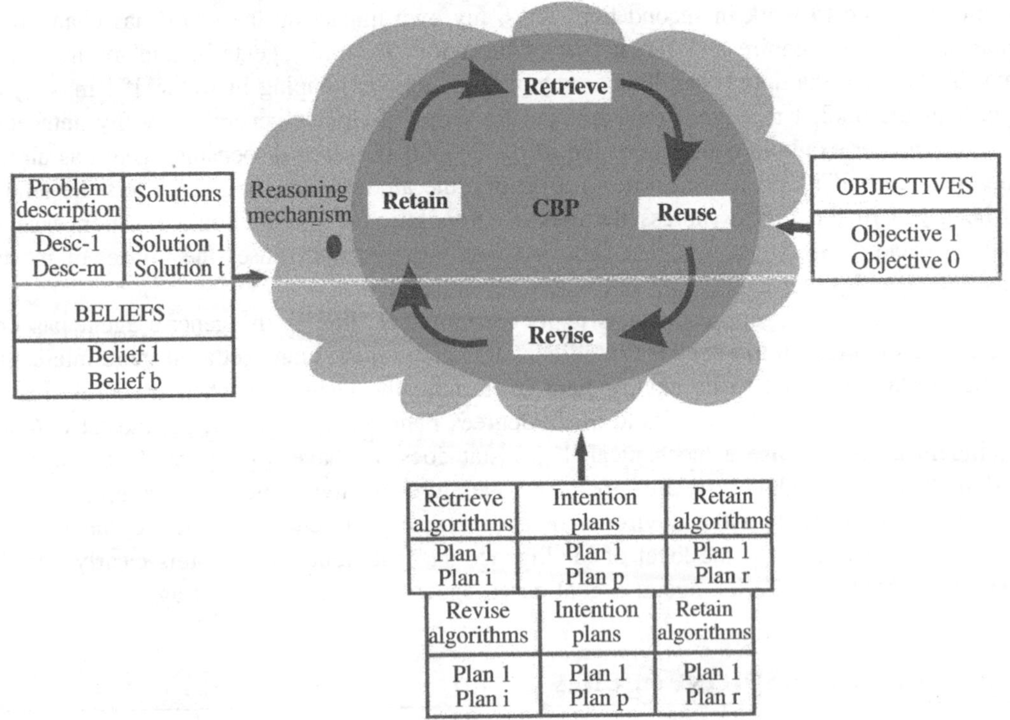

Figure 2.12 Nature inspired agent.

2.6 Planning Agent

An agent which responds to goals through forming plans to achieve them and then enacting these plans through interaction within a domain can be termed as a planning agent. The agent first generates a goal to achieve, and then constructs a plan to achieve it from its current state. Once it has a plan, it keeps executing it until the plan is finished, then begins again with a new goal. Figure 2.13 shows the cycle of operations in a planning agent.

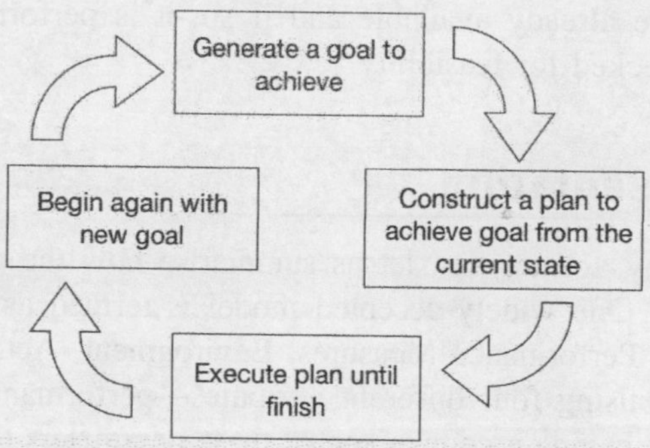

Figure 2.13 Planning agent cycle.

For planning agents to work, there are various assumptions. Primarily the result of an action should be well-defined. There should not be an uncertainty. Each and every action in the plan should be a result of an agent acting on it. And no concurrent actions are allowed. Agent should always have the complete knowledge of the world.

Planning being the oldest subfield of artificial intelligence, has a very active subfield which is open to exploration. Planning agents work in a bottom-up manner and O is the theories in planning. Such approaches have always led to many valuable insights which provided researchers new areas for research. The major aim of any planning is to find out the most suitable path to the very nearest solution. Plans based on beliefs may fail, but a plan based on theories never goes in vain. The agent always makes sure that the developed plan has the most feasible solution to it which is achievable in the shortest time spam.

Planning agents can be partial planning agents or partial-order planning agents. In the earlier one, the operator works on the short-term goals and once it is finished it keeps on adding the next job. It is similar to a building task. Whenever new tasks are added the agent runs a feasibility check and if finds issues, then that particular job is replaced by another suitable one. This always provides a medium for debugging. In the case of partial ordering planing agent, there will be some decisions open for the agent to take action according to the environment. Once an action is taken then rollback is not possible in partial order planning.

Truck route planning can be seen as a typical example where planning agents work with cent percent accuracy. Truck route planning is a skilled task, requiring a great deal of local knowledge about roads, traffic and restrictions. It can also be a time-consuming task if done manually. In an ideal world, the transport manager would carry out a debrief once the driver finishes all of his deliveries. The driver will provide feedback on the planned route or explain the reason for a late delivery or detour to input into the next day's truck route planning exercise. In reality, it is rarely possible for a transport manager to debrief all the drivers immediately upon return and record all of the information formally. The feedback process becomes adhoc and dependent upon the planner remembering what a driver has said to help improve future route plans.

An agent comes to help in such a scenario, whenever there is a change in the environment, the agent registers it. Planning agents will have a associative memory which keeps track of the events happening for similar action. When such a situation arises again, agent checks whether any feasible solutions are already available and if so, it is performed else a new chain of possible solutions are checked for feasibility.

2.7 PEAS Representation

As we have learned many agent types, let us summarise how the agents can be represented in the most easiest way. One widely accepted model is termed as the PEAS representation model. PEAS stands for Performance Measures, Environment, Actuators, and Sensors. Here any agent is represented using four different attributes—performance measure, environment, actuators and sensors. By having a comparison of these parameters the agent quality/reliability can be checked.

Figure 2.14 shows an autonomous car which acts as an intelligent agent which perceives input through the cameras and processes the signals to have a safe ride.

Figure 2.14 Autonomous car.

TABLE 2.1 PEAS Attribute Representation of a Vacuum Cleaner

Agent	*Performance*	*Environment*	*Actuator*	*Sensor*
Vacuum cleaner	Cleaning Efficiency Battery life	Room Furniture Car Other surfaces	Sheet Brush Suction pipes	Dirt sensor Obstacle sensor Topology sensor
Part picking robot	Percentage of parts in correct bins	Conveyor belt with parts Bin	Joint hand Arm	Camera Joint angle sensor

Performance

The performance factors for a self-driven car are the speed, safety while driving (both of the car and user), time taken to drive to a particular location, comfort of the user, etc.

Environment

The road on which the car is being driven, other cars present on the road, pedestrians, crossings, road signs, traffic signals, etc., all act as its environment.

Actuators

All those devices through which control of the car is handled, are actuators of the car. For example, the steering, accelerator, breaks, horn, music system, etc.

Sensors

All those devices through which the car gets an estimate about its surroundings and it can draw certain perceptions out of those are its sensors. For example, camera, speedometer, GPS, odometer, sonar, etc.

How PEAS representation is carried out for some of the real-world examples. Table 2.1 shows the PEAS attribute representation of a vacuum cleaner and part picking robot. Another example which explains the PEAS representation that can be quoted from real-world situation is the use of online medium for purchasing. Internet shopping agent, which can be the most relevant example in the current era when 65% of buying happens through online medium. Here what are the performance measures? The price, quality of the product, efficiency and appropriateness of the items are major things the customer looks for. Environment belongs to the vendors of shops or shipping agencies which deliver the item. Actuators can be the fill-in-forms, and sensors are the HTML page where the shops are listed.

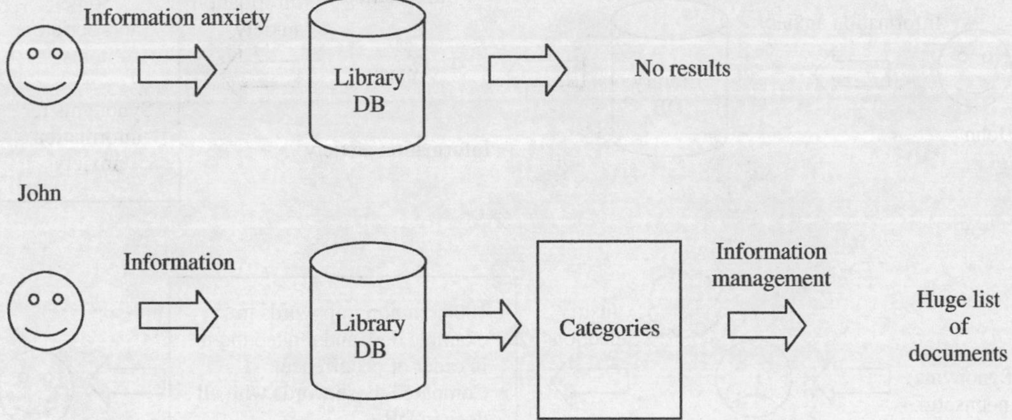

Figure 2.15 Search synonyms for 'Information Anxiety' and 'Information'.

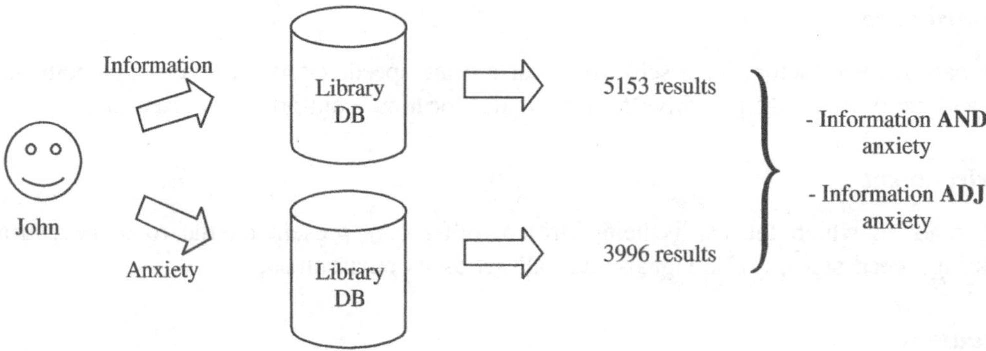

Figure 2.16 Search synonyms for 'Information Anxiety.'

2.8 Intelligent Database

Earlier, databases were viewed as computer systems that stored record oriented and business type data such as manufacturing inventories, bank records, sales transactions, etc. A database system was not expected to merge numeric data with text, images, or multimedia information and it expected automatically notice patterns in the data are stored. Algebra search can be initiated in the database through SQL queries. Traditional database search is shown in Figure 2.15 and Figure 2.16.

The concept of an 'Intelligent Database' views database as a system that manages information (rather than data) in a way that appears natural to users and which goes beyond simple record keeping. The concept of intelligent database is: Intelligent database system = Traditional database system + Intelligent system. A search word, synonyms for 'information anxiety' can be operated in different ways in a database (Figure 2.17). Database stores the history of search that can normally be used in future (Figure 2.18).

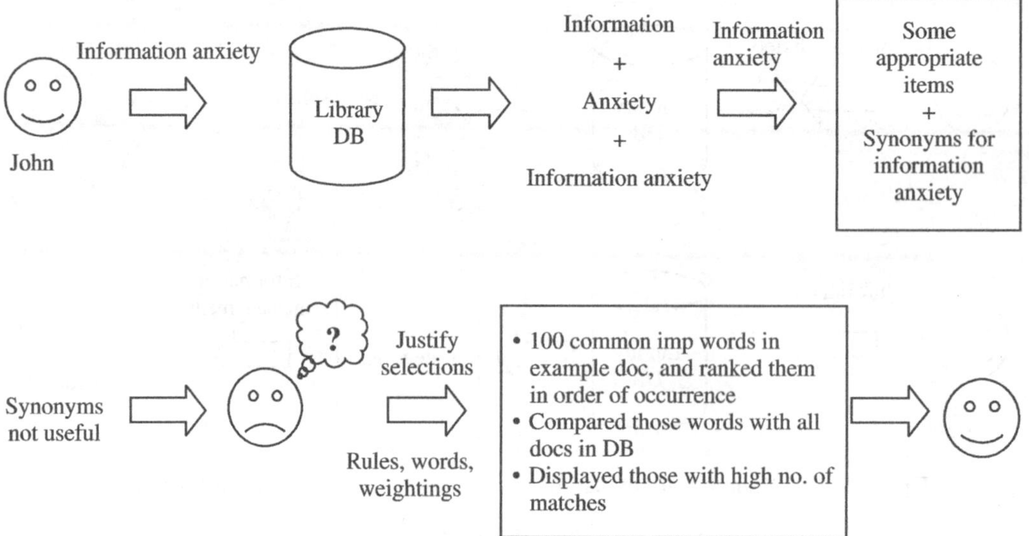

Figure 2.17 Search synonyms for 'Information anxiety'.

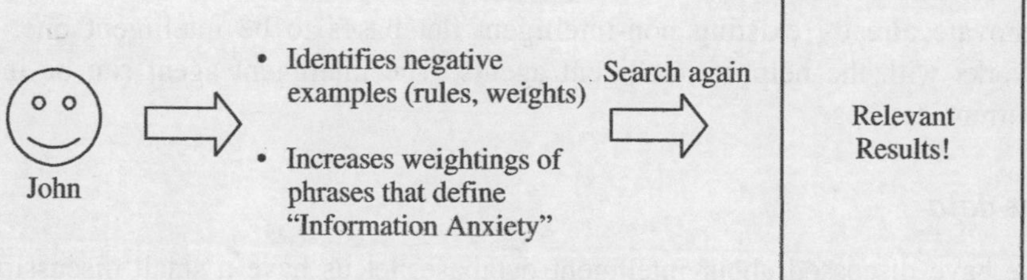

Figure 2.18 Database stores history of the search.

According to John Hopkins Institute, online glossary, an intelligent database is a full-text database that employs artificial intelligence, interacting with users to ensure that returned items (hits) contain most relevant information as possible.

The concept of intelligent database assumes three levels of intelligence for such systems: high-level tools, user interface and database engine. High-level tools manage data quality and automatically discover relevant patterns in the data with a process called data mining. This layer often relies on the use of artificial intelligence techniques. The user interface uses hypermedia in a form that uniformly manages text, images and numeric data. The intelligent database engine supports other two layers, often merging relational database techniques with object orientation.

Intelligent databases have now become widespread. For example, hospital databases can now call up patient histories consisting of charts, text and x-ray images just with a few mouse clicks, and many corporate databases include decision support tools based on sales pattern analysis, etc.

In this context, a research on the topic to make databases more intelligent that intelligence should be proper. A proper planning is required to concentrate more on all the three above-mentioned levels of a system. That is, to develop more efficient tools to manage data quality and knowledge discovery. To develop a better and efficient user interface, a better database engine is highly essential.

Problems need to be identifying for intelligent database design. However, as a primary problem, a planning is required to concentrate on a student database managed in the college/university. Secondary problems focussed to other databases like in business, research, images or such large databases.

In students database, all necessary and sufficient data related to the students are maintained. Optimisation can be applied here to minimise the storage required.

Parents can be informed about the ward in case of emergency, automatically by making use of the data in database. In addition, useful patterns/knowledge can be retrieved from the database by means of the tools, automatically (For example, patterns regarding students who are weak in programming skills from the marks stored in the database). This can be extended to other details regarding the students extracurricular activities, achievements, career, etc.

The intelligence given to a database is through machine learning techniques either supervised or unsupervised terminologies. Major component in the design is natural language processing followed by intelligent prediction. Intelligence database can apply to other databases also. Intelligent database system can make relational schema automatic and hence it becomes more effective. The problem can be extended to build any and/or all databases intelligent. So, we can renovate already existing non-intelligent databases to be intelligent one. Intelligent database works with the help of intelligent agents. The intelligent agent can be incorporated with the current database.

Intelligent data

As now we have discussed about intelligent database, let us have a small discussion on what an intelligent data is. This era is witnessing an unusual data exploration in each and every stream of life. Let it be a daily profit earning company or be a multi-million international company, the amount of data it consumes and produces is in the range of many TB. While producing such immense data, it is always a question on how we can analyse and generate conclusions from them.

Intelligent data or data intelligence finds its relevance in this scenario is the analysis of various forms of data in such a way that it can be used by companies to expand their services or investments. Data intelligence can also be referred to companies' use of internal data to analyse their own operations or workforce to make better decisions in the future. Business performance, data mining, online analytics, and event processing are all types of data that companies gather and use for data intelligence purposes. It is normally mistaken for business intelligence but there is a clear difference in the mechanism. Privacy concerns can sometimes arise as a result of data intelligence gathering. Customers or clients may not want the companies they support to be eavesdropping on their personal online habits or get information about them from social networking sites.

There are places where data is intelligently processed even without our knowledge. For example, there are applications wherein once we give our photo it generates a picture saying how we will be looking after 10 years or 20 years. There are photo edit application which fits an extra sun glass or hat to our original image. We are just excited seeing these applications and leave it there. But does that data end there? No. Such images are data for various other applications like face recognition and prediction. They use such warped images to train internally to produce more accurate results. Thus, even a data which is not worthy in a user's eye will be a piece of valuable information to others.

2.9 Exercises

1. Differentiate between utility function and performance measure.
2. Why is autonomy an essential characteristic of agents in multi-agent systems?
3. What are types and applications of intelligent agents?
4. What is the difference between agent function and agent program?
5. What is the difference between framework, architecture, and model?
6. Discuss social intelligent agents?
7. What kind of behaviour does the stochastic environment posses?

8. How can we define an 'unethical intelligent agent' in a cooperative multi-agent environment?
9. What are the limits or drawbacks of Multi-Agents Systems (MAS)?
10. Explain the various characteristics of agents?
11. Assume that agent A requires the square root of 4.5, and that agent B can solve this. Explain with a worked example how agent C might broker this request.
12. What is a simple reflex agent?
13. Describe intelligent agents on web.
14. How does a simple reflex agent differ from other types of intelligent agents?
15. How does the 'condition-action rule' work with a simple reflex agent?
16. What are the properties of task environments?
17. Differentiate between real-time and reinforced agents.
18. How can deep learning offer better training to agent environment? Comment.
19. Describe a scenario in which a goal-based agent would act rationally but a simple reflex agent would not. Explain why this is the case. Provide enough details so that your conclusion can be supported.
20. Consider an agent that checks its user's e-mails and notifies the user. If a regular e-mail arrives and the user is in a meeting, then the agent will only highlight the e-mail on screen. If the user is not in a meeting, then the agent will beep. However, if the e-mail is urgent, then the agent will always play some music. Describe the components that such an agent will have.
21. Pick an Artificial Intelligence system that has appeared in the news or on TV recently and work out PAGE (i.e. Percepts, Actions, Goals, Environment) descriptions for it. How do these relate to their PEAS descriptions?
22. The definition of an "agent" given in the lectures is quite broad. Can everything be described as an agent? What is an example of a non-agent? What about clocks – in what sense are they agents? Does the distinction between agents and non-agents really make any sense?
23. What are the four basic types of agent program in any intelligent system? Explain how did you convert them into learning agents?

CHAPTER 3

Problem-Solving

Problems are different in nature. The mathematical or statistical problems are entirely different from the artificial intelligence problem. We have find best solutions to the problems by using easy techniques. Some of the real-world problems in such classes are given in Figure 3.1. For example, let us take magic square of size 10 × 10 magic square. If we use naive search space, the computation becomes $100^{100} = 10^{200}$. It may better with permutations by $100! \approx 10^{158}$. If we consider a 400 × 400 magic square, search space = $160000! \approx 10^{31}$. We will see methods which can solve 400 × 400 magic square in less than one hour CPU-time.

In psychology, problem-solving refers to a state of desire for reaching a definite goal from a present condition that is either not directly moving toward the goal, is far from it, or needs more complex logic for finding a missing description of conditions or steps toward the goal. In computer science and in the part of artificial intelligence that deals with algorithms, problem-solving incorporates a number of methods known as algorithms, heuristics, root cause analysis, etc.

To understand artificial intelligence, we can define some common problems encountered in our day to day life. Almost all the problems stated in artificial intelligence commonly use the term 'state'. It defines the state of the solution for the given problem at that particular step. In short, we can summarise the solution of a problem to be a collection of the problem states. The problem-solving procedure used is, to apply an operator to a state to get the next state. Then another operator is applied to the new state to get the next. The process of deriving new state from the current state by applying the operator is continued till the desired state is reached. This form of problem-solving is termed 'state space' approach.

The subject area of artificial intelligence spans a broad prospect. The study of artificial intelligence involves various knowledge representation schemes, intelligent search methods, techniques for automating machine learning and much more. The various application fields include Expert systems, Image recognition, Game-playing, Theorem-proving, Natural language processing, Robotics and many other disciplines. Artificial intelligence spreads over a wide discipline spectrum of knowledge from Cognitive science, Computer science, Mathematics, Philosophy, Psychology, and Engineering. We need to discuss some common questions regarding artificial intelligence as follows:

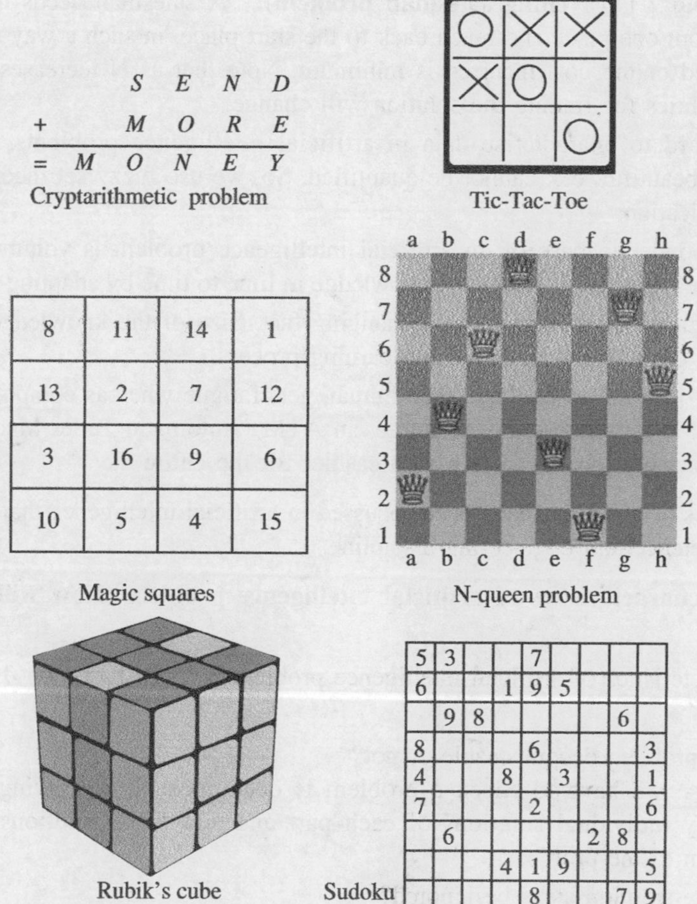

Figure 3.1 Real-world mathematical puzzles.

What are artificial intelligence problems? How are artificial intelligence problems different from others?

The answers can be summarised as follows:

1. If a solution to the problem needs symbolic representation in computers, then we can consider it as an artificial intelligence problem.

   Example: Understanding of English sentences by a computer when they are represented in the symbolic form.

2. If there is combinatorial explosion in outputting the result of a traditional problem, then we can consider such problems as artificial intelligence problem.

   Example 1 (8-Queens problem): Given a chess board of size 8 × 8, we have to place 8 queens on the chess board such that no queens contradict each other. There are 96 possible solutions.

Example 2 (Travelling salesman problem): A salesman needs to visit N cities, start from one place and reach back to the start place, in such a way that the distance travelled or the cost incurred is minimum. Note that as N increases, the number of possibilities for finding the solution will change.

3. It is hard to characterise data in artificial intelligence problems. Adjectives like white, beautiful, etc. cannot be quantified. So, we use fuzzy set theory to solve such quantification.
4. The knowledge base of an artificial intelligence problem is voluminous. This size increase happens by change of knowledge in time to time by adapting new knowledge.
5. The data or knowledge base is changing fast. Size of the knowledge base increases day by day. This is achieved by learning process.
6. If dirty jobs are to be done, then human gets fatigue whereas computers can perform without any tiredness. For example, an ATM (Automatic Teller Machine) of a bank is performing jobs as a clerk cum cashier for the entire day.

These types of questions have been focussed in artificial intelligence that are wide spread in the area of science and engineering discipline.

What are the characteristics of artificial intelligence problem (How will you analyse a problem?)

The characteristics of artificial intelligence problem is listed based on the analysis of its characteristics.

1. Is the problem decomposable or not?
   Suppose we have to check a problem is decomposable by diving into sub units. Identify individual solutions of each part and add these solutions that gave final solution to the problem.
2. Can the solution steps be ignored?
   In theorem proving, the solution step can be ignored. In bridge or chess, the solution steps are not ignorable. Also they are not recoverable. But in 8-puzzle problem, the solution steps are recoverable.
3. Is the universe predictable?
   In 8-puzzle problem or any scientific problem, the universe (all in the state space representation) is certain whereas in playing cards, the outcome is not certain. That is, the universe is not predictable.
4. Is solution to the goal absolute or relative?
   In 8-puzzle problem any path leading to the solution may be taken since the aim is to obtain goal state, and here the solution to the goal is absolute. In travelling salesman problem, we have to specify the path rather than the goal. Here the solution to the goal is relative.
5. Is the knowledge base consistent or not? If a knowledge base contains some fact A and ¬A then it is not consistent.
6. The role of knowledge.
   In 8-puzzle, chess, etc., we got a solution with the help of a knowledge base. In the

election of a politician, collecting knowledge base from newspaper or from any other resources will not be helpful to decide his winning or losing. So, in this context, knowledge base does not play any role.

7. Is the interaction with the computer necessary?
   Some problems may need human interaction. For example, for playing computer chess, computer bridge, etc., solving scientific problems through programs, human interaction is not necessary.

This chapter discusses techniques in artificial intelligence that can be used to solve real-world problems. These problems shows different characteristics and solution methodologies. These methods involve the process of defining the search space, defining start and goal states and the different paths from source to goal states. A set of rules, called production rules, solves the problem in a search space from the start state to goal state. The best path from start state to goal state is obtained by the set of production rules. At step defines the artificial intelligence problem itself followed by defining a state space with possible configurations of the relevant objects. From the initial state, the solution process may start by specifying one or more states in the state space. Considering these states that would be acceptable as solutions to the problem or would lead to a solution are considered. If it is a solution, then the state is called goal state. The set of production rules describes the action on operators. The following section discusses the production system that can solve an artificial intelligence problem.

3.1 Production Systems

The production system is a model of computation that has proved particularly important in artificial intelligence, both for implementing search algorithms and for modelling human problem-solving. A production system consists of following components (Figure 3.2).

- Global database (working memory)
- A set of rules (production rules)
- A goal

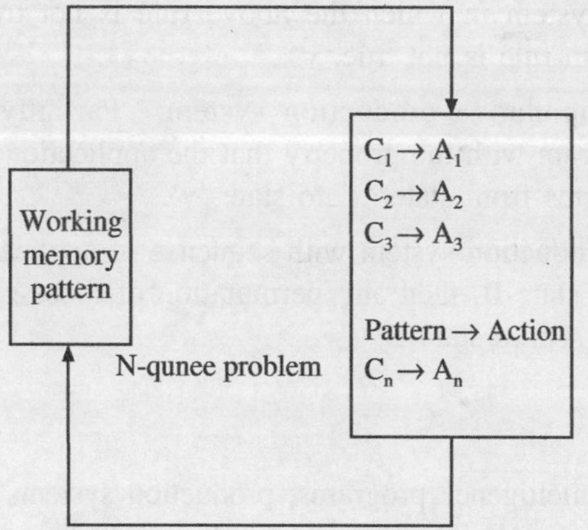

Figure 3.2 Production system.

Production rule set is of the form A→B that consists of current problem state in the left hand side and next problem in the right hand side. That is, B is the result or generated output state from state A. The working memory is obtained from knowledgebase of the production system which contains no control or programming information. During the execution how many rules need to be activated? In real-world problem-solving several rules may be activated in the working memory. A control strategy will reduce this problem by well written hierarchical arrangement. For each cycle of the control program, an exhaustive search is carried out in the production rule set. Remember, the simple production system will not store results of any computations in future use. So, there is no learning process required. An ideal production system has no conflicts in the production rules. It should ensure that no conflict will occur when new production rules are added into the knowledgebase. When a conflict occurs in the rule with existing rules while adding into the knowledgebase, it should be resolved in the following ways:

- Rules are created with priority stamps and fire a rule based on obtained priority value. This is a simple technique that have some order of firing frequency. These technique is adapted by many of the expert systems because of its simplicity.
- Fire a rule with longest matching strategy. This is obtained by counting the number of matching constraints. This is used by the strict conditions that have discrimination power provided in the knowledgebase.
- Fire a rule that is recently used which follows the path of greatest activity.

There are following four basic types of production systems:

1. A monotonic production system: Monotonic Production System (MPS) is a program in which the application of a rule never prevents later application of the another rule that could also have been applied at the time when first rule was selected.
2. A non-monotonic production system: Non-monotonic production system is a program in which the application of a rule prevents the later application of another rule which may not have been applied at the time when first rule was selected, that is, it is a system in which the above rule is not true, that is, the monotonic production system rule is not true.
3. A partially commutative production system: Partially commutative production system is a program with the property that the application of those rules is allowable and also transforms from state 'x' to state 'y'.
4. It is a type of production system with sequence of application rules. The rules are in the form A into state B, then any permutation of those rules that is acceptable also transforms state A into state B.

Advantages

For structuring artificial intelligence programs, production systems work as an excellent tool. This technique is highly modular, as individual rules could be added, modified or removed independently. The production rules are expressed in a natural form while it is used in problem-solving. So, the statements contained in the knowledgebase should be recorded for an expert thinking out loud.

Disadvantages

The fact that may be very difficult to analyse the flow of control within a production system is due to the individual rules that do not call each other. This is a major drawback of the production system building. The production systems always describe the sort of operations that are being performed while searching solution to a problem. Several production systems exist for any type of problems. The efficiency varies from problem to problem. In a wide sense we may not feel that there is any relation between the problem and production system. But it is not so. For ignorable problems, we make use of partially commutative and monotonic production systems. These systems are important for humans from the viewpoint of implementation. As these systems are implemented without the ability of backtracking to previous states, efficiency can be increased because the system will not keep track of the changes made in the search process. In systems like puzzle games where the changes can be reversed and the order of operation is not critical, we can make use of the monotonic partially commutative systems. Production systems which are not partially commutative are useful for many problems in which irreversible changes occur, such as chemical analysis. In such systems the order of operation is very important and the correct decisions need to be taken at the first encounter itself.

Production system can be illustrated with some examples discussed in the successive subsections.

3.1.1 Travelling Salesman Problem

Enumerate the global database for the following problem and find a solution:

A salesman must visit each of the five cities as shown in Figure 3.3(a). There is a road between every pair of cities and the corresponding distances are given. The problem is to find a minimum path that he visits each of the cities only once and returns to the starting city.

The goal is to find the minimum path. Rule is to branch out from a node which has minimum distance from the root node.

Solution

Let A is selected as starting node. We can specify the global database in the form of a graph. The minimum path is A→C→D→E→B→A and Cost = 34. Figure 3.3(b) shows the global database for the travelling salesman problem.

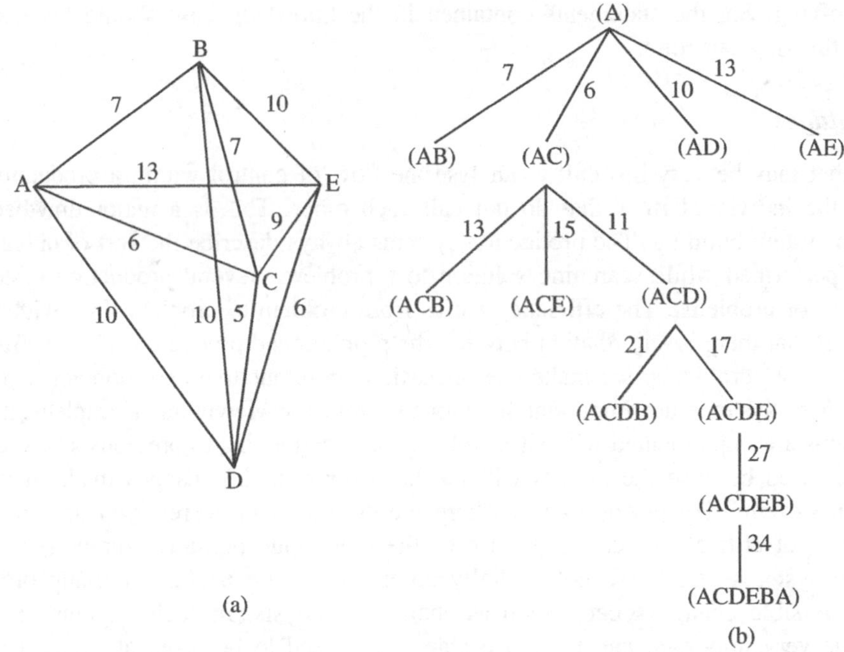

Figure 3.3 (a) 5 Cities connected (b) Global database.

3.1.2 Water Jug Problem

Specify the global database, set of rules for the following water-jug problem:

There are two jugs of 4 litre and 3 litre capacities. Neither has any maskings made on it. There is a pump to fill the jugs with water. How to obtain exactly two litres of water in the 4 litre jug.

Solution

Rules of the problem are:

1. We can fully empty the jug onto the ground or from one jug to the other.
2. We should not fill up a jug or empty a jug half way or fractionally from the pump or onto the ground.

The global database can be represented by an ordered part of number (x, y) for $x = 0, 1, 2, 3, 4$ and $y = 0, 1, 2, 3$ which represent the amount of water in 4 litre jug and 3 litre jug respectively.

The applicable rules of the production system are:

1. Fill 4 litre jug if it is partially filled.
   $((x, y) / x < 4)) \rightarrow (4, y)$
2. Fill 3 litre jug if it is partially filled.
   $((x, y) / y < 3)) \rightarrow (x, 3)$
3. Pour d litres of water from 4 litre jug onto the ground.
   $((x, y) / x > 0)) \rightarrow (x - d, y)$

4. Pour d litres of water from the 3 litre jug onto the ground.
   $((x, y) / y > 0)) \rightarrow (x, y - d)$
5. Empty 4 litre jug onto the ground.
   $((x, y) / x > 0)) \rightarrow (0, y)$
6. Empty 3 litre jug onto the ground.
   $((x, y) / y > 0)) \rightarrow (x, 0)$
7. Pour water from 3 litre jug into 4 litre jug until it is full.
   $((x, y) / x + y \geq 4)$ and $(y > 0)) \rightarrow (4, 4 - (4 - x)$
8. Pour water from 4 litre jug into 3 litre jug until it is full.
   $(x, y) / (x + y \geq 3)$ and $(x > 0)) \rightarrow (x - (3 - y), 3)$
9. Pour all water from 3 litre to 4 litre.
   $((x, y) / (x + y \leq 4)$ and $(y > 0)) \rightarrow (x + y, 0)$
10. Pour all water from 4 litre to 3 litre.
   $((x, y) / (x + y \leq 3)$ and $(x > 0)) \rightarrow (0, x + y)$

Table 3.1 shows the solution for water jug problem derived from the set of rules discussed above.

TABLE 3.1 Solution to the water jug problem

Rule no.	*Result*	
	A	B
Initial State	0	0
1	4	0
8	1	3
6	1	0
10	0	1
1	4	1
8	2	3

3.1.3 The Knight's Tour Problem

In chess game, a knight can move two squares either horizontally or vertically followed by one square in an orthogonal direction as long as it does not move off the board. Find out the series of legal moves in which the knight lands on each square of a chess board exactly once. Given a 3 × 3 chess board.

Solution

The 16 production rules are rule number, condition and action given as follows:

1. Knight on square 1 → Move Knight to 8
2. Knight on square 1 → Move Knight to 6
3. Knight on square 2 → Move Knight to 9
4. Knight on square 2 → Move Knight to 7

5. Knight on square 3 → Move Knight to 4
6. Knight on square 3 → Move Knight to 8
7. Knight on square 4 → Move Knight to 9
8. Knight on square 4 → Move Knight to 3
9. Knight on square 6 → Move Knight to 1
10. Knight on square 6 → Move Knight to 7
11. Knight on square 7 → Move Knight to 2
12. Knight on square 7 → Move Knight to 6
13. Knight on square 8 → Move Knight to 3
14. Knight on square 8 → Move Knight to 1
15. Knight on square 9 → Move Knight to 2
16. Knight on square 9 → Move Knight to 4

The goal is to place the knights in the 3 × 3 board.
A production system solution to the 3 × knight's tour problem.

The solution to Knight tour problem is illustrated in Table 3.2.

TABLE 3.2 Solution to the Knight tour problem

Iteration no.	*Current square*	*Goal square*	*Conflict rule no.*	*Fire rule*
0	1	2	1, 2	1
1	8	2	13, 14	13
2	3	2	5, 6	5
3	4	2	7, 8	7
4	9	2	15, 16	15
5	2	2	Halt	

3.1.4 Missionaries and Cannibals Problem

Three missionaries and three cannibals are needed to cross a river. They made an agreement that all would like to get to the other side. But there is no surety for missionaries that the cannibals have agreed to. To manage the trip missionaries want to cross the river in such a way that number of missionaries on either side of the river is never less than the number of cannibals who are on the same side. There is only one boat available for their trip that holds two people at a time. How can everyone get across the river without the missionaries risking being eaten?

The solution is to find a sequence of travelling path from one bank to other using a boat through the river. We can represent 'M' as Missionary and 'C' as Cannibal while satisfying the following production rules:

1. (0, M) : One missionary travelling the boat from bank-1 to bank-2
2. (M, 0) : One missionary travelling the boat from bank-2 to bank-1

3. (M, M) : Two missionaries travelling the boat from bank-1 to bank-2
4. (M, M) : Two missionaries travelling the boat from bank-2 to bank-1
5. (M, C) : One missionary and one Cannibal travelling the boat from bank-1 to bank-2
6. (C, M) : One missionary and one Cannibal travelling the boat from bank-2 to bank-1
7. (C, C) : Two Cannibals travelling the boat from bank-1 to bank-2
8. (C, C) : Two Cannibals travelling the boat from bank-2 to bank-1
9. (0, C) : One Cannibal travelling the boat from bank-1 to bank-2
10. (C, 0) : One Cannibal travelling the boat from bank-2 to bank-1

Applying a sequence of production rules we have reached a solution to the problem. Table 3.3 shows the solution.

TABLE 3.3 Rules applied and their sequence in missionaries and cannibals problem

After application of rule	*Persons in the river bank-1*	*Persons in the river bank-2*	*Boat position*
Start state	M, M, M, C, C, C	0	bank-1
5	M, M, C, C	M, C	bank-2
2	M, M, C, C, M	C	bank-1
7	M, M, M	C, C, C	bank-2
10	M, M, M, C	C, C	bank-1
3	M, C	C, C, M, M	bank-2
6	M, C, C, M	C, M	bank-1
3	C, C	C, M, M, M	bank-2
10	C, C, C	M, M, M	bank-1
7	C	M, M, M, C, C	bank-2
10	C, C	M, M, M, C	bank-1
7	0	M, M, M, C, C, C	bank-2

3.2 State Space Representation

We know that a production system consists of a global database, a set of rules and a goal. The initial state can be represented in a way in which the computer can understand. This representation is known as state space representation. By taking up an applicable rule, we can derive another state in the solution path of the problem. The same has to be represented in the computer in a form in which it can understand. Thus, all intermediate configurations which are obtained by applying the rules in a production system is known as set of states in the global database (space), which can be represented in the form in which computer can understand. Following sections discuss problem examples using state space representation.

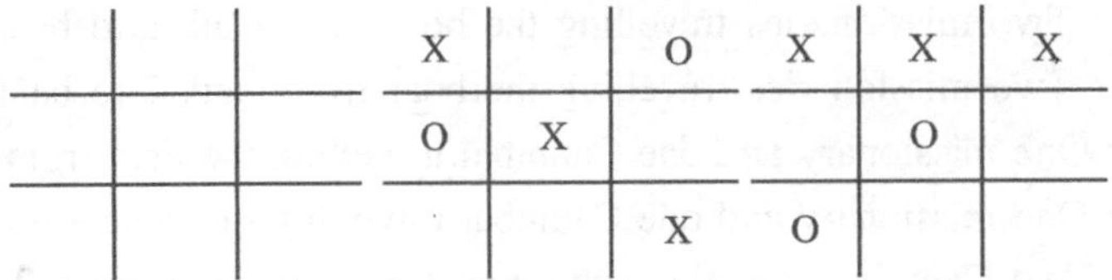

Figure 3.4 Initial state and two different goal states.

3.2.1 State Space Search

Problem in a state space is represented in terms of states and operators that change states. A state space consists of the following:

1. Representation of a state in the system can be in the form of data structure. For example, in a board game the board represents the current state of the game.
2. Set of operators that can change state to state. In a board game, the operators are the legal moves from any given state. Often the operators are represented as programs that change a state representation to represent the new state.
3. An initial state.
4. Set of final states that may be desirable in some cases and undesirable in others. This set is often represented implicitly by a program that detects terminal states.

3.2.2 Tic-Tac-Toe as a State Space

A good representation of state space model can be illustrated through games such as Tic-Tac-Toe. Each state of a game can be described through the contents of board and the next turn of the player. The tic-tac-toe board has an array of 9 cells, each of which may contain an * or o or be empty. Figure 3.4 represents the initial state and two different goal states.

State:
Next move of a player: * or o.
Configuration of the board: Board configuration may change according to each move of the player.
Operators: Change an empty cell to * or o.
Start state: Initially board is empty followed by *'s turn.
Terminal states: Three *'s in a row; Three o's in a row; All cells full.

The starting state of the tic-tac-toe can be represented as {0,0,0,0,0,0,0,0,0}. The goal state or winning combination will be board position having 'O' or 'X' separately in the combination of {1,2,3}, {4,5,6}, {7,8,9}, {1,4,7}, {2,5,8}, {3,6,9}, {1,5,9}, {3,5,7} element values (since these are tile positions). Hence, two goal states represented in Figure 3.4 can be represented as {2,0,1,1,2,0,0,0,2} and {2,2,2,0,1,0,1,0,0}.

When board position is satisfying the winning condition the game enters into a stop condition in the state space. The valid transitions of this problem are simply putting '1' or '2' in any of the element positions containing 0. Normally, all the valid moves are defined and stored and selection of move is from the store only. Tic-tac-toe having 3^9 valid transitions in the store.

3.2.3 Missionaries and Cannibals Problem

Recall Missionaries and Cannibals problem again. We can solve this problem in state space representation.

State representation

1. BOAT position: Original (T) or final (NIL) side of the river.
2. Number of Missionaries and Cannibals on the original side of the river.
3. Start is (T 3 3); Goal is (NIL 0 0).

Operators

1. (MM 2 0) Two Missionaries cross the river.
2. (MC 1 1) One Missionary and one Cannibal.
3. (CC 0 2) Two Cannibals.
4. (M 1 0) One Missionary.
5. (C 0 1) One Cannibal.

Missionaries and Cannibals search graph is shown in Figure 3.5.

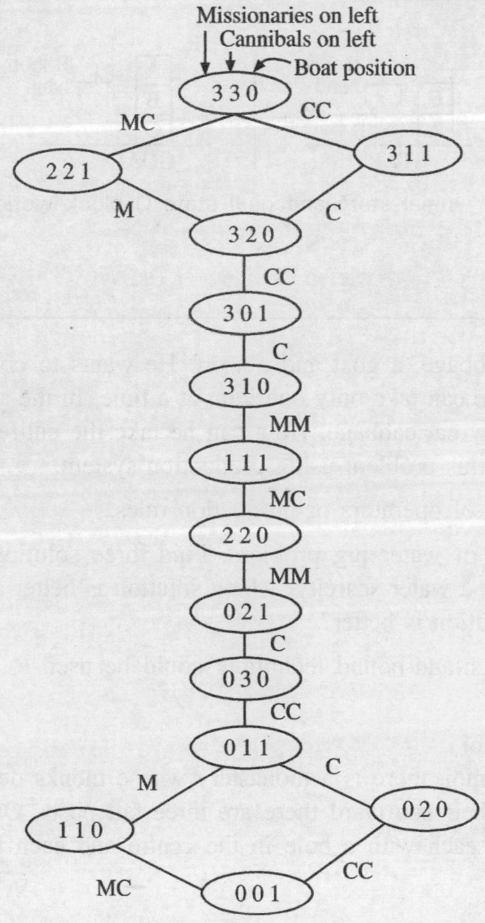

Figure 3.5 Missionaries/Cannibals search graph.

3.2.4 Block World Problem

The problem can be represented as, "There are some cubic blocks, out of which, each is placed either on the table or on other block forming a particular configuration. We have to move the blocks to arrange them to form some other given configuration by applying minimum number of moves. The requirement of moving a block is that only the top block from a group can be shifted in a move. It can be placed either on the table or on the top of any other block".

Solution

The problem can be represented in a state space representation. The containers are placed one over the other in several groups. To load these blocks in ship or train, a crane or a robotic arm is used to arrange these containers in different groups. Start and goal states are given in Figure 3.6. The problem can be solved by moving the blocks from one location to the other. Sequence of actions are given as follows:

1. Grab C	2. Pickup C	3. Place on table C
4. Grab B	5. Pickup B	6. Stack B on C
7. Grab A	8. Pickup A	9. Stack A on B

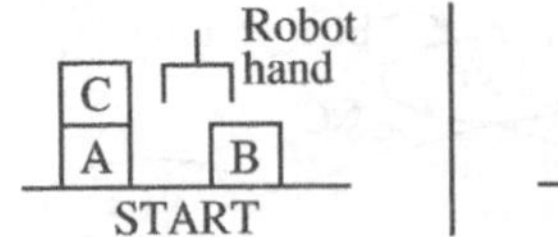

Figure 3.6 Initial state and goal state O block world problem.

3.3 Exercises

1. A man carries a cabbage, a goat and a wolf. He wants to cross a river and there is a boat available but he can take only one item at a time. In the absence of man, wolf may eat goat or goat may eat cabbage. How can he take the entire three into the other side of the river? Solve this problem using production system.
2. Explain the concept of operators or production rules.
3. Define the concept of water jug problem. Find three solutions to water jug problem. Compare, if there is a water scarcity, which solution is better and if there is no scarcity of water, which solution is better?
4. Describe how branch-and-bound technique could be used to find the shortest path in water jug problem?
5. **The Tower of Hanoi**
   Somewhere near Hanoi, there is a monastery whose monks devote their lives to a very important task. In their courtyard there are three tall posts. One of these posts is a set of sixty four disks, each with a hole in the centre and each having a different radius.

When the monastery was established, all of the disks were on one of the posts, each disk resting on the one just larger than it. The monks task is to move all the other disks that must be on one of the poles. Only one disk may be moved at a time and all the other disks are to be placed on top of a smaller disk. The third pole can of course, be used as a temporary resting place for the disks. What is the quickest way for the monks to accomplish their mission?

6. **Missionaries and Cannibals problem**
   Three missionaries and three cannibals are all together on one side of a river. There is one boat that can carry either one or two persons. How can they all cross the river such that there are never less missionaries than cannibals on either side of the river? How would you represent a state of that search problem? Which abstraction did you use in order to obtain an adequate representation?
7. **The Monkey and bananas problems**
   A hungry monkey finds himself in a room in which a bunch of bananas is hanging from the ceiling. The monkey unfortunately, cannot reach the bananas. However, in the room there are also a chair and a stick. The ceiling is just at the right height so that a monkey standing on a chair could knock the bananas down with the stick. The monkey knows how to move around, carry other things around, reach for the bananas, and wave a stick in the air. What is the best sequence of actions for the monkey to take to acquire lunch?
8. **The 8 Queens problem**
   The goal of Eight Queens problems is to place 8 Queens on a chess board such that no queen attacks any other queen (A queen attacks any piece in the same row, column or diagonal).
9. Consider 4 queens problem, where four queens are positioned on a 4 × 4 chess board such that no queen threatens any other queen.
   (a) A backtracking procedure positions the queens row by row in the leftmost non-threatening column. Which states of the search space are generated until a solution is reached?
   (b) The procedure is enhanced by the following heuristics: Let $D(i; j)$ be the length of the larger diagonal for the position $(i; j)$. In every row i, the column j is chosen, for which $D(i; j)$ is minimum. Which states of the search space are generated until a solution is reached?
10. **Block World Problems**
    There are some cubic blocks, out of which, each is placed either on the table or on other block forming a particular configuration. We have to move the blocks to arrange them to form some other given configuration by applying minimum number of moves. The requirement of moving a block is that only the top block from a group can be shifted in a move. It can be placed either on the table or on top of any other block.
11. **Magic Square**
    It is given a square of same number of rows and columns, where consecutive numbers are filled in blank spaces of each row and column such that numbers in each row, column and two diagonals, add up to the same number. The numbers have to be consecutive and one number can be used only once.

CHAPTER 4

Heuristic Search Techniques

Many of the problems that fall within the area of artificial intelligence are too complex to solve by direct techniques. These problems must be attacked by appropriate search methods armed with powerful direct techniques that are available to guide the search. These methods are varieties of heuristic search. These methods are independent of any particular task or problem domain. Their efficiency highly depends on the domain-specific knowledge rather than all applications. Their process is so vulnerable. That these technique are so-called *weak methods*.

A utility that is mapped from problem state descriptions to channel of desirability is called heuristic function. Usually, it is represented as a number that guides the most gainful direction. That is, heuristic is a technique that improves the efficiency of the search process. This chapter discusses the various heuristic methods used in artificial intelligence.

4.1 Generate and Test

This is a common technique used to solve problems. For some problem, generate a possible solution and test whether it is a solution to that problem. Generate and test can also operate by generating solution randomly, but there is no guarantee that a solution will ever be found. Procedure for generate and test is as follows:

1. Generate a possible solution which can either lead to the problem space or a path from the initial state.
2. Test to see if the generated solution is a real solution by a comparison to the state reached with the set of goal states.
3. Return, if it is a real solution. Otherwise repeat from step 1.

Generate and Test is not a powerful technique and is not used to solve complex problems. This is because it makes some guesses for a solution. We can use powerful techniques for complex problems.

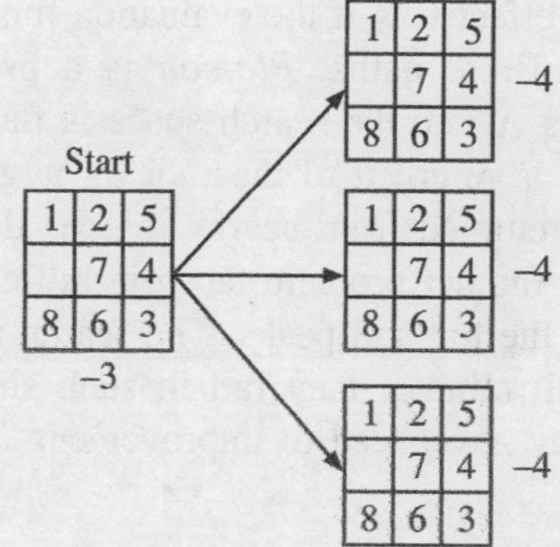

Figure 4.1 Local maxima values.

4.2 Hill Climbing

New techniques are generated when a high quality heuristic function is available for evaluating states. Hill climbing is such a problem-solving technique that is used to solve artificial intelligence problems. Most of the time no other useful knowledge is available while solving the problem. Hill-climbing strategies expand current state in the search and evaluate its children. For further expansion, the best child is selected, then are removed its siblings and its parent. The search process is terminated when it reaches a state that is better than any of its siblings. A major problem of hill climbing strategy is its tendency to stop at local maxima (Figure 4.1). The algorithm halts when it reaches a state that has a better evaluation than any of its children. Suppose this is not a goal but just a local maxima, the algorithm fails to find a solution. The best solution will be that state space where objective function has maximum value (global maximum). The attainment of local maxima and global maxima in hill climbing technique is given in Figure 4.2.

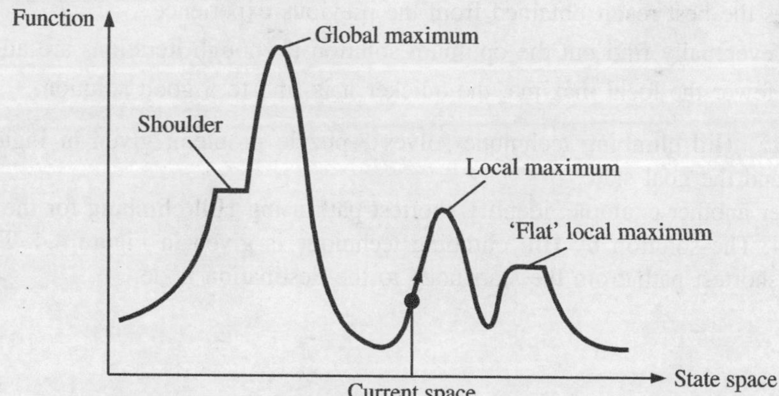

Figure 4.2 Local maxima and global maxima.

Hill climbing can be used effectively if the evaluation function is sufficiently informative to avoid local maxima and infinite paths. *Plateau* is a problem faced in hill climbing technique that sometimes occurs. When the search space is flat, or sufficiently flat, a *plateau* is encountered. Then the value is returned to the root by target function. This is impossible to differentiate from the value returned from nearby regions due to the accuracy used by the machine to represent its value. Another problem faced in hill climbing is ridges, which means no steep sloping sides between the top and peaks. The search makes little progress unless the top is directly reached. The hill climber may fail in such situations to determine in which direction it should step, and may never lead to improvement.

Function hill climbing

```
inputs: problem, a problem
local variables: current, a node
          next, a mode
current ← Make-Node(Initial-State[problem])
loop do
          next ← a highest-valued successor of current
          if Value[next] ← Value[current] then return current
          current ← next
returns a solution state
end
```

Random restart hill climbing is a remedy that improves the hill climbing technique. The improvement techniques are as follows:

- Generates different starting points when no progress can be made from the previous point.
- Saves the best result obtained from the previous experience.
- Can eventually find out the optimum solution if enough iterations are allowed.
- The fewer the local maxima, the quicker it is find to a good solution.

Example 4.1: Hill climbing technique solves 8-puzzle problem, given in Figure 4.3, with install state and the goal state.

Consider another example, identify shortest path using Hill climbing for the graph given in Figure 4.4. The solution by Hill climbing technique is given in Figure 4.4. The dark line leads to the shortest path from the start node to the destination node.

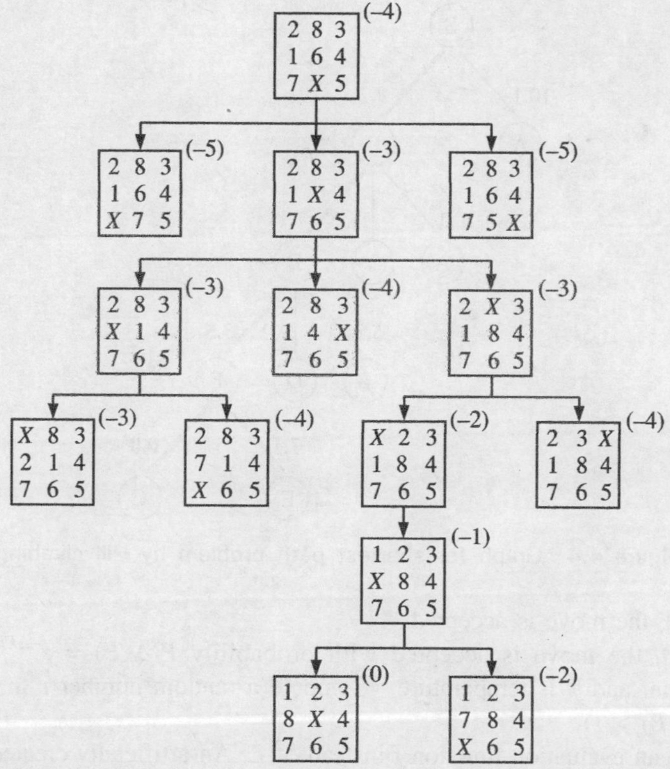

Figure 4.3 Solution to 8-puzzle problem using Hill climbing.

4.3 Simulated Annealing

Annealing technique is a thermal process for obtaining low-energy state of a solid in a heat bath. The process consists of increasing temperature (solid melts by increasing the temperature of heat bath to a maximum value) and decreasing temperature (particles arrange themselves in the ground state of the solid by decreasing the temperature of heat bath). The particles are arranged in an extremely structured lattice, for which the corresponding energy is minimum in the ground state of solids. The ground state is obtained by sufficient slow cooling and the maximum value of sufficiently high temperature.

In order to improve the hill climbing, we can incorporate a modification in this algorithm using simulated annealing or tabu search. The technique by which a metal is heated to a very high temperature, and later allowed to cool slowly, thereby relieving the internal stresses, is called annealing. Kirkpatrick proposed this method in 1983. Simulated annealing is analogous to hill climbing in which it is possible to make some moves that are locally non-optimum. By analogy with statistical thermodynamics, it is assumed that the goal is to minimise the 'energy' E of a system. If move causes a change ΔE,

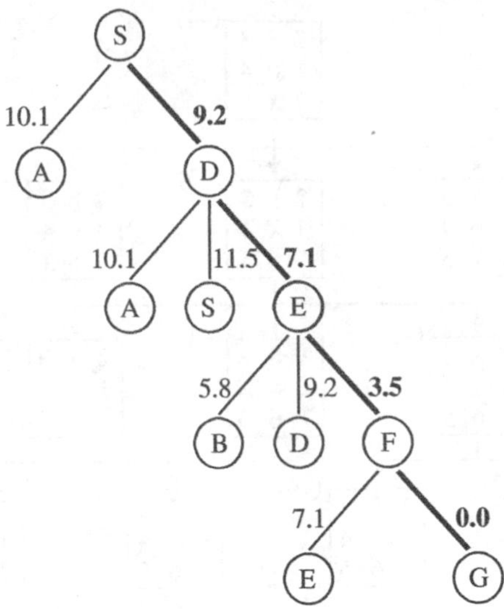

Figure 4.4 Graph for shortest path problem by Hill climbing.

If $\Delta E \leq 0$, the move is accepted.

If $\Delta E > 0$, the move is accepted with probability $P(\Delta E) = e^{-\Delta EkT}$ where k is the Boltzmann constant and T is temperature. (Compute a random number r in [0, 1] and accept the move if $P(\Delta E) > r$)

We can use an evaluation function f instead of E. An artificially created 'temperature' is high enough for many initial moves. This initial temperature is accepted and gradually reduced to zero. The steps involved in simulated annealing are as follows:

1. **Initialise:** Start with a random initial placement, means initialise a very high 'temperature'.
2. **Move:** Disturb the placement through a defined move.
3. **Calculate score:** Calculate change in the score due to the move made.
4. **Choose:** Accept or reject the move with respect to the change in score. The acceptance probability depending on the present 'temperature'.
5. **Update and repeat:** Modify the temperature value by its lowering. Go back to step 2.

When the 'freezing point' is reached, process is terminated. Simulated annealing procedure is given as follows:

Function simulated-annealing (problem, schedule)

Current ← Make-Node(Initial-State[problem])

For t ← 1 to ∞

  T = schedule[t] // T goes downwards.

  if T = 0 then **return** current

```
    next ← Random-Successor(current)
    ΔE ← f-Value[next] - f-Value[current]
    if ΔE > 0 then current ← next
    else current ← with probability e^(ΔE/T)
End
```

Simulated annealing is a well-suited technique for problems that are struggled with some inertia like:

- A problem that required very large memory with number of variables and possible states. This makes in-feasibility in the search process.
- The continuous process makes a 'frustration' due to its impossibility to optimise all cost measures concurrently.
- Significant improvements from a random starting position are possible.
- It is possible to arrive numerous fine near-optimum solutions.
- The hill climbing technique gets confused by local maxima.

Computational complexity of this algorithm is proportional to N or a small power of N, while finding the exact optimum NP is often complete. Simulated annealing is used to solve travelling salesman problem, graph partitioning, matching problem, quadratic assignment, quadratic assignment, linear arrangement and scheduling problems. In engineering problems, simulated annealing is applied in VLSI: placement and routing, facilities layout, image processing and code design.

Example 4.2: Maximise the continuous function: $f(x) = x^3 - 60x^2 + 900x + 100$

A solution x is represented as a string of 5 bits. The initial solution is 10011 ($x = 19$, $f(x) = 2399$) with two testing scenarios like initial temperature $T_0 = 500$ and $T_0 = 100$. The cooling constant, $T = 0.9$. The search is stored for finding out the best solution from the current solution. While searching, parameters used are temperature and the number of iterations performed at each temperature.

For $T = 500$ and initial solution (10011), output is in Table 4.1.

TABLE 4.1 First scenario for $T = 500$

T	*Move*	*Solution*	*f*	*Δf*	*Move*	*New neighboring solution*
500.0	1	00011	2287	112	yes	00011
450.0	3	00111	3803	< 0	yes	00111
405.0	5	00110	3556	247	yes	00110
364.5	2	01110	3684	< 0	yes	01110
328.0	4	01100	3998	< 0	yes	01100
295.2	3	01000	3972	16	yes	01000
265.7	4	01010	4100	< 0	yes	01010
239.1	5	01011	4071	29	yes	01011
215.2	1	11011	341	3728	no	01011

TABLE 4.2 Second scenario for $T = 100$

T	*Move*	*Solution*	*f*	Δf	*Move*	*New neighboring solution*
100	1	00011	2287	112	no	10011
90.0	3	10111	1227	1172	no	10011
81.0	5	10010	2692	< 0	yes	10010
72.9	2	11010	516	2176	no	10010
65.6	4	10000	3236	< 0	yes	10000
59	3	10100	2100	1136	yes	10000

For $T = 500$ and initial solution (10011), output is in Table 4.2.
When temperature is not high enough, algorithm gets stuck.

4.4 Search Techniques

Many search techniques are used in artificial intelligence operating on graphs and major search techniques are discussed in this section. Evaluation functions are keys in the search technique. Some of the evaluation functions are:

$h^*(n)$ = Cost of the shortest path between node n and a goal node (usually not known).

$h(n)$ = Estimated cost to get to a nearest goal from node n.

$g(n)$ = Cost to get to node n from the start node via the lowest-cost path found so far.

These functions or their extracts are used for identifying an item in a problem domain. Let us have a discussion about graphs, which leads to search algorithms.

A graph can be defined as $G = (V, E)$, where

V : Set of vertices

E : Set of edges connecting the vertices

An edge $e = (u, v)$ is a pair of vertices

In a directed graph:

The in-degree of a node v is the number of distinct edges $(w, v) \in E$.

The out-degree of a node v is the number of distinct edges $(v, w) \in E$.

A node with in-degree 0 is a root.

There are two types of representation for graphs—Adjacency list and Adjacency matrix. Adjacency matrix M with entries for all pairs of vertices, $M[i, j]$ = True if there is an edge (i, j) in the graph and $M[i, j]$ = False if there is no edge (i, j) in the graph. Space complexity of adjacency representation is $O(|V|^2)$. The Adjacency list of a vertex v: a sequence of vertices adjacent to v that represents the graph by adjacency lists of all its vertices. Space complexity of this representation is $O(|V + E|)$. The graph representation is highly dependable in the implementation time. Figure 4.5 shows an example for adjacency matrix and adjacency list representation respectively.

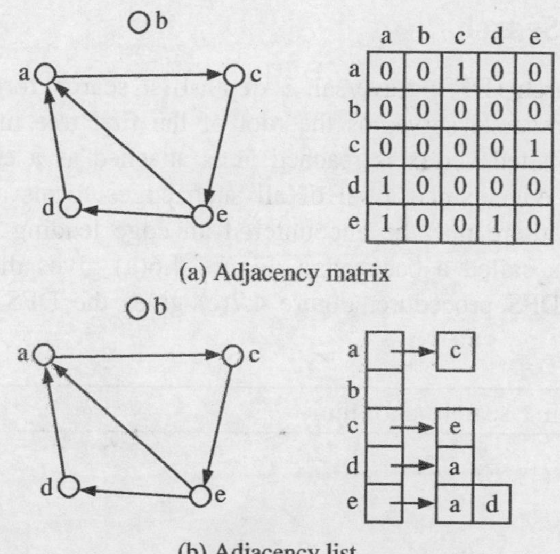

Figure 4.5 (a) Adjacency matrix representation (b) Adjacency list representation.

4.4.1 Graph Search

A large variety of problems can be represented by a graph and its solutions can be considered as defining specific nodes. Solving the problem is reduced to searching the graph for those nodes starting from an initial node and each transition in the graph corresponds to a possible action that ends when reaches a final solution node. For a problem, it is defined by

1. An initial state
2. A successor function $S(X)$ = Set of action-state pairs
3. A set of specific nodes: the goals
4. A path cost (additive)

A solution is the sequence of actions leading from the initial state to a goal. Basic graph search algorithm is given in Algorithm 1. The set of all possible paths can be represented as a tree. Graphs are turned into trees by duplicating nodes and breaking cyclic paths.

Algorithm 1 Graph search algorithm

Input : a graph $G = (V, E)$
set of start nodes
Boolean procedure goal (n) that tests if n is goal node.
frontier = {< *s* >: sisastartnode}
while *frontier* is not empty:
  select and remove path < n_o,...,n_k > for frontier;
    if goal (n_k)
      return < n_o, ..., n_k >
    for every neighbour n of n_k
    add < n_o, ..., n_k > to *frontier*
endwhile

4.4.2 Depth First Search

During Depth First Search (DFS) traversal, a depth-first search forest is also constructed. Starting vertex of the traversal serves as the root of the first tree in such a forest. For the first time, if a new unvisited vertex is reached, it is attached as a child to the vertex. This edge is called the tree edge, since a set of all such edges forms a forest. Other than its immediate predecessor, there may be encountered an edge leading to a previously visited vertex. Such an edge is called a back edge. Figure 4.6(a) gives the DFS movement, and Algorithm 2 gives the DFS procedure. Figure 4.7(c) gives the DFS movement for a graph shown in Figure 4.7(a).

Algorithm 2 Depth first search algorithm

```
Input : Graph G = (V, E)
dfs(G)
begin
  count = 0;
  for (each vertex v in V)
     if (v is marked with a 0)
        dfs(v);
dfs(v);
Visits recursively all the unvisited vertices connected to a vertex v and
assigns them a number in the order they are encountered.
count = count + 1;
for (each vertex w in V adjacent to v)
  if (w is marked with 0))
     dfs(w);
end
```

This algorithm takes the time proportional to the size of the data structure used for representing the graph. Thus, if adjacency matrix is used, the time taken by this algorithm is $O(|V^2|)$. If adjacency list is used it becomes $O(|V| + |E|)$. Where V is the number of vertices and E is number of edges in the graph.

4.4.3 Breadth First Search

Instead of depth first search tree, Breadth First Search (BFS) forest is generated. Here also, a tree edge is generated as before. If an edge leads to a previously visited vertex other than its immediate predecessor the edge is called a cross edge. BFS movement is given in Figure 4.6(b).

New states in the order of their distance from the initial state is generated by breadth-first search. Before examining the states at level $i + 1$, all states at level i are examined. Major advantages of BFS include a guarantee to find an optimum solution (in terms of smallest number of steps to reach the goal) and it can always find a goal node if one exists (complete). One of the major disadvantages is high storage requirement (exponential with tree depth). The time complexity of algorithm is same as depth first search.

Graphically we can represent the control moves in DFS and BFS shown in Figure 4.6(a) and (b) respectively. BFS algorithm is listed in Algorithm 3. The BFS move for a graph (Figure 4.7(a)) is given in Figure 4.7(b).

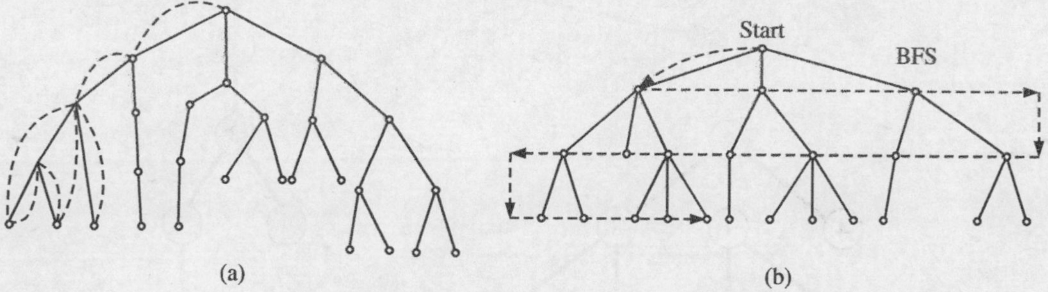

Figure 4.6 (a) DFS moves (b) BFS moves.

Algorithm 3 Breadth first search algorithm

```
Input : Graph G = (V, E)
bfs(G)
begin
  count = 0;
  for (each vertex v in V)
    if (v is marked with a 0)
      bfs(v);
bfs(v);
  Visits all the unvisited vertices connected to a vertex v and assigns
  them a number in the order they are encountered.
count = count +1; mark v with count and initialize a queue with v;
while (the queue is not empty)
  for (each vertex w in V adjacent to the fronts vertex v)
    if (w is marked with 0)
      count = count + 1; add w to queue;
    remove vertex v from the front of queue;
end
```

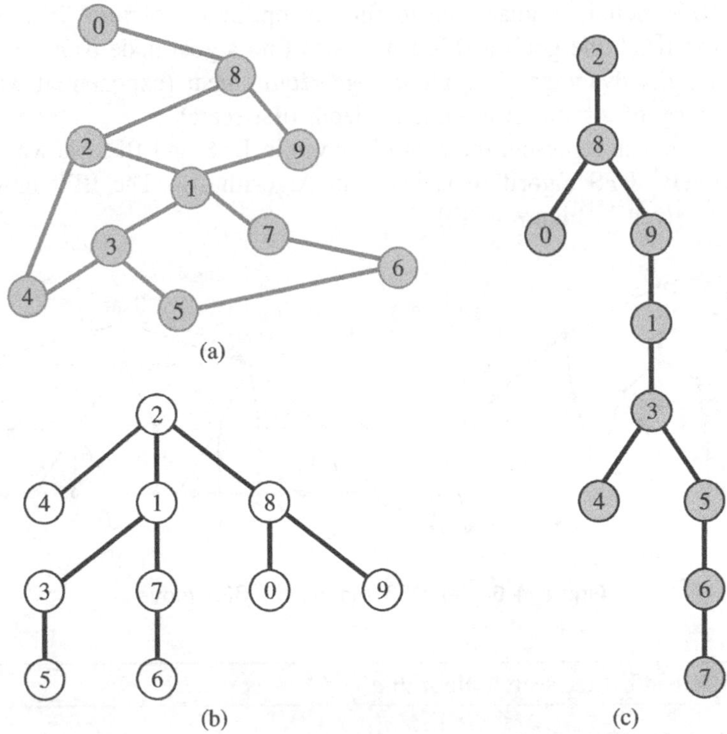

Figure 4.7 (a) A graph (b) BFS tree (c) DFS tree.

4.4.4 Iterative Deepening Search

Iterative Deepening Search (IDS) or Iterative Deepening Depth First Search (IDDFS) increases the efficiency by combining space efficiency of depth first search and fastness of breadth first search (for nodes closer to root). Iterative deepening search procedure is given in Algorithm 4. The time complexity of algorithm is same as depth first search.

For example, consider the graph in Figure 4.7(a). Consider node 0 is the source and node 6 is the target then IDS move is

Depth 0: 0

Depth 1: 0, 8

Depth 2: 0, 8, 2, 9

Depth 3: 0, 8, 2, 9, 1, 4

Depth 4: 0, 8, 2, 9, 1, 4, 3, 7

Depth 5: 0, 8, 2, 9, 1, 4, 3, 7, 5, 6

Algorithm 4 Means end analysis algorithm

```
IDS(source, target, max-depth)
  for limit from 0 to max-depth
    if DLS(src, target, limit) = true
      return true
  return false

Procedure DLS(source, target, limit)
  if (source = target)
    return true;
  //if reached the maximum depth,
  //stop recursing.
  if (limit ≤ 0)
    return false;
  if reach adjacent i of source
    if DLS(i, target, limit)
      return true
  return false
```

4.4.5 Uniform Cost Search

Uniform cost search algorithm is different from BFS and DFS because of the costs coming into play. That is, when a different cost is available for each edge of the graph, the algorithm traversing via different edges might not have the same cost and its goal is to find a path where the cumulative sum of costs is least. Starting from the root node, uniform cost search expands towards other nodes. This algorithm is useful to solve any graph/tree when the optimum cost is in demand. Using priority queue, uniform cost algorithm is implemented that gives maximum priority to the lowest cumulative cost. This algorithm becomes a BFS algorithm when the path cost of all edges is the same. Uniform cost search is optimum because at every state the path with least cost is chosen. Algorithm may enter infinite loop when the number of steps involved in searching and only concerned about path cost. Uniform cost search procedure is given in Algorithm 5.

Algorithm 5 Uniform cost search algorithm

```
1. insert the root into queue
2. while queue is not empty
   Dequeue the maximum priority element from the queue
   (If priorities are same, alphabetically smaller path is chosen)
   if the path is ending in the goal state,
     print the path and exit
   else
     Insert all the children of dequeued element, with the cumulative costs as priority
```

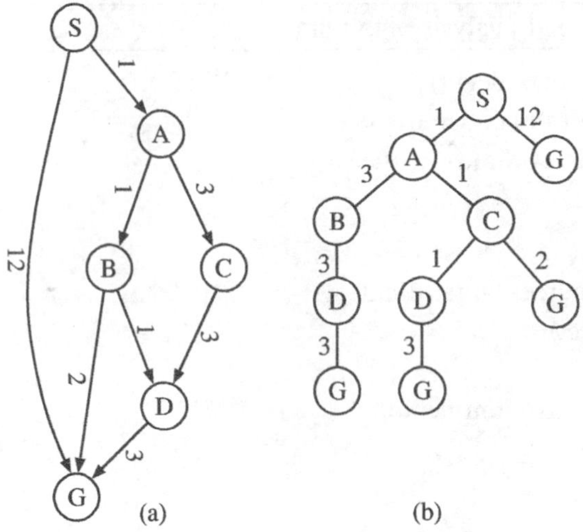

Figure 4.8 (a) A graph (b) Search tree.

For example, consider a search graph in Figure 4.8(a). The uniform cost search procedure is performed as shown in Figure 4.8(b). Applying the algorithm we will get the iterations and reach to the final goal with the priority queue as follows:

Initialisation : {[S, 0]}

Iteration 1 : {[S → A,1], [S → G, 12]}

Iteration 2 : {[S → A → C, 2], [S → A → B, 4], [S → G, 12]}

Iteration 3 : {[S → A → C → D, 3], [S → A → B, 4], [S → A → C → G, 4], [S → G, 12]}

Iteration 4 : {[S → A → B, 4], [S → A → C → G, 4], [S → A → C → D → G, 6], [S → G, 12]}

Iteration 5 : {[S → A → C → G, 4], [S → A → C → D → G, 6], [S → A → B → D, 7], [S → G, 12]}

Iteration 6 gives the final output as S → A → C → G.

4.4.6 Greedy Method

Greedy method suggests constructing a solution through a sequence of steps, each expanding a partially constructing solution obtained so far, until a complete solution to the problem is reached. On each step the choice made must be:

- Feasible, that is, it has to satisfy the problem's constraints.
- Locally optimum, that is, it has to be the best local choice among all feasible choices available on that step.
- Irrevocable, that is, once made, it cannot be changed in the subsequent steps of algorithm.

These requirements explain on each step, that it suggests a 'greedy' grasp of the best substitute available in the trust. Thus, locally optimum choices of sequence will yield a globally optimum solution to the whole problem.

Greedy search basically 'minimise estimated cost to reach a goal'. This algorithm uses a heuristic function that calculates such cost estimates

h(*n*) = *Estimated cost of the cheapest path from the state at node n to a goal state.*

Major algorithms that use greedy techniques are Prim's algorithm, Kruskal's algorithm, Dijksta's algorithm and Best first search. To identify minimum spanning tree from a graph Prim's algorithm and Kruskal's algorithm are used. Dijksta's algorithm is used for identifying shortest path and best first search is a graph search algorithm.

4.4.7 Best First Search

Best first search is a combination of depth first and breadth first searches. Best first search is a greedy method used to solve problems. We may reach a solution while using depth first algorithm without computing all nodes. Best fit search is one of the finest search methods which uses a combination of Depth First Search (DFS) and Breadth First Search (BFS). As DFS is used, the solution could be found out easily without computing all nodes. BFS gives advantage by making sure that the search does not get trapped in dead ends. Best fit search switches between BFS and DFS thus gains the advantage of both. At each step the most prominent node is selected. Whenever the node selected is not prominent it can switch the nodes by standing on the same level as that of the tree (that is, a change from DFS to BFS). On the analysis of procedure, these are no better than the previous unexpanded node and branch that are not reached. Two lists are maintained by the algorithm. One of the lists contains a set of candidates yet to explore and the other has the list of visited nodes. The best first procedure is given in Algorithm 6.

Algorithm 6 Best first search

1. Place starting node on a queue.
2. If the queue is empty, declare failure and stop.
3. if first element in the queue is goal node, declare success and stop otherwise,
4. Remove the first element from the queue, expand it, and compute estimated goal distances for each child, place the children on queue (at either end of the queue) and arrange all the queue elements in ascending order corresponding to the goal distances from the start node of the queue.
5. **return** to step (2)

Example 4.3: Consider the following graph (Figure 4.9).

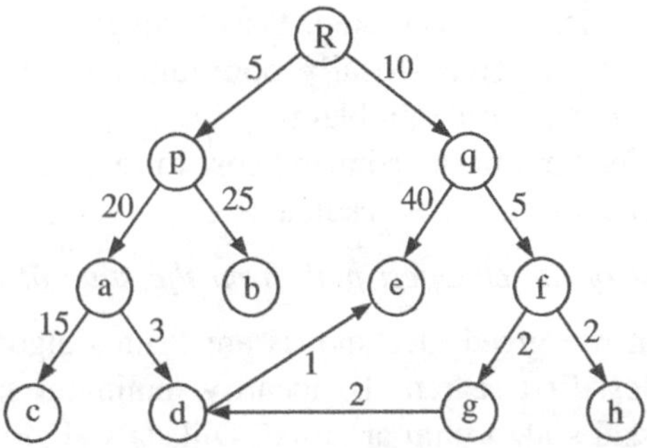

Figure 4.9 A directed graph which shows distance between a pair of nodes.

The goal node is *e*, and the cost is given.
The following steps are carryed out by Best-First search:

1. ((R, 00))
2. ((Rp, 24) (Rq, 40))
3. ((Rpa, 4) (Rpb, ∞)(Rq, 40))
4. ((Rpac, ∞) (Rpad, 1)(Rph, ∞) (Rq, 40))
5. ((Rpade, 0)(Rq, 40)(Rpac, ∞) (Rpb, ∞))

The cost is 29 units and the path is R → p → a → d → e.

4.4.8 Beam Search

Beam search is very close to breadth first technique and it explores the search tree level by level. But it keeps only the best *w* nodes from each level. This algorithm is very competent since it explores only *w* nodes at each level but it may fail due to local minima. The beam search procedure is given in Algorithm 7.

Algorithm 7 Beam search

1. Form a one-element queue consisting of the root node.
2. Until the queue is empty or the goal has been reached, determine if the first element in the queue is goal node.
   if the first element is not the goal node, remove the first element from the queue, find the first element's children, if any.
   Sort all the children found so far, by estimated remaining distance.
   Add the best *w* children to the back of the queue.

Sort all the parts in the queue.

4.4.9 Branch and Bound Search

Branch and bound technique uses the costs of already formed partial paths to guide the search. With the minimum cost, expands the partial path and tries to find a goal state. Keeping the expansion process of the paths by which costs are greater than or equal to the minimum cost of the solution path, even if the goal is found. This is branching process to the solution. The procedure is given in Algorithm 8.

Algorithm 8 Branch and bound search

1. Form a queue of partial paths. Let the initial queue consists of the zero-length, zero-step path from the root node to nowhere.
2. Until the queue is empty or the goal has been reached, determine if the first element in the queue is goal node.
   if the first path does not reach the goal node
      Remove the first path from the queue.
      Form new paths from the removed path by extending one step.
      Add the new paths to the queue
      Sort the queue by increasing cost accumulated so far.

Pure branch and bound uses information of the search that has been obtained and uses information about the cost of remaining steps to the goal. From a given node, cost of the path to goal is estimated, and total cost of a path from the start through current node to the goal can be estimated. The estimated cost of the solution path is equal to sum of cost from the start to present node and estimated cost of the remaining path to the goal. The path with the smallest such estimate to expand next path is selected. The estimated total cost is $f(n) = g(n) + h(n)$.

4.4.10 A* Algorithm

Consider descendants of nodes in a graph. Based on lowest estimated total cost of a path through the node, a heuristic search chooses the next node to expand. Then there exist functions that can find two different costs.

Estimated total cost $f(n) = g(n) + h(n)$

Known cost $g(n)$ = Cost from start to n

Estimated cost $h(n)$ = Cost from n to goal

The heuristic function h, estimates the cost of getting from a given node n to the goal. For a good heuristic function, the search is highly efficient and directed. As an example, road distance is an excellent heuristic distance estimator for route finding between cities. For poor heuristic function having pathologies, inclusion of the known cost g keeps the search from getting stopped or going too far from target.

For finding the most prominent node we make use of a heuristic function say f'. Where f' is an approximation to f. It is a combination of g and h'. Where g is the cost for going from initial state to current node, here in this context, the number of arcs traversed, and h' is an estimate of the initial cost of getting from the present node to the designation node. Thus, function f can be considered as the estimate or approximate value of getting from initial to goal state. The function f' is the approximate estimate or value of getting from the initial state to the goal state. Both g and h' are positive valued variables. Simplified form of the Best First algorithm is A* algorithm. We have noted that $f' = g + h'$ in A* algorithm. Where g is a unit measure in terms of travel time taken from the initial node to the current node. Also h' is an estimate of the time taken for solution from the present node. Thus, f' is a computation estimate from the initial node to the solution node. A* procedure is given in Algorithm 9.

Algorithm 9 A* Algorithm

1. Repeat the following steps until all the nodes in search place are examined.
2. Check the head of list of paths to find whether it leads to goal state (if the goal node is present). If it does report success and stop. Otherwise,
3. Expand the node at head of the list of paths by one level. //This will lead to the creation of many more new paths.
4. Sort all the parts in the list, by estimate of their remaining cost, so that most promising path will be head of the list.

Note: The remaining cost to the goal node may lead to the exact minimum cost. Instead we calculate the remaining cost to the leaf node from a given node, which will be only an estimate.

For example, consider the graph in Figure 4.9. The following are the steps carryied out in the search for node *e* using *A** algorithm:

1. ((R, ∞))
2. ((Rp, 24),(Rq, 9))
3. ((Rq, 9),(Rp, 24)
4. ((Rqf, 4), (Rp, 24))
5. ((Rq fg, 3), (Rp, 24))
6. ((Rqfg, 3), (Rp, 24))
7. ((Rqfgd, 1), (Rp, 24))
8. ((Rqfgde, 0), (Rp, 24))

The cost is 19 units and the path is R → q → f → g → d → e.

The detailed steps that illustrate A* algorithm with the example graph are given in Figure 4.10. We have seen how to make a solution using the Hill climbing technique. We can see how A* algorithm gives a solution, such as shortest path from node A to node G. Step by step solution is plotted as Step 1 through Step 6 in Figure 4.10.

4.5 Problem Reduction

Sometimes, it is very hard to solve certain problems. Such a hard problem can be reduced to a number of simple subproblems. Each of these simple problems are solved, and then combined into one resulted solution to the hard problem. This is the basic insight behind the method of problem reduction.

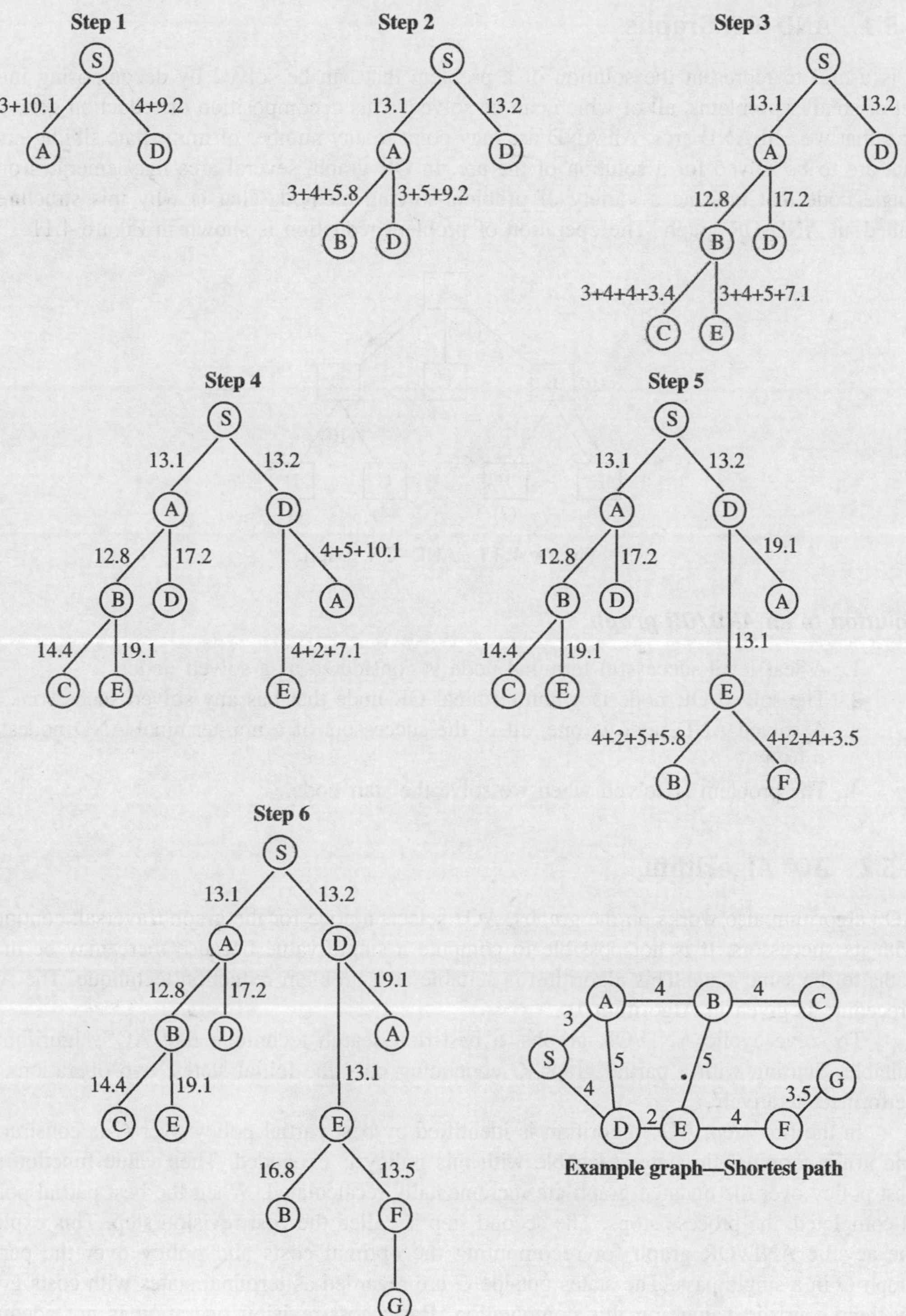

Figure 4.10 A* solution to the shortest path problem.

4.5.1 AND-OR Graphs

It is useful to represent the solution of a problem that can be solved by decomposing into a set of smaller problems, all of which can be solved. This decomposition or reduction generates arcs that we call AND arcs. An AND arc may point to any number of immediate sibling nodes that are to be solved for a solution of the arc. In OR graph, several arcs may emerge from a single node that indicate a variety of problem-solving method. That is why this structure is called an AND-OR graph. The operation of problem reduction is shown in Figure 4.11.

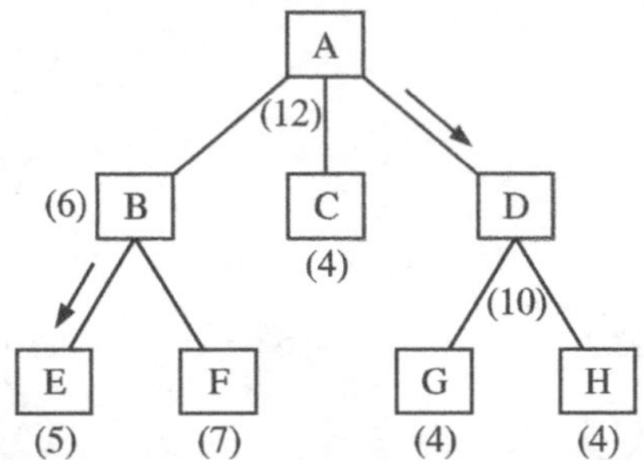

Figure 4.11 AND-OR graph.

Solution to an AND/OR graph

1. A leaf level successful terminal node is considered as a solved node.
2. The solved OR node is a non-terminal OR node that has any solved successors.
3. A solved AND node is one, all of the successors of a non-terminal AND nodes are solved.
4. The problem is solved when we solve the start node.

4.5.2 AO* Algorithm

AO* algorithm also works on the graphs. AO* selects a node for the graph traversal, continued with its successors. It is not possible to compute a single value because there may be many paths to the same state. This algorithm is suitable as a problem reduction technique. The AO* procedure is given in Algorithm 10.

To solve cyclic AND/OR graphs, a best-first search technique like AO* algorithm is suitable. Starting with a partial graph *G* containing only the initial states, two operations are performed iteratively.

In the first step, AO* algorithm is identified by best partial policy over *G* is constructed and a non-terminal tip state reachable with this policy is expanded. Then value function and best policy over the updated graph are incrementally recalculated. When the best partial policy is completed, the process stops. The second step is called the cost revision step. This exploits the acyclic AND/OR graph for recomputing the optimal costs and policy over the partial graph *G* in a single pass. The states outside *G* are regarded as terminal states with costs given by their heuristic values on this computation. Basic cost revision operation is not adequate when AND/OR graph contain cycles. Backward propagation is shown in Figure 4.12. Some examples are discussed in the following sections:

4.5.3 Towers of Hanoi Problem

We can trace back the towers of Hanoi problem to the ancient era. In the great temple of Brahma in Benares, on a brass plate under the dome marks centre of the world, There are 64 gold disks that the priest always carries one at a time between these diamond needles. The Brahma's immutable law states that, no disk can be placed on top of a smaller disk. It is considered that during the beginning of the world all the 64 disks formed a tower in the needle. Now, the process of transfer of disks is taking place from one needle to another. It is believed that when the last disk also comes in place, by again forming the Tower of Brahma in another needle, it marks the end of world.

Algorithm 10 AO* Algorithm

1. Initialise the graph into initial node.
2. Traverse the graph by any graph traversal algorithm.
   Using current path accumulate nodes that have not yet been solved or expanded.
3. Select any of the selected nodes, expand it and if it has no successors call this value SUCC, otherwise calculate each successor's f' value.
4. If f' is zero then mark the node as FINISH.
5. Using backpropagation, change the value of f' for the newly created node to reflect its successors.
6. Mark the parent node as FINISH if its siblings are FINISH. Use the most promising routes for node mining.
7. When the starting node is FINISH or the value is greater than SUCC then stop, else repeat step 2.

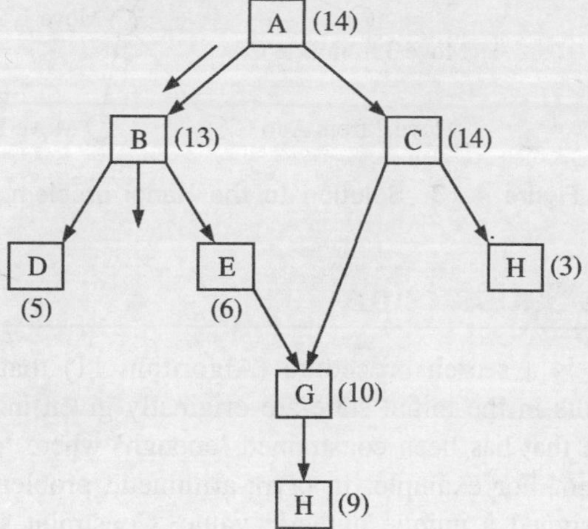

Figure 4.12 A necessary backward propagation (The last node H (9) comes then value of *G* becomes (10). Before G has value (5)).

Simple problem reduction technique is useful for solving the 'Towers of Hanoi' problem. One way of solving the real 'Towers of Hanoi' problem is by moving a tower of *n* disks from pole **A** to pole **B** by using an intermediate pole **C**. The solution involves the following set of simple reasoning:

1. To move all the disks to pole **B**, we must certainly move the biggest disk there. Here we need to empty pole **B** just prior to moving the biggest disk to it.
2. Looking for the initial configuration, we must remove all other disks then move the biggest disk. In addition, the other disks had better position, not to be moved to pole **B**, since we would not be able to move the biggest disk into the pole. So, we should first move all other disks to pole **C**.
3. Then complete the key step of moving the biggest disk from pole **A** to pole **B** and go on to solve the problem.

To solve this problem, the first step is to move the largest block from pole **A** to pole **C**. This will leave 'Towers of Hanoi' problem of size 3, where the aim is to move the disks from pole **B** to pole **C**. Because the disk that is on pole **C** is the largest disk, any other disk can be placed on top of it, and because it is in its final position, it can effectively be ignored. In this way, a Towers of Hanoi problem of any size *n* can be solved by first moving the largest disk to pole **C**, and then applying the 'Towers of Hanoi' solution to the remaining disks, but swapping pole **A** and pole **B**. We have to reduce the problem of moving a tower to the one of moving a tower with height one less than that of moving the biggest disk. This solution can be most successfully rendered as a recursive procedure, that is, a procedure that calls itself, by definition. Figure 4.13 gives a solution to the Hanoi problem for four disks (numbered 1, 2, 3 and 4) and poles (Named A, B and C).

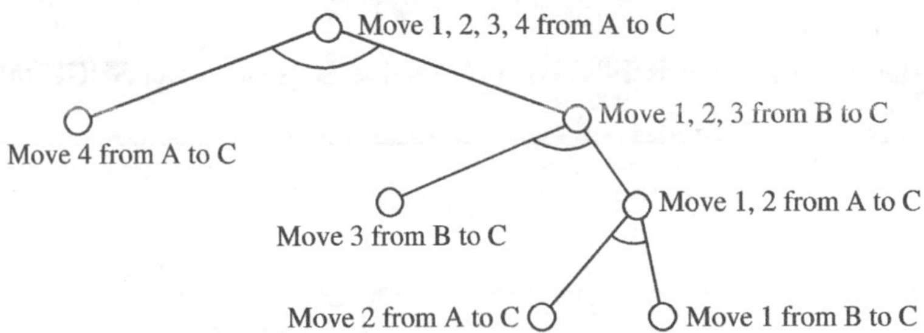

Figure 4.13 Solution to the Hanoi problem.

4.6 Constraints Satisfaction

Constraints satisfaction is a search procedure (Algorithm 11) that operates in a space of constraint sets. Constraints in the initial state are originally given in the problem description. Any state is a goal state that has been constrained 'enough' where 'enough' must be defined for each problem domain. For example, in crypt-arithmetic problems, 'enough' means that each letter has been assigned a unique numeric value. Constraint satisfaction is a two-step process. First, constraints are discovered and propagated as far as possible through the system.

If there is no solution reached then search begins. A guess is made and added as a new constraint propagation that can occur with this new constraint. The step is repeated again and so forth. The solution process continues in cycles, at which two significant things are done on each cycle.

1. Propagate the constraints using rules that correspond to the arithmetic properties.
2. Guess a value for some letter whose value is not yet determined.

Algorithm 11 Constraints satisfaction algorithm

Propagate available constraints:

All objects should be opened and must be assigned values in a complete solution.

Repeat the succeeding steps until inconsistency or all objects are assigned valid values:

Select an object and strengthen this object as much as possible by a set of constraints that is applied to the object.

If set of constraints is not matched from previous set then

Open all objects that share any of these constraints.

Remove selected object.

Return solution, if union of constraints exposed above defines a solution.

Otherwise return failure, if union of constraints discovered above defines a contradiction.

Explain constraint satisfaction while solving the crypt-arithmetic problem

Many problems in artificial intelligence can be viewed as problems of constraints satisfaction in which the aim is to discover some problem state which satisfies a given set of constraints. The procedure for constraint satisfaction is given below.

Until a complete solution is found we work out the following steps:

1. From the search graph, select an unexpected node.
2. Generate all possible new constraints by applying a constraint inference rule to the selected node.
3. If a contradiction is arrived in the set of constraint then report failure.
4. If a complete solution is met by the set of constraints then report success.
5. Apply the rules to new problem state, if neither a constraint nor a solution has been found. Insert these new states in the search graph.

The problem is to plot integers from 0 to 9 for the alphabet specified in the crypt-arithmetic problems with following constraints:

1. No two alphabets should have the same integer value.
2. Having allotted the different values for different alphabets, we have to perform the arithmetic operations specified in the problem.

Example 4.4: Consider

```
 F O R T Y
   + T E N
   + T E N
 ---------
   S I X T Y
```

Here the variables are {*E, F, I, N, R, S, T, X, Y*} and domain contains {0, ..., 9} for the letters and {0, 1} for the carries.

Step 1: Here N + N + Y = Y; so N is either 0 or 5

Also E + E + T = T; so E is either 0 or 5

Assume E is 5 and N is 0

So $c_1 = 0$, $c_2 = 1$

Step 2: 1 + R + T + T = X with a carry c_3 and $c_3 = 2$

O + 2 = I

Here O must be a higher number.

O = 9 and I = 1

Hence $c_4 = 1$

I + F = S

T and R must be higher numbers.

Then only we get a carry $c_3 = 2$.

So, T = 8 and R = 7.

Step 3: Just assume

F = 2 then S = 3

also Y = 6

So

```
 F O R T Y        2 9 7 8 6
   + T E N          + 8 5 0
   + T E N          + 8 5 0
 ---------        ---------
   S I X T Y        3 1 4 8 6
```

Similarly we can calculate the values.

Example 4.5:

```
   S E N D        8 5 4 2        9 5 6 7
 + M O R E      + 0 9 1 5      + 1 0 8 5
 ---------      ---------      ---------
 M O N E Y      0 9 4 5 7      1 0 6 5 2
```

How will this result arrive?

We can solve this problem another way, by rules. In this problem, variables are {D, E, M, N, O, R, S, Y, R_1, R_2, R_3, R_4 }, domain is 4{0,...,9} for the letters and {0, 1} for the carries. Constraints: all different (S, E, N, D, M, O, R, Y) where $S \neq 0$ and $M \neq 0$. 5 constraints for columns or one single constraint.

$D + E = Y + 10 \times R_1$
$R_1 + N + R = E + 10 \times R_2$
$R_2 + E + O = N + 10 \times R_3$
$R_3 + S + M = O + 10 \times R_4$
$R_4 = M$

$1000 \times S + 100 \times E + 10 \times N + D$
$+ 1000 \times M + 100 \times O + 10 \times R + E$
$= 10000 \times M + 1000 \times O + 100 \times N + 10 \times E + Y$

Performing the way as shown above we will get the result. Other examples for cryptarithmetic problem are as follow:

```
  DONALD        526485
 +GERALD       +197485
 -------       -------
  ROBERT        723970

  CROSS          96233
 +ROADS         +62513
 ------         ------
 DANGER         158746

   TWO            765
  +TWO           +765
  ----           ----
  FOUR           1530
```

According to Naoyuki Tamura, there are following seven solutions to the problem (TWO + TWO = FOUR):

938 + 938 = 1876
928 + 928 = 1856
867 + 867 = 1734
846 + 846 = 1692
836 + 836 = 1672
765 + 765 = 1530
734 + 734 = 1468

4.7 Means-ends Analysis

Means-ends Analysis, a technique that was first used as General Problem Solver (GPS) by Newell and Simon. It is a problem-solving method in which the current state is compared to the goal state. The distinction between them is divided up into subgoals in order to achieve the goal state by the use of the available operators. Means-end analysis is one of the weak search methods that have been utilised in both cognitive architectures and more general artificial intelligence research.

A simple idea is hidden in the Means-ends analysis technique. It is a planner that has an ability to compare two problem states. This technique resolve many ways for a problem state that differ from other problem state. The human behaviour shows a strong evidence for matching and difference reduction mechanism referred to as means-ends analysis. In means-ends analysis, the difference reduction methods (means) were strongly linked to the specific differences to be reduced (ends).

The operators for difference reduction were indexed by differences they could reduce. One of the artificial intelligence programs to exploit means-ends analysis is general problem solver (GPS). Means-ends analysis method uses a set of rules that can transfer one problem state into another. These rules are usually not represented with complete state descriptions on each side instead, they are represented in the left side. This describes the set of conditions that must meet applicable rules (these conditions are called rule's preconditions) and a right side that describes those aspects of a problem state that will be changed by the application rules. A separate data structure called difference table indexes rules by difference that can be used in reduction purpose.

Moves of a household robot is a suitable example to explain means-ends analysis. Suppose, the robot used in the domain was given the problem of moving a desk with two things on it from one room to another. The objects on top must also needs move. The main difference between the start and goal states would be the location of desk. The available operators are given in Table 4.3. Either PUSH or CARRY could be chosen for reduce the difference. If we select CARRY first, its preconditions can meet. These results in two more differences that must be reduced the location of robot and the size of desk. The location of robot can be handled by applying WALK, but there are no operators that can change the size of an object, so this path leads to dead end. Following the other branch, we attempt to apply PUSH. The difference table for this household robot can be represent in Table 4.4. The means-end analysis procedure is given in Algorithm 12.

Algorithm 12 Means-end analysis algorithm

Until the goal is reached or no more procedures are available:
  Calculate difference between the current state and the goal state.
  Use the difference that describes a procedure that will hopefully get closer to goal.
Use the procedure and update current state.
Report *success*, if the goal state is reached otherwise report *fail*.

TABLE 4.3 Operators of a household robot

Operator	*Preconditions*	*Results*
PUSH(obj, loc)	at(robot, obj)& large (robot, obj)& clear(robot, obj)& armempty(robot)	at (obj, loc)
CARRY (obj, loc)	at (robot,obj)& small (obj)	at (obj, loc)

(Contd.)

Operator	Preconditions	Results
WALK(loc)	none	at (robot, loc)
PICKUP(obj)	at (robot,obj)	holding (obj)
PUTDOWN(obj)	holding (obj)	¬holding (obj)
PLACE (obj1, obj2)	at (robot,obj2)& holding (obj)	on(obj1, obj2)

TABLE 4.4 Difference table for a household robot

	Push	*Carry*	*Walk*	*Pick up*	*Put down*	*Place*
Move object	*	*				
Move robot			*			
Clear object				*		
Get obj on obj						*
Get arm empty					*	*
Be holding object				*		

4.8 Exercises

1. What do you know about Heuristic search?
2. Explain the concept of Heuristic search techniques.
3. Differentiate problem trees from problem graphs.
4. Why are Heuristic search techniques called week methods? Explain the same by using nearest neighbour algorithm of travelling sales man problem.
5. What are the different data structures used to implement the open list in BFS, DFS, Best first search?
6. Discuss Hill climbing with steepest ascent heuristic search techniques. Explain algorithm with the help of an example.
7. Explain the concept of Best first search procedure with an implementation of A* algorithm.
8. What do you know about OR arcs in A* algorithm? Explain in detail.
9. Explain the concept of problem reduction method with an implementation of AO* algorithm.
10. Differentiate between A* algorithm and AO* algorithm? Which one is selected for a problem?
11. What do you know about constraint satisfaction procedure? Explain the same with Constraint inference rules by using search trees.
12. What are the ways of using Heuristic information?

13. How to define a problem as state space search? Discuss it with the help of an example.

14. Solve the arithmetic problem using constraints satisfaction search procedure,

NOON + SOON + MOON = JUNE.

15. Solve the arithmetic problem using constraints satisfaction search procedure,

FOUR + FOUR = EIGHT.

16. Missionaries and Cannibals problem

Three missionaries and three cannibals are all together on one side of a river. There is one boat that can carry either one or two persons. How can they all cross the river such that there are never less missionaries than cannibals on either side of the river?

(a) Draw the complete search tree for that problem

Hint: Do not further expand the tree for states that occur for the second time on one path of the tree.

(b) Define an admissible heuristic function h for the Missionaries and Cannibals problem. Explain why you have chosen this function.

(c) Show that your function h fulfils the triangle inequality.

(d) Explain graphically how an A*-search procedure traverses the search-space.

17. Set of cities and their connecting roads are given in the following table (Meaning of the entries: from city A there is a road to B of length 3 km, a road to C of length 2 km, and one to D of length 8 km, etc.):

City	Connecting roads (to-city, length)
A	(B, 2) (C, 2) (D, 8)
B	(A, 3) (C, 2) (F, 2)
C	(A, 2) (B, 2) (E, 9)
D	(A, 8) (E, 3) (F, 2)
E	(C, 9) (D, 3) (F, 6) (G, 4)
F	(B, 2) (D, 2) (E, 6) (G, 10)
G	(E, 4) (F, 10)

We are looking for the shortest route to go from A to G. The evaluation function h should assume that the cities are aligned in alphabetical order at a respective distance of 1 km, that is, $h(A) = 6$; $h(B) = 5$; : : : ; $h(F) = 1$.

(a) What is the shape of the search tree after having expanded 4 nodes using the uniform-cost strategy? Indicate in which order the nodes are expanded.

(b) What could a Hill-climbing search for that problem look like? Assume the initial state is the route A-B-C-E-F-G. The state transitions consist of adding, removing or replacing exactly one city. Which route could the Hill-climbing strategy find? Which problems can occur?

(c) Describe the (complete) search tree for the greedy search strategy based on the evaluation function h. Indicate in which order the nodes are expanded.

(d) Describe the (complete) search tree for A* algorithm using the evaluation function h. Indicate in which order the nodes are expanded. Indicate how the (complete) search tree would look like when using an evaluation function h_0 which assumes the cities are at 5 km distance from one another, that is, $h'(A) = 30; : : : ; h'(F) = 5$). Compare the obtained solutions. How does this depend distance on h'?

Hint: Note that A* algorithm terminates only when selecting a goal node for expansion, and not when generating that goal-node.

18. The 8-puzzle is a 3 × 3 square tray in which are placed 8 square tiles as shown in Figure 4.14(a). See, remaining ninth square is free. Each tile is numbered. A tile that is adjacent to the blank space can be moved into that space. There is a starting and a goal position in the game. The goal is to transform the starting position into the goal position by moving the tiles around.

19. Considering DFS, BFS and IDS algorithms, which algorithm gives the best solution for a non-weighted graph?

20. An undirected graph is given in Figure 4.14(b). Identify DFS and BFS traversal paths. Are they same?

21. Let us revisit the Sudoku problem (Figure 4.14(c)). What would an admissible heuristic function look like? What would a good heuristic function look like?

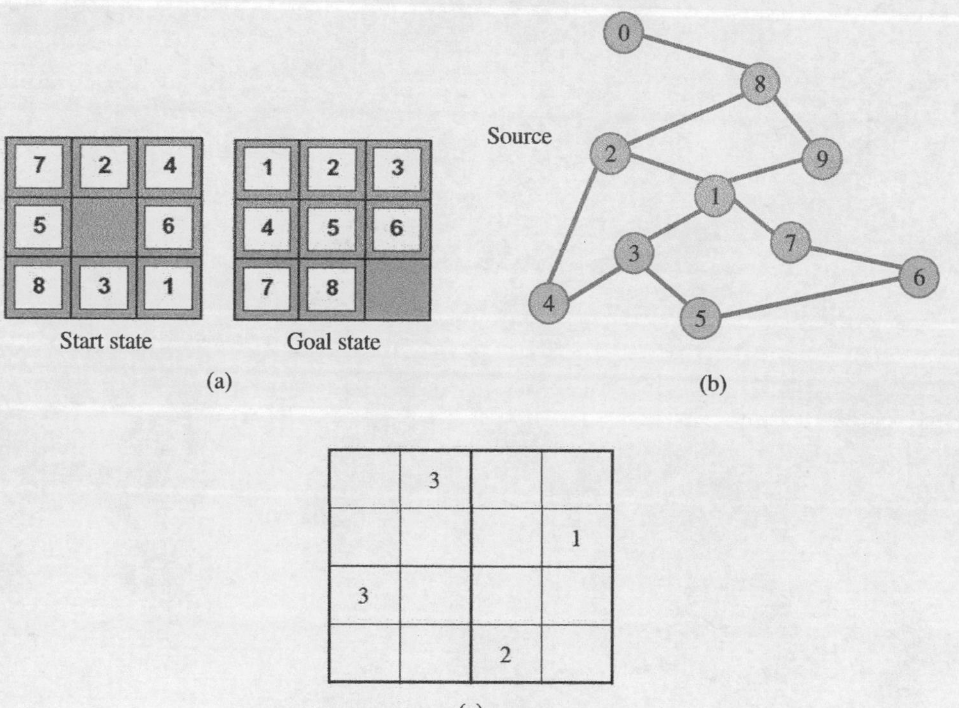

Figure 4.14 (a) 8 puzzle example (b) Graph example (c) Sudoku example.

22. Consider the state space in Figure 4.15(a). G is the goal state. Note that the values of $g(n)$ are given on the arcs. Is $h(n)$ admissible? What is the optimum solution path? What is the cost of this path?

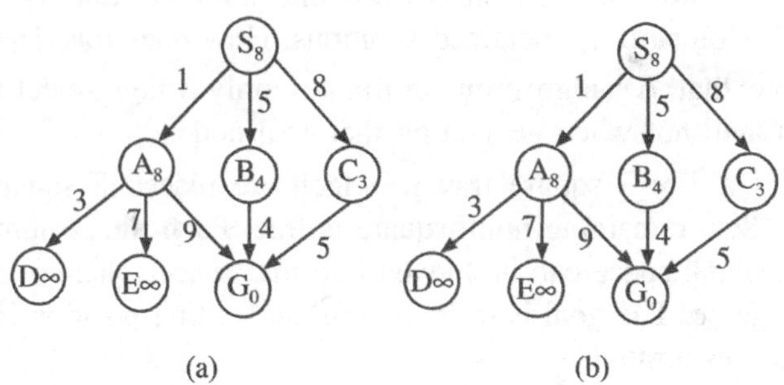

Figure 4.15 (a) Graph example-1 (b) Graph example-2

23. Consider the state space in Figure 4.15(b), G_1 and G_2 are goal states. Is $h(n)$ admissible? What is the optimum solution path? What is the cost of this path?
24. How an algorithm's performance is evaluated? Compare different uninformed search strategies in terms of the four evaluation criteria.
25. A* algorithm is based on which search method?

CHAPTER 5

Knowledge Representation

Data represents unorganised and unprocessed facts and figures. Usually data is static in nature with a set of discrete facts about events. It simply exists and has no significance beyond its existence. For example, "The price of crude oil is $80 per barrel". From this sentence one may get raw data such as price of crude oil. We cannot say, this is the current price of the crude oil. So, we need an informative phase. Information is processed data. Data that has been interpreted so that it has meaning for the user. Information can be considered as an aggregation of data which makes decision making easier.

For example, "The price of crude oil rises from $80 to $120". This is an information which shows that, at present there is hike in the crude oil by $40.

By the *Oxford dictionary* definition, knowledge is defined as expertise and skills acquired by a person through experience or education, the theoretical or practical understanding of a subject.

Knowledge is human understanding of a subject matter that has been acquired through proper study and experiences. Knowledge is usually built by learning, thinking and proper understanding of the problem area. It can be considered as the integration of human perceptive process that helps them to draw meaningful conclusion. Figure 5.1 shows relation between data, information and knowledge. To explain knowledge, consider the following sentence:

"When crude oil prices go up by $10 per barrel, it is likely that petrol price will rise by 50p per litre", which extends to knowledge.

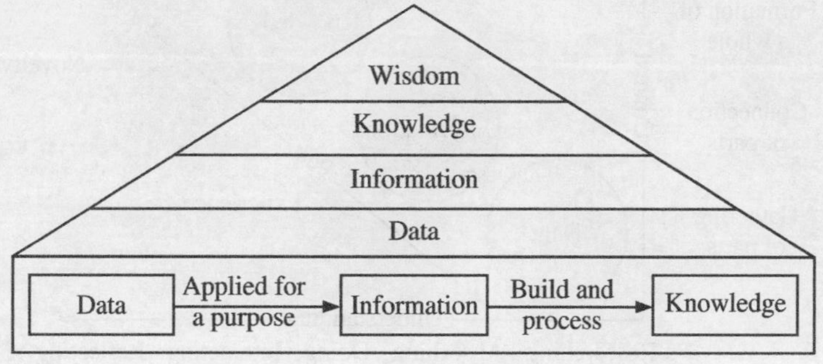

Figure 5.1 Data, information and knowledge.

Wisdom is the ability of human being to judge between right and wrong or good and bad. Wisdom is an understanding and realising of people, things, events or situations, resulting in the capability to choose or find an alternative to constantly create the best results within a minimum time.

The kind of knowledge which needs to be represented in artificial intelligence systems are objects, events, performance, meta-knowledge, facts and knowledgebase. The objects are facts in a domain of real-world (For example, 'a car has wheels'). The actions which occur in our world are events (For example, 'drive a car'). Knowledge about how to do things is understood from performance (For example, 'good mileage for the car'). Meta-knowledge is knowledge about a preselected knowledge or what we know (For example, 'Jimny is an upcoming car from Suzuki'). The truth about real-world representation is called as facts (For example, 'Rolls-Royce is a luxurious car'). Finally, knowledgebase is a group of sentences which gives agent based information. The above example sentences are grouped information of car, which is stored in the knowledgebase. This chapter explains a detailed view of knowledge and its operations.

5.1 Knowledge Management

Knowledge management is a set of strategies and practices used in an organisation to identify, create and represent, distribute and enable adoption of insights and experience. Knowledge management emerged as a scientific discipline in 1990s.

What is knowledge management? It is desirable to address this question after developing some concepts regarding 'knowledge'. Consider the following observation made by Neil Fleming from the DIKW hierarchy in Figure 5.2.

- Collection of data is not information.
- Collection of information is not knowledge.
- Collection of knowledge is not wisdom.
- Collection of wisdom is not truth.

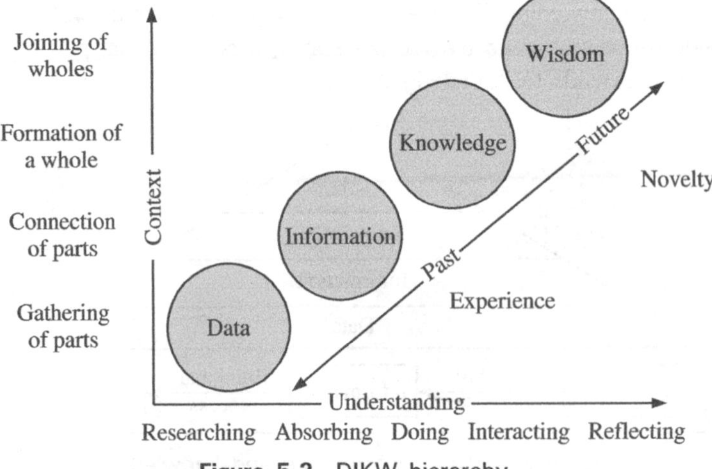

Figure 5.2 DIKW hierarchy.

The basic concept was to view information, knowledge and wisdom not as simple collections but to view as a whole because together it represents more ideas and helps to make more perfect conclusions.

Let us consider an example on stock market and relate the terms data, information, knowledge and wisdom to capital, share value, and profit percentage.

Data: While using numbers like 1000, 2000 or while using 20% or 50% in a situation completely out of context of stock market scenario, these are just pieces of data. The capital, share value, rate and profit percentage can also be termed as data as each of these provides separate meaning according to the context in which they are used.

Information: Data turns to valid information when used in the correct context. The terms capital or share or dividend represent a particular meaning when used in a share market environment. The user can gather more information and draw better conclusion. Here in this situation the following terms are interpreted as:

Capital: means their cash plus the financial assets they have invested.

Investment rate: factor used to compute the extent of investment by a particular company. It can either be a fixed rate or a fixed increase from its previous rate.

Consider a sentence, "Today the share value of Indian oil increased by 1%". This is information for the share holders who invested in the Indian oil company.

Knowledge: Let the capital that an investor possess be rupees 1000. He buys a share for rupees 50. After 2 years the share value increases to 500. Now, the investor sells his share in the market and he gets a profit of rupees 450 from a single share. This now represents a piece of knowledge. By this representation we could easily spot on how the process is evolved and how the output was obtained. By looking at the progressive growth, investor can set the investment rate and can buy the share of companies which has steady and progressive growth. More share the investor possesses larger the chance to get more profit (Considering the fact that the share value is increasing).

With respect to above context, consider the following sentence: "The crude oil usage increases in the Indian market". From this information we have a part of knowledge obtained. "The share value of Indian Oil will increase".

Wisdom: Creating wisdom from knowledge is a big effort. This is the front line principle in developing almost all system principles. The main idea focussed at any action which produces a result that encourages the system to have more of the same action produces. Growth is an everlastingly emergent characteristic for faster or later expansion that runs into limits. Analysis and study of single patterns in the scope will never help in discovering new patterns and creating new learning. Process needs to be a continuous one, which when evolved over time discovers better conclusions and findings.

For example, consider the following sentences:

"The number of vehicles on the Indian roads increases day to day". "In future government subsidies of fuels will be withdrawn".

Knowledge is obtained as "Oil price and share value of Indian Oil Company will increase". From this part of knowledge a wisdom thought can be obtained as "If I purchase more shares of Indian Oil Company, which will be a beneficiary for my future".

5.1.1 Value of Knowledge Management

From an organisational viewpoint, data is anything that represents facts or values of results. Data never form knowledge; the relation between data or the pattern of relationship is the basics of knowledge formation. These relationships need to be represented in a very simple manner so that they are easy to be interpreted. When understood these representations become information or knowledge. This is then updated in the user knowledgebase and stored for future retrieval.

Without associations and related data, it is hard to understand the situations. By associations we are able to perceive the apt meaning. For example, Consider the share of a company ABC, let the initial share value was rupees 500. If we say that the market value of ABC had increased by 50% than the last quarter, then we can easily calculate its present share value. The value could be calculated as we know all the attributes are clearly mentioned. But if we try to answer a question like, whether the share value will shoot up in the next quarter, then we are helpless. This is because although we have data and information, but we lack the knowledge. This is one of the major pitfalls, we are not able to predict the trend of the market. For that, we need more information. We must study the sales trend dealt by the company. Study on the company's immediate competitors, the current products, and the demand of the products is helpful in data analysis. These factors affect the growth of a company and which in-turn reflects in the share value. Various other elements like customer satisfaction, probability of extending the current production area also matters for the same. If we have a full data set like this we can have a data analysis and predictions could be made with most accuracy. Anything less, will create a weak system that has data and information but no knowledge.

From the last example, we can see that past and current data has been used to create or predict the future of share value. This is what knowledge management is about. It basically deals with data analysis of related or associated data for creating a prediction system. They capture, hold and reuse the information and are analysed together to convey meaningful results to a person.

One way of gathering knowledge is through Internet. Knowledge nodes are computers placed in different parts of the world, connected in a large communication network. This knowledge acquisition is done by search engines. Web browsers such as Internet Explorer, Mozilla Firefox, Google Chrome or Opera are major players in this area. The information retrieval is done by crawlers or spiders on the internet which identify and retrieve information from each of the knowledge resources sited in different locations.

5.1.2 Categories of Knowledge

Knowledge can be categorised into the following different ways:

1. Symbolic, sub-symbolic and hybrid

   Symbolic: represented as discrete unit (explicit) expressed in formal languages.

   Sub-symbolic: cannot be represented by single symbols (implicit), stored in a form of information.

   Hybrid: systems involving both types of knowledge representation.

2. Declarative and procedural
   Declarative: Knowledge are facts (objects, events) that we observe in the environment. Can describe current state of a problem.
   Example: I am a human or the sun shining.
   Procedural: knowledge represents the skills or rules we use to operate with and reason using facts.
   Example: IF John is a human THEN John has two legs.

5.2 Types of Knowledge

We cannot say there are different types of knowledge. But we are calling two classes of knowledge—declarative knowledge and procedural knowledge. There are three kinds of declarative knowledge—relational knowledge, inheritable knowledge and inferential knowledge. These knowledge is conceptual only and represented in different ways.

5.2.1 Declarative Knowledge

Declarative Knowledge is to know about something with concepts, facts, and objects. The knowledge is represented independently from the mechanisms that will use it as declarative knowledge. This gives a control in adequate use of the knowledge performed by means of general purpose heuristics that determine the best way to use the knowledge. This means the use of knowledge about the control that guides solving mechanism.

Simple relational knowledge

Declarative fact is to represent them as relations which is a simplest way to represent. The same mechanism is used to store in databases too. Databases are intended to provide support for relational knowledge. Even though it gives little opportunity for inference, it may be used as the knowledge basis for inference engines. Simple relational knowledge can be represent in the following ways.

- Store the facts in simple way.
- Set out each fact about a set of objects systematically in columns.
- Little opportunity for inference.
- Knowledge basis for inference engines.

The easy way to represent declarative facts is to use relations that can be represented as tables (as in databases). For example, information about the students in a department (Table 5.1). This table gives the information of three students who study in different classes with their marks for subject/total.

TABLE 5.1 Simple relational knowledge

Name	*Class*	*Marks*
C. Vinod	S3 CS	564
J. Prakash	S5 CS	430
R. Hari	S5 EC	390

Major problem is that there is not much knowledge represented by this method. So, we need some mechanisms to improve the representation. The inference engine obtains new knowledge from the representations.

Inheritable knowledge

We have seen that relational knowledge is made up of objects consisting of attributes and corresponding associated values. We can extend the base more by allowing property inheritance mechanisms. Elements that inherit values from members of a class and data organised into a hierarchy of classes is an inheritance property. A possibility occurs in the augment for basic representation with inference mechanism that operates on a structure of representation. This is effective when the designed structure is used by inference mechanism.

It is useful to structure the knowledge hierarchically (taxonomical hierarchy). The goal is to represent the knowledge as a graph or a tree and to use generalisation and specialisation goal (Figure 5.3). The nodes are concepts/classes, the edges are relations with is-a(class-class) relation or instance-of (class-instance) relation. The reasoning mechanism is the inheritance of properties and values of simple/multiple inheritance or default values.

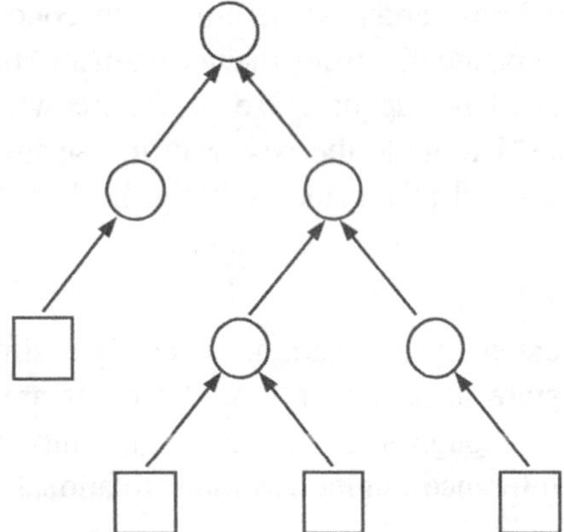

Figure 5.3 Inheritable knowledge.

Inferential knowledge

Inheritance is a powerful form of inference, but it is not the only useful form. Sometimes all the power of traditional logic is necessary to describe the inferences that are needed. The knowledge is described using logic. The semantics of the logic connectives and inference mechanisms (For example, Modus Ponens rule) can be used to obtain new knowledge

$$\forall\ x;\ y : \text{person}(x) \land \neg \text{under age}(x) \land \neg\ \text{job}(x;\ y) \rightarrow \text{unemployed}(x)$$

The inference mechanisms for first order logic are the different algorithms from automatic theorem proving. They are used for problem-solving that describes relationships between various ideas such as kind of, part of, and grouping of something. They also gives relationship that exists between concepts or objects.

5.2.2 Procedural Knowledge

Procedural knowledge can be represented as programs in many ways. The machine uses this knowledge when it executes the code to perform a task. This knowledge includes control information in its representation and/or specific procedures needed to use it. The knowledge is defined as algorithms that allow to infer new knowledge from facts. This kind of knowledge is computationally more efficient, but it is more difficult to obtain new knowledge from it and it is complex to acquire/modify.

For example, Date of birth = DD-MM-YYYY; function Age (Date of birth: date)

Procedural knowledge usage has some advantage as follows:

- Domain specific or heuristic can be represented.
- Facilitates extended logical inferences, such as default reasoning.
- Models a side effect of actions that are used in procedural knowledge. The generated rules may become false in time and it is tricky to keep track of this rule in large systems.

Some problems using procedural knowledge are based on certain procedure as follows:

1. Completeness: We cannot represent all cases.
2. Consistency: All deductions are not correct.
   For Example, If we know that *Tweety is a bird* we might deduce that *Tweety can fly.*
   Later we might discover that *Tweety is an emu.*
3. Modularity: Which is sacrificed. Changes in knowledgebase might have far-reaching effects.
4. Cumbersome: This is controlling information.

5.3 Knowledge Representation

In order to solve a complex problem in artificial intelligence, one needs both large amount of knowledge and some mechanism for manipulating that knowledge to create solution to new problems. A variety of ways to represent knowledge (facts) have been exploited in artificial intelligence programs.

First we need to finalise on the types of knowledge that need to be represented in artificial intelligence systems as follows:

Objects: Facts about the real-time entities in our world domain. For example, "Guitar has strings, trumpets are brass instruments".

Events: Actions that occur in our world. For example, "John played the guitar in 13AD's band".

Performance: It represents a behaviour. For example, "Playing the guitar", it involves knowledge about how to do it.

Meta-knowledge: It is the knowledge about a knowledge. This knowledge can read straight signs along the way to find out where it is. For example, "Jain's robot who plans a trip".

Thus, for problem solving in artificial intelligence, we need to represent knowledge. Two entities deals in the knowledge representation.

Facts: What we need to represent to the truths about the real-world. This can be regarded as the knowledge level for a domain. We have to manipulate the representation of the facts for a sentence or a knowledge. This is regarded as the symbol level since we usually define the representation in terms of symbols that can be translated by programs.

5.3.1 Approaches to Knowledge Representation

Basically four properties are needed to process a good system for the representation of knowledge in a particular domain.

1. **Representational adequacy:** Ability to represent all kinds of knowledge that are needed in a domain. It is how facts about the world can be represented in data structures. If representation is too weak, you cannot get at necessary discriminations in the representational adequacy.
   Consider the following facts:

   "John believes no-one likes Brussels sprout".

   "John will have to finish his assignment before he can start working on his project".

   These all can be represented as a string! But it is harder than to manipulate and draw conclusions.

2. **Inferential adequacy:** It is an ability to formulate the representational structures in such a way as to derive a new structure corresponding to new knowledge inferred from old. Representing knowledge is not very interesting unless you can use it to make inferences. That is, drawing new conclusions from existing facts.

   "If its raining, John never goes out"

   "It is raining today so."

   For this, come up with solutions to complex problems, using the represented knowledge. Inferential adequacy refers to how easy it is to draw inferences using represented knowledge. Representing everything as natural language strings has good representational adequacy and naturalness, but very poor inferential adequacy.

3. **Inferential efficiency:** Ability to incorporate knowledge structure with additional information that can be used to focus the attention of inference mechanism in most promising directions. You may be able to make complex deductions given knowledge represented in a sophisticated language.
   But it may be just too inefficient. Generally, more complex the possible deductions, the less efficient will be the reasoner which needs representation and inference system sufficient for the task, without being hopelessly inefficient.

4. **Acquisitional efficiency:** This is an ability to acquire new knowledge using automatic methods wherever possible rather than confidence on human intervention. It must allow to incorporate new knowledge to the representation easily. Ideally an intelligent agent should be able to obtain new information autonomously and incorporate it to the representation.

5.3.2 Issues in Knowledge Representation

A large amount of knowledge and some mechanism for manipulating that knowledge is the key of problem-solving. The basic goal of knowledge representation is to assist inference (conclusions) from knowledge. The issues arise while many knowledgebase techniques are used in a single problem. Some of these are explained as follows:

1. **Important attributes:** There are two attributes that are of very general significance. These attributes are important because they support property inheritance. There are two attributes "instance" and "isa", that are of general significance.
2. **Relationships among attributes:** The attributes that are used to describe objects as entities that we represent. There are four such properties as follows:
   (a) Inverse—When checking consistency, while a value is added to one attribute.
   (b) An 'isa' hierarchy of attribute—It is a generalisation vs specialisation. For example, height of a person is a specialisation of general attributes like physical size.
   (c) Techniques for reasoning about value—Reasoning values of attributes will not be given explicitly. Information values like height, weight, age, etc.
   (d) Single-valued attribute—It is a unique value that replaces knowledge. For example, president of India.
3. **Choosing the granularity of representation:** The level at which the knowledge needs to be represented and what are the primitives. Primitives are fundamental concepts such as seeing, playing and holding. English is a very rich language with over half a million words available in its dictionary. It is clear that we will find difficulty in deciding upon which words to choose as our primitives in a sequence of situations.

   There are quite a few arguments against the exercise of low-level primitives. First, a lot of storage when broken down into primitives is one of the simple high-level facts. Second, a related problem in knowledge is initially presented to the system in a moderately high-level form, such as in English, then considerable work must be done to reduce the knowledge into primitive form. Third, problems with the low-level primitives are in use in many domains. It is not at all clear what should be the primitives to be used.

   Example of Granularity:
   Suppose we are interested in following facts:
   "John spotted Maya"
   This could be represented as
   Spotted (agent(John), object (Maya))
   Such a representation would make it easy to answer questions such as:
   "Who spotted Maya?"
   Suppose we want to know:
   "Did John see Maya?"
   Given only one fact, we cannot discover that answer.
   We can add other facts, such as

Spotted (x, y) → saw (x, y)
We can now infer answer to the question.

4. **Representing sets of objects:** There are several reasons to represent a set of objects. First, there are some properties that are true for sets and that are not true for the individual members of the set. There are three obvious ways in which sets may be represented. The simplest way is 'name' representation. There are two ways to state a definition of a set and its elements. First, list members of the set. Such a specification is called an extensional definition. Second, is to provide a rule that when particular object is evaluated, return true or false depending on whether the object is a set or not.

   Example: Consider the assertion made in the sentences:

   "In Australia, there are more sheeps than people", and
   "English speakers are found all over the world."

   To describe these facts, the only way is to attach assertion to the sets representing people, sheep, and English.

5. **Find the right structure:** For describing the right structure of a particular situation, we can form certain requisites. This requires, selecting an initial structure and then revising the choice. It is necessary to solve the following problems for doing the above requirements:

   - How to carry out a preliminary selection of the most suitable structure?
   - How to fill in suitable details from the present situations?
   - How to find an improved structure if one chosen at first turns out not to be suitable?
   - What to do if none of the accessible structures is suitable?
   - When to produce and keep in mind a new structure?

   There is no good, general purpose method for solving all these problems. Some knowledge representation techniques solve some of them. It is very important in the structure of a sentence. Consider the following example:

   "John went to Ambrozia last night. He ordered a large rare pizza. He paid the bill and left".
   "Did John eat dinner last night?"

   The answer is 'yes', for the above question. Notice that nowhere in the story telling John is eating. The knowledge representation solves such problems.

5.4 Knowledgebase

A knowledgebase is a type of database for knowledge management. The knowledgebase contains means for information to be collected, organised, shared, searched and utilized. The knowledgebase has facts, rules and its connected relations. The facts are specifics to the world. For example, Consider the following facts:

"VIT is a deemed University"

"The first thing I did at the party was talk to John".
The rules (axioms) in the knowledgebase that describe ways to infer new facts from existing facts. For example, consider the following sentences:
"All triangles have three sides"
"All elephants are grey"

There are general facts which are known to all. Such facts are stored in knowledgebase. The facts and rules are stated in a formal language by which it can connect each other when problem-solving encounters. This is generally some form of logic (predicate calculus). For a practical side, in order to perform their desired tasks, medical experts require access to knowledge about their domains. Figure 5.4 shows an example of relationship—facts and representation schemes with respect to knowledgebase.

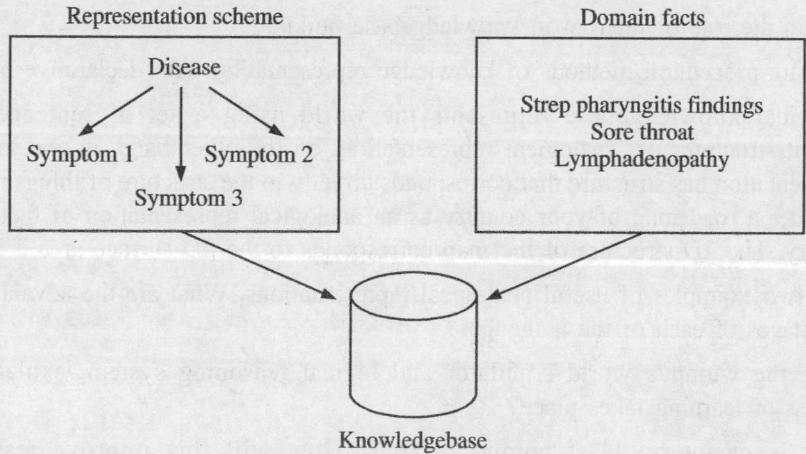

Figure 5.4 Knowledgebase representation—Medical facts.

There are mainly two types of knowledgebases that are essentially closed or open information repositories as follows:

Machine-readable knowledgebases: These store knowledge in the form of a computer readable attributes. Automated deductive reasoning is the major concern used in this storage. This representation contains a set of data, often in the form of rules that describe the knowledge in a logically consistent manner. Logical operators (conjunction, disjunction), implication and negation are used to build relationship between bricks of facts.

Human-readable knowledgebases: These are designed to allow people to retrieve and use the knowledge. These are implemented in the form of helpdesks that share information among people within a domain. They may store troubleshooting information, articles, user manuals, knowledge tags, white papers, or answers to regularly asked questions. A web-search engine can be used to locate information in the system, or users may browse through a scheme of classification.

5.5 Exercises

1. What is the difference between data, database and database management system?
2. What are the major differences between data, information and knowledge?
3. What is difference between database and knowledgebase?
4. How will you differentiate procedural knowledge from declarative knowledge?
5. What are the major issues in knowledge representation?
6. How does knowledge representation work?
7. How will you store knowledge?
8. How is knowledge management helpful in problem-solving?
9. Explain the role of internet in knowledgebase building.
10. Compare procedural methods of knowledge representation with declarative methods.
11. A logical knowledgebase represents the world using a set of sentences with no explicit structure. An analogical representation, on the other hand, is one in which the representation has structure that corresponds directly to the structure of things represented. Consider a road map of your country as an analogical representation of facts about the country. The 2D structure of the map corresponds to the 2D surface area.
12. Give two examples of useful analogical representations. What are the advantages and is advantages of each of the languages?
13. Using the wumpus world simulator and logical reasoning system, explain how the process of learning takes place?
14. What is meant by least commit search? How will this improve learning in a knowledgebase?
15. With the help of electronic circuit problem, discuss knowledge engineering.
16. Explain why you need flip-flops? Give an order-of-magnitude estimate of how many gates and flip-flops you need to build the knowledge-base. Illustrate with an example.
17. How question answering system helpful in knowledge representation?
18. What are knowledge acquisition techniques? How to build a knowledge base for a medical treatment plan by knowledge acquisition technique?
19. Consider a task requiring knowledge, like baking a cake. Examine a recipe and state what needs to be known to follow the recipe.
20. Describe some ways in which the first-hand knowledge we have of some topic goes beyond what we are able to write down in a language. What accounts for our inability to express this knowledge?

CHAPTER 6

Knowledge Representation Structures

The previous chapter shows, there are many problems with respect to representing knowledge in a computer. There are some methods to store knowledge in the computer memory. Most common knowledge representation schemes fall into one of the following four categories:

1. Structured knowledge representation

Structural representations give the 'packaging' of knowledge into well-defined pieces with higher levels of organisation. The following are examples of such type of structural representations:

1. Frames
2. Data Base Management Systems (DBMS)

Frames are complex data structures that contain, information about the concept being described. Procedural information detail on how the frame may change over time.
For example, the concept 'shopping' may be represented as:

Concept: Shopping
Location: Textiles
Actions: Item selection (procedure)
Paying (procedure)

Later sections detail on frame based knowledge representation. There are mainly two types of databases which are frequently found in clinical settings, viz., Relational databases, Object-oriented databases. Relational databases are based on a record structure in which there exist a number of fields in each record.

For example, a disease record might have the following fields as shown in Figure 6.1:

Disease name: Primary field
Organ system: Affected
Gender: Affected
Diagnostic: Test

Example of disease record table			
Disease name	Organ system	Tests	Gender
Pneumonia	Respiratory	X-ray	Both
Hypothyroidism	Endocrine	TSH	Both

Figure 6.1 Diseases record representation.

Usually records are collected together in tables. Each row in the table represents a unique record and each column in the table represents a feature of the record.

2. Logic-based knowledge representation

Logic-based knowledge representation is in the form of propositional logic statements about the world that are either true or false (propositions). For example, consider two sentences that indicate in medical terms as follows:

P = MCV is decreased in iron deficiency anaemia

Q = MCV is increased in pernicious anaemia.

Here when a patient is anaemic, the MCV (Mean Corpuscular Volume) measurement allows classification as either a microcytic anaemia (MCV below normal range), or normocytic anaemia (MCV within normal range) or macrocytic anaemia (MCV above normal range).

Propositional logic provides rules for manipulating statements like P and Q, P or Q, P and not (Q).

Using predicate calculus, the anaemia propositions may be rewritten as:

MCV (increased, pernicious anaemia)

MCV (decreased, iron deficiency)

In this form, questions can be asked of the following type:

MCV (x, iron deficiency)

Steps in a logic based system for the diagnosis of anaemia associated with 'Increased' MCV. This will search through all the 'MCV' logic predicates and find those that have 'Increased' as an object.

All matching predicates would then be returned from the knowledgebase. In our case, there is only one such predicate, i.e. pernicious anaemia.

3. Procedural-based knowledge representation

The procedural systems include a 'process' to aid in making the diagnosis. Process statements are provided in the form of rules. Rule-based systems are prototypical procedural representations. A procedural recasting of the anaemia facts would be:

IF MCV is increased THEN conclude pernicious anaemia

IF MCV is decreased THEN conclude iron deficiency anaemia.

Procedural representation shows rules for a production system.

4. Graph/network knowledge representation

Graphical or network structures are used to represent knowledge structures, which consist of the following:

Nodes—representing facts, events, objects, processes, etc.

Arcs—links between the nodes.

The significance of the network model is the capacity of networks to capture causal, temporal and other hard-to-model knowledge quite readily. The major knowledge structures are as follows:

- Frame networks
- Semantic networks
- Causal networks
- Decision trees
- Artificial neural networks

Let us have a look into first order logic, a powerful knowledge representation method for natural sentence.

6.1 First Order Logic

First order logic is a collection of formal systems used in Computer science, Mathematics, Philosophy and Linguistics. First order logic is one of the knowledge representation methods in artificial intelligence. It is an extension to propositional logic which communicates to represent the natural language statements in a short way. Consider the following sentence:

"John is my friend"

Here 'John' is the subject and 'is my friend' is the predicate.

Consider two facts as follows:

'P' represents the fact "John likes chocolate"

'Q' represents the fact "John has chocolate"

How will you represent "John likes chocolate and he does not have any" in simple form?

Simply, $P \wedge \neg Q$

So, what is the meaning of $P \rightarrow Q$?

"If John likes chocolate then John has chocolate". Is it correct meaning of the symbolic representation?

It is a very general system of logic which accurately expresses a huge variety of assertions and modes of reasoning. Any sentence in a language is explained with syntax and semantics. To explain syntax we need to look the legitimate expressions. The legitimate expression in predicate calculus is known as well-formed formula (*wff*). The components of an expression are predicate, symbols, objects, variable symbols, function symbols and constant symbols.

- Examples of objects are people, houses, numbers, theories, colours, baseball game etc.

- Examples of relations are an ordered pair, for example, brother of, bigger than, inside, part of, etc.
- Examples of properties are round, prime, multistoried, etc.
- Examples of functions are best friend, father of, mother of, etc.

User defines these primitives:

- Constant symbols (That is, the 'individual entities' in the world)
  Example, John, 3
- Function symbols (mapping between individuals)
  Example, father-of(Reji) = John, colour-of(moon) = White
- Predicate symbols (mapping from individuals to truth values)
  Example, greater(5,3), white(moon), colour(moon, white)
- A predicate is a relation which has objects as its arguments.
  Example, "John likes football".
  This sentence can be represented as: likes(John, football)

In predicate calculus, certain connectives and quantifiers are used for creating *wff*. These symbolic representations are the following:

Connectives

$\wedge$	and
$\vee$	or
$\neg$ or ~	not
$\rightarrow$	implication
$\Leftrightarrow$	iff

Quantifiers

First order logic contains two standard quantifiers, called Universal and Existential.

Universal quantifiers

Universal quantification corresponds to conjunction ('and') in which $(\forall x)P(x)$ means that P holds for all values of x in the domain associated with that variable.

Example, $\forall$x: penguin(x) $\rightarrow$ birds(x) means "All penguins are birds".

Universal quantifiers are usually form 'if-then rules' by using 'implies' symbol.

Example, $\forall$x: computer-science-student(x) $\rightarrow$ smart(x) means "All computer science students are smart."

It is rarely used universal quantification to make cover statements about every individual in the world such as:

∀x: computer-science-student(x) ∧ smart(x) meaning that "everyone in the world is a computer science student and is smart".

Existential quantifiers

Existential quantification corresponds to disjunction ("or") in which (∀x)P(x) means that P holds for some value of x in the domain associated with that variable.
Example, ∃x: bird(x) ∧ lives-Antarctica(x)

Existential quantifiers are usually used with "and" to specify a list of properties or facts about an individual.
Example, ∃x: computer-science-student(x) ∧ smart(x) means "there is a computer-science student who is smart".

A common mistake is to represent this English sentence as the first order logic sentence: ∃x: computer-science-student(x) → smart(x)

But consider what happens when there is a person who is NOT a computer-science-student.

In predicates, the coefficient having connective quantifiers, true and false values form the basis of predicate calculus. Sentences are built up from terms and atoms. Table 6.1 gives priority based operations of symbols and connectives.

TABLE 6.1 Connectives and symbols in decreasing order of operation priority

Connective	*Symbols*	*Read as*
Assertion	p	"p is true"
Negation	¬p	"p is false"
Conjunction	p ∧ q	"both p and q are true"
Disjunction	p ∨ q	"either p is true, or q is true, or both "
Implication	p → q	"if p is true, then q is true", "p implies q"
Equivalence	p ↔ q	if and only if "p and q are either both true or both false"

A term which denotes a real-world individual can be a constant or variable symbol. It can also be an *n*-place function with n terms. For example, x and $f(x_1, ..., x_n)$ are terms, where each x_i is a term.

An atom is an *n*-place predicate of *n* terms. It can have value of either true or false. If P and Q are atoms, then ¬P, P ∨ Q, P ∧ Q, P → Q, P ⇔ Q are atoms.

If a sentence does not contain any "free variables" then it is termed as a well-formed *wff*. That is, all variables are either "bound" by universal or existential quantifiers. For example, ∀x: P(x, y) has x bound as a universally quantified variable, but y is free.

6.1.1 Basic Predicate Representations

Convert the following sentences into first order logic:

1. John lives in a yellow house.
   lives(John, house) $\wedge$ colour(house, yellow)
2. John plays centre field or shortstop.
   plays (John, centre-field) $\vee$ plays(John, short-stop)
3. If the car belongs to John then it is green.
   belongs(John, car) $\rightarrow$ colour (car, green)
4. Voltaire did not write computer chess.
   $\neg$ write(voltaire, computer-chess)
5. All elephants are gray.
   $\forall$x: elephant(x) $\rightarrow$ colour(x, gray)
6. There is a person who wrote computer chess.
   $\exists$x person(x) $\wedge$ wrote(x, Computer chess)
7. Every gardener likes the sun.
   $\forall$x: gardener(x) $\rightarrow$ likes(x, sun)
8. Not every gardener likes the sun.
   $\neg$ ($\forall$x: gardener(x) $\rightarrow$ likes(x, sun))
9. You can fool some of the people all of the time.
   $\exists$x: (person(x) $\wedge\forall$ t(time(t) $\rightarrow$ can-fool(x, t)))
10. You can fool all of the people some of the time.
    $\forall$x (person(x) $\rightarrow$ $\forall$t (time(t) $\wedge$ can-fool(x, t)))
11. You cannot fool all of the people all of the time.
    $\neg$ ($\forall$x:$\forall$t: (person(x) $\wedge$ time(t) $\rightarrow$ can-be-fooled(x, t)))
12. All violet mushrooms are healthy.
    $\forall$x: (mushroom(x) $\wedge$ purple(x)) $\rightarrow$ healthy(x)
13. No violet mushroom is healthy.
    $\neg\exists$x: violet(x) $\wedge$ mushroom(x) $\wedge$ healthy(x)
    or, equivalently,
    $\forall$x: (mushroom(x) $\wedge$ violet(x)) $\rightarrow$ $\neg$healthy(x)
14. There are exactly two violet mushrooms.
    $\exists$x: $\exists$y: mushroom(x) $\wedge$ violet(x) $\wedge$ mushroom(y) $\wedge$ violet(y) $\wedge\neg$(x = y) $\wedge$ ($\forall$z) (mushroom(z) $\wedge$ violet(z)) $\rightarrow$ ((x = z) $\vee$ (y = z))
15. John is not tall.
    $\neg$tall(John)
16. X is above Y if X is directly on top of Y or else there is a pile of one or more other objects directly on top of one another starting with X and ending with Y.
    $\forall$x: $\forall$y: above(x, y) $\Leftrightarrow$ (on(x, y) $\vee$ ($\exists$z) (on(x, z) $\wedge$ above(z, y)))

6.1.2 Conversion of WFF to Clause Form

Clause form represents the logical database as a set of disjunctions of literals. A literal is an atomic expression or the negation of an atomic expression.

Algorithm

With the following example, we can convert a *wff* into the equivalent clause form:

$\forall x: \{P(x) \rightarrow \{\forall y[P(y) \rightarrow P(f(x, y))] \wedge \neg \forall y[Q(x, y) \rightarrow P(y)]\}\}$

1. Eliminate implication symbol using $A \rightarrow B$ by $\neg A \vee B$.

   $\forall x:\{\neg P(x) \vee \{\forall y[\neg P(y) \vee P(f(x, y))] \wedge \neg \forall y[\neg Q(x, y) \vee P(y)]\}\}$

2. Reduce the scope of each negation symbol by distributive laws (deMorgan's laws).
   $\forall x:\{\neg P(x) \vee \{\forall y[\neg P(y) \vee P(f(x, y))] \wedge \exists y[\neg Q(x, y) \vee P(y)]\}\}$

3. Standardise variables so that each quantifier binds a unique variable.

   $\forall x:\{\neg P(x) \vee \{\forall y[\neg P(y) \vee P(f(x, y))] \wedge \exists w[\neg Q(x, w) \vee P(w)]\}\}$

4. Move all quantifiers to the left of the formula without changing the relative order.

   $\forall x:\forall y \exists w:\{\neg P(x) \vee \{[\neg P(y) \vee P(f(x, y))] \wedge [\neg Q(x, w) \vee P(w)]\}\}$

5. Eliminate the existential quantifiers by Skolem constants.
   Every existentially quantified variable can be replaced by a unique Skolem function whose arguments are all the universally quantified variables on which the existential depends, without changing first order logic.

   Examples

   "Everybody likes something"

   $\forall x\ \exists y: [Person(x) \wedge Likes(x, y)]$
   $\forall x: [Person(x) \wedge Likes(x, S_1(x))]$
   Where $S_1(x)$ = "that which x likes"

   Another example

   "Every philosopher writes at least one book"

   $\forall x: \exists(y)[Philosopher(x) \wedge Book(y)) \rightarrow Write(x, y)]$
   $\forall x: [(Philosopher(x) \wedge Book(S_2(x))) \rightarrow Write(x, S_2(x))]$

   In our problem, after eliminating existential quantifier
   $\forall x:\forall y:\{\neg P(x) \vee \{[\neg P(y) \vee P(f(x, y))] \wedge [\neg Q(x, g(x)) \vee P(g(x))]\}\}$

6. Drop the prefix

   $\{\neg P(x) \vee \{[\neg P(y) \vee P(f(x, y))] \wedge [\neg Q(x, g(x)) \vee P(g(x))]\}\}$

7. Convert the matrix into a conjunction of disjunct.

   $(a \wedge b) \vee c \equiv (a \vee c) \wedge (b \vee c)$

   $\{\neg P(x) \vee \neg P(y) \vee P(f(x, y))\} \wedge \{\neg P(x) \vee \neg Q(x, g(x)) \vee P(g(x))\}$

8. Create a separate corresponding clause for each conjunct.
   (a) $\neg P(x) \vee \neg P(y) \vee P(f(x, y))$
   (b) $\neg P(x) \vee \neg Q(x, g(x)) \vee P(g(x))$
9. Standardise a part of the variable in the set of clauses generated in the previous step, that is, rename the variables so that no two clauses may be reference to the same variable.
   (a) $\neg P(x_1) \vee \neg P(y) \vee P(f(x_1, y))$
   (b) $\neg P(x_2) \vee \neg Q(x_2, g(x_2)) \vee P(g(x_2))$

6.1.3 Resolution

Resolution is a technique for proving theorems in propositional or predicate calculus that has been a part of problem-solving. Resolution proves a theorem by negating statement to be proved and adding this negated goal to a set of axioms that are known to be true. The steps involved in resolution technique are given in Algorithm 13.

Algorithm 13 Propositional resolution

Convert all the propositions of P to clause form.
Negate P and convert the result to clause form. Add it to the set of clauses obtained in step 1.
repeat
  (a) Select two clauses and call their parent clauses.
  (b) Resolve them together, then the resolvent will be the disjunction
      of all of the literals of both of the parent clauses with the following exception: if there are any pairs of literals L and ¬ L such that one of the parent clauses contains L and the other contains ¬L, then select one such pair and eliminate both L and ¬L from the resolvent.
  If the resolvent is the empty clause, then a contradiction has been found.
    If it is not found, then add in the set of clauses available to the procedure.
until either a non-progress condition or a contradiction is found

Example 6.1: Consider the following set of axioms in proportional logic.

1. P
2. $(P \wedge Q) \rightarrow R$
3. $(S \vee T) \rightarrow Q$
4. T

Corresponding clause form is

1. P
2. $\neg P \vee \neg Q \vee R$

3. ¬ S ∨ Q
4. ¬ T ∨ Q
5. T

Using resolution by refutation we can solve R (Figure 6.2). In the principle of resolution by refutation we negate the goal sentence and try to look at the given set of sentences for a match. That is, a predicate which is existing in the goal sentence. The matching will take place between predicates of the form as follows:

L and ¬ L

After matching, the variables will attain constant values from either of the predicates. This process is known as substitution. When a substitution occurs within the predicate of the form L and ¬ L, we say that unification has been carried out within L and ¬ L.

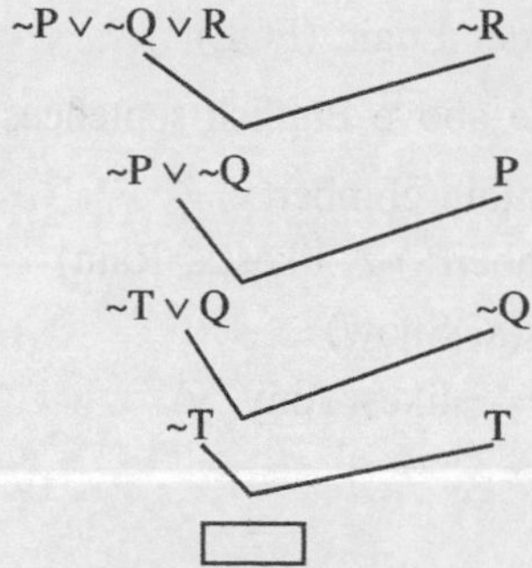

Figure 6.2 Refutation tree.

Resolution examples

1. Construct the following sentences into clause form and extract the answer.
   "If fido goes wherever John goes and if John is at school, where is fido?"
   The equivalent *wff*

   (a) ∀x: at (John, x) → ¬ at (fido, x)
   (b) at (John, school)

   Goal: at (fido, x)
   The equivalent clause form is

   (a) ¬ at (John, x) ∨ at (fido, x)
   (b) at(John, school)

   By resolution, we need negation of the goal statement ¬ at (fido, x). To carry out answer extraction process we shall consider both the goal and the negation. Resolution tree is shown in Figure 6.3.

2. Convert the sentences into clause form and answer the question:
   Tony, Tensing and Hilari belong to the MountainClub. Every member of the MountainClub is either a skier or a mountain climber or both. No mountain climber

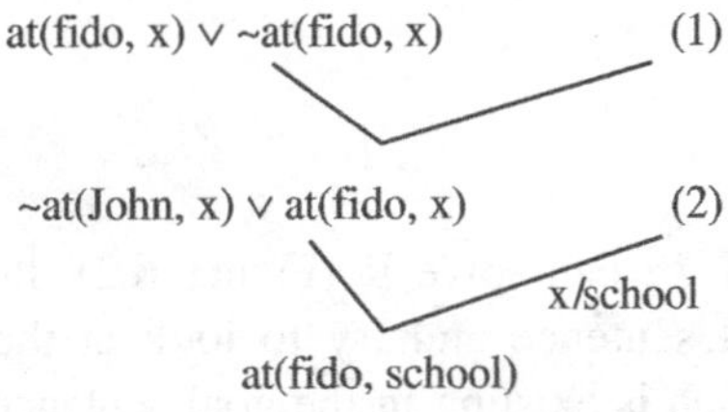

Figure 6.3 Resolution tree.

likes rain, and all skiers like snow. Ellen dislikes whatever Tony likes and likes whatever Tony dislikes. Tony likes rain and snow.

Question: Is there a member of the MountainClub who is a mountain climber but not a skier?

Translation into First Order Logic (FOL)

We can now translate the above English sentences into the following FOL *wffs*:

(a) ∀x: Skier(x) ∨ mountain-climber(x)
(b) ¬ ∃x: mountain-climber(x) ∧ likes(x, Rain)
(c) ∀x: skier(x) → likes(x, Snow)
(d) ∀y: likes(Ellen, y) ⇔ ¬likes(Tony, y)
(e) likes(Tony, Rain)
(f) likes(Tony, Snow)
(g) Query: ∀x mountain-climber(x) ∧ ¬skier(x)
(h) Negation of the Query: ¬∃x mountain-climber(x) ∧ ¬skier(x)

Conversion to clause form

(a) skier(x1)∨ mountain-climber(x1)
(b) ¬mountain-climber(x2) ∨ ¬likes(x2, Rain)
(c) ¬S(x3) ∨ likes(x3, Snow)
(d) ¬likes(Tony, x4) ∨ ¬likes(Hilari, x4)
(e) likes(Tony, x5) ∨ likes(Hilari, x5)
(f) likes(Tony, Rain)
(g) likes(Tony, Snow)
(h) Negation of the Query: ¬mountain-climber(x7) ∨ skier(x7)

Resolution refutation proof

Clause 1	Clause 2	Resolvent	
8	1	9. skier(x1)	x7/x1
9	3	10. likes(x1, Snow)	x3/x1
10	4	11. ¬likes(Tony, Snow)	x4/Snow, x1/Hilari
11	7	12. False	{}

Answer extraction:
The refutation tree is shown in Figure 6.4.

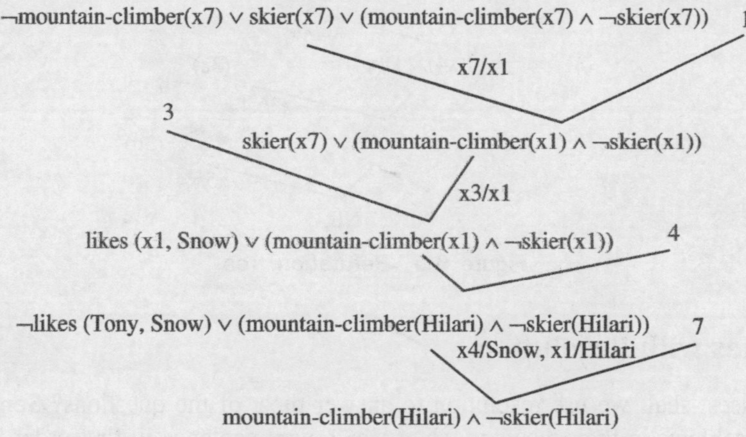

Figure 6.4 Refutation tree.

Answer to the Query:
Hilari

3. Using resolution by refutation answer the goal sentence.

   (a) Whoever can read is a literate.
   (b) Dolphins are not literate.
   (c) Some dolphins are intelligent.

   Goal sentence: 'Some who are intelligent can not read' Solution:

   Classical form can be

   R : Read L : Literate I : Intelligent D : Dolphin

Then wff:	Clause form
$R(x) \rightarrow L(x)$	1. $\neg R(x1) \vee L(x1)$
$D(x) \rightarrow \neg L(x)$	2. $\neg D(x2) \vee \neg L(x2)$
$D(x) \rightarrow I(x)$	3. (a) $D(x3)$ (b) $I(x4)$

   Goal is: $\neg I(x5)$ vee $R(x5)$

   Negation of goal sentence is wrong. So, goal sentence will be true, that is, some who are intelligent can not read. The refutation tree is given in Figure 6.5.

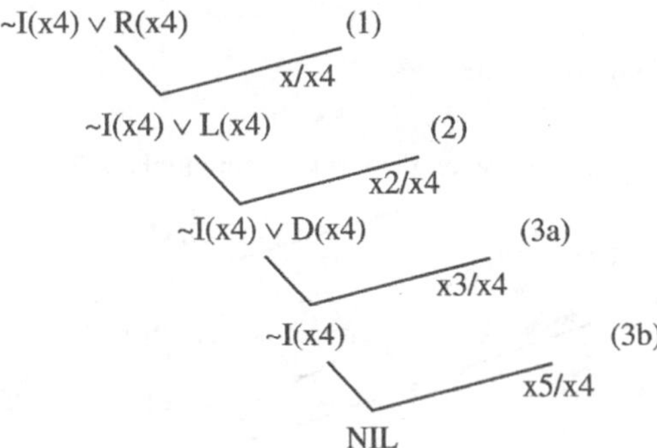

Figure 6.5 Refutation tree.

6.1.4 Issues with Resolution

A question arises, shall we use resolution to answer most of the questions? Sometimes clause forms result problems in their meaning when we convert sentence in first order logic to clause form. The following are certain issues with resolution techniques:

- Requires full formal representation in first order logic (for conversion to clause form).
- Resolution defines a search space (which clauses will be resolved against which others define the operators in the space) → search method required.
- In the worst case, resolution is exponential in the number of clauses to resolve. Actual: exponential in average resolvable set (= branching factor).
- Can we define heuristics to guide search for Best first search, or A* or B*? (Not in the general case)?

6.2 Unification Algorithm

The unification algorithm is in first order logic with first element is the name of a predicate and the remaining elements are arguments. In other words, unification tries to identify two symbolic expressions by replacing certain subexpression variables with other expressions. It is an automated reasoning technology using first order logics. Unification procedure is given in Algorithm 14. Before applying unification algorithm some precondition should be checked. Different predicate symbols cannot be unified so check if those symbols, atoms or expression are same. Number of arguments in both expressions must be identical. The algorithm fails if there are two similar variables present in the same expression. Using a substitution process, two different atomic expressions are made as identical. For example, S_1 = Professor (X) and S_2 = Professor (John). Then Unify (S_1, S_2) = Unify (Professor (X), Professor (John)).

Algorithm 14 Unification algorithm

Function unify (E_1, E_2)
Case:
1. Both E_1 and E_2 are constant or the empty list
   if $E_1 = E_2$ then return
   Else **return** FAIL
2. E_1 is a variable:
   If E_1 occurs in E_2 then return FAIL
   else return $\{E_1/E_2\}$
3. E_2 is a variable:
   if E_2 occurs in E_1 then return FAIL
   else return $\{E_1/E_2\}$
4. Either E_1 or E_2 are empty then return FAIL

Otherwise:
  $H\ E_1$ = First element of E_1
  $H\ E_2$ = First element of E_2
  $SU\ BS_1$ = unify ($H\ E_1$, $H\ E_2$)
  if $SU\ BS_1$ = FAIL then return FAIL
  $T\ E_1$ = apply ($SU\ BS_1$, rest of E_1)
  $T\ E_2$ = apply ($SU\ BS_1$, rest of E_2)
  $SU\ BS_2$ = unify ($T\ E_1$, $T\ E_2$)
  if $SU\ BS_2$ = FAIL then return FAIL;
  else return composition ($SU\ BS_1$, $SU\ BS_2$)

For example:
Unify ((parents x (father x) (mother bill)), (parents bill (father bill) y))
The step by step illustration of unification algorithm is given in Figure 6.6.

6.3 Frames

Frames were introduced by Mervin Minsky in 1975, as a data structure to represent a mental model of a stereotyped situation, such as, driving a car, attending a meeting, eating in a restaurant, etc. Knowledge about an item is stored in a single unit called a frame. When a new situation is encountered, an appropriate frame is selected for reasoning to the situation.

Frames are general record structures which consist of a collection of slots and slot values. Slots contain names, values and such fields called links. The links may also have names and any number of values. The slots in a frame specify a character of an entity for which the frame is being represented. Some slots will also include instructions for manipulations like entering default values, a few conditions for filling a slot for procedures such as:

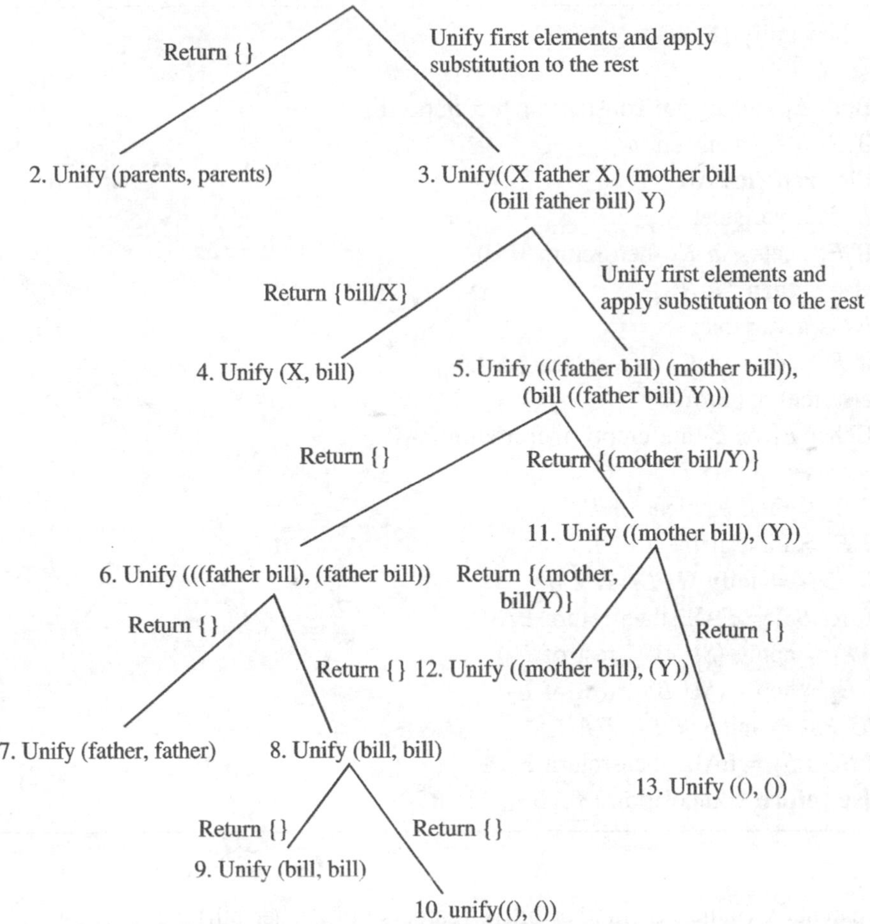

Figure 6.6 Trace of unify ((parents X (father X) (mother bill)) and (parents bill (father bill) Y).

If needed, If added, If removed, etc.

The slots are similar to traditional record that contains information relevant to stereotyped entities. The slot in the frame contains information such as:

- Frame identification
- Relationship of this frame with other frames
- Descriptors of requirements for a frame
- Procedural information used in the structure described.
- New instance information

Frame system supports class inheritance. The slots and default values of a class frame are inverted across the class/subclass and class/member hierarchy.

Example: A frame that describes lecturer in Operating system by Prof. John is shown in Figure 6.7.

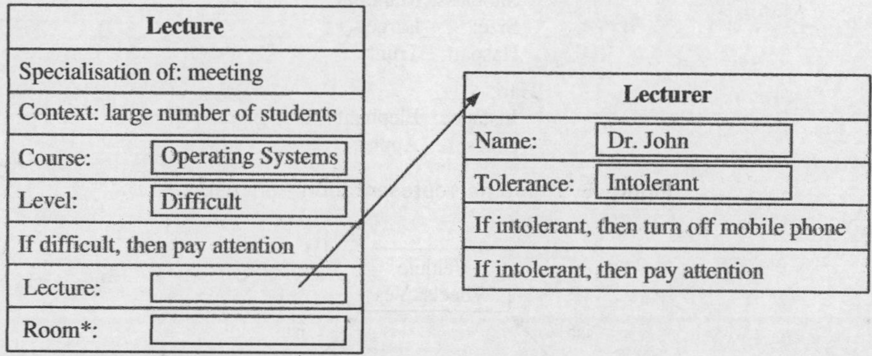

Figure 6.7 Frame representation—lecturer.

Example: A frame description of a hotel room is shown in Figure 6.8. This frame describes specialisation of a hotel room which has a bed with king size and other furniture in that room.

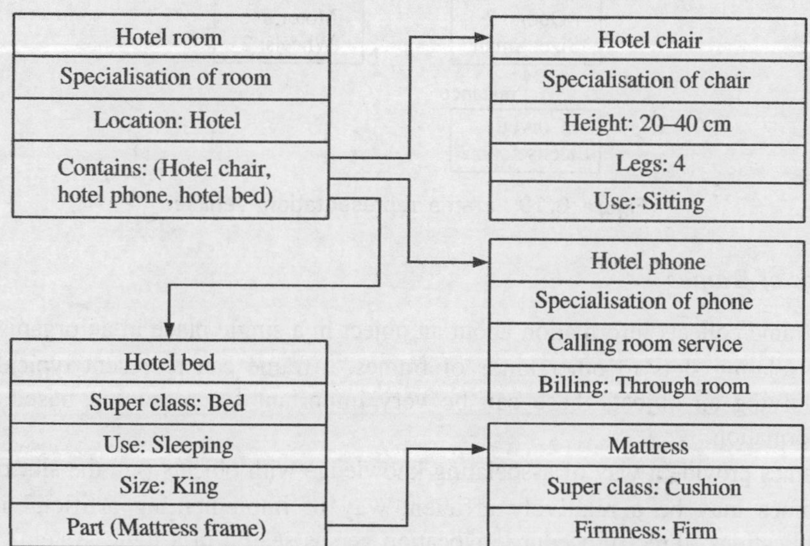

Figure 6.8 Frame representation for a hotel room.

Another example of a frame 'animal' is given in Figure 6.9. Here we have slot-value (elephant, size, large), instance (Hari, elephant), value (obj, slot, V) : instance (obj, class), slot-value (class, slot, V). Frames also allow multiple inheritance (Hari is an elephant and is a circus animal). Another example is given in Figure 6.10 regarding vehicle frame.

Mammal:
  Subclass: Animal

Elephant:
  Subclass: Mammal
  Size: Large
  Haspart: Trunk

Hari:
  Instance: Elephant
  Likes: Apples

Figure 6.9 Frame representation—animal.

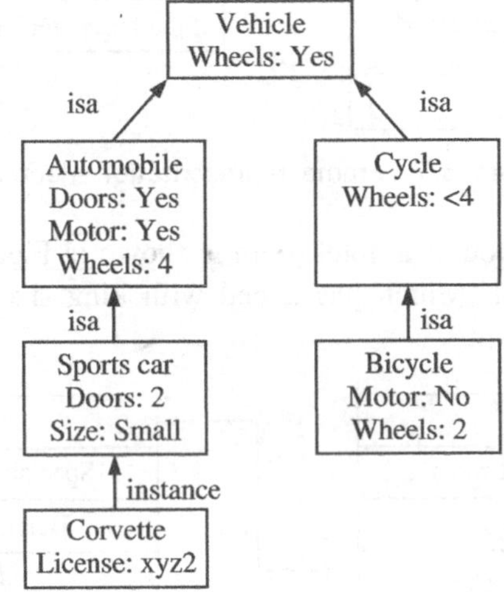

Figure 6.10 Frame representation-vehicle.

Advantages of frames

- A frame collects information about an object in a single place in an organised fashion.
- By relating slots to other kinds of frames, a frame can represent typical structures involving an object; these can be very important for reasoning based on limited information.
- Frames provide a way of associating knowledge with objects (via the slot procedures).
- Frames may be a relatively efficient way of implementing artificial intelligence applications (direct procedure invocation versus search in a logic system).
- Frames allow data that are stored and computed to be treated in a uniform manner. (Example, AGE might be stored, or might be computed from BIRTHDAY.)

Object-oriented programming has much in common with frames. Frames are good for applications in which the structure of the data and the available information are relatively fixed. The procedures (methods) associated with slots provide a good way to associate knowledge with a class of objects. Frames are not particularly good for applications in which deep deductions are made from a few principles (as in theorem proving).

Disadvantages of frames

Frames encourage baroque representations; little guide to good structuring of a domain. All problems with respect to the object-oriented systems are also applicable for frames.

Some things that can be represented in logic cannot be represented well, or at all, in frames.

- Slot fillers must be "real" data. For example, it is not possible to say that John is a butcher or a baker, since there is no way to deal with a disjunction in a slot filler.
- It is not possible to quantify over slots. For example, there is no way to represent "Some student made 100 on the exam".
- It is necessary to repeat the same information to make it usable from different viewpoints, since methods are associated with slots or particular object types. For example, if there are knowledge stores regarding John, it may be easy to answer "whom does John love?" but hard to answer "who loves Mary?".

In summary, frames are instance of structured representation (schemes). Frames posses static data-structure representing stereotyped situation. Frames provide object-oriented system properties. Frames offer which things were just typical of a class, and which were definitional for knowledge representation and could not be overridden. Using an *asterix* (*) to denote typical values.

6.4 Conceptual Dependency

It is a theory of how to represent a kind of knowledge about event, that is, usually contained in natural sentences. The goal is to represent the knowledge in a way that facilitates drawing inference from the sentences and which is independent of the language in which the sentences were originally stated.

The Conceptual Dependency (CD) theory was proposed by Schank and Ridge in 1973, offering a set of four primitive conceptualisations from which the world of meaning is built. They are equal and independent.

1. ACTs actions
2. PPs objects (Picture producers)
3. AAs modifiers of action (Action aiders)
4. PAs modifiers of objects (Picture aiders)
5. T (Times)
6. LOC (Locations)

A primitive is a set of idea (words) that maps to an object. There are some primitives as follows:

1. ATRANS—Transfer a relationship (example, give)
2. PTRANS—Transfer physical location of an object (example, go)
3. MOVE—Move body part by owner (example, kick)
4. ATTEND—Focus sense organ (example, listen)

5. MTRANS—Transfer mental information (example, tell)
6. SPEAK—Utter a sound (example, say)
7. GRASP—Actor grasping a thing (example, clutch)
8. INGEST—Actor ingesting a thing (example, eat)
9. EXPEL—Actor getting rid of a thing from body
10. PROPEL—Apply physical force to an object (example, push)

Conceptual dependency possesses certain symbols, which gives interconnection between two objects. These symbols aid relations between each objects. The symbols used in CD are as follows:

1. $\leftarrow$: Indicates the direction of dependency
2. $\Leftrightarrow$: Indicates the agent-verb relationship
3. P Indicates the past tense
4. O Indicates the object relation
5. D Indicates the direction of the object in the action
6. C Indicates the condition

Example:

Consider the following sentence:

"John throws the ball"

$$\text{John} \Leftrightarrow \text{PROPEL} \overset{O}{\leftarrow} \text{Ball}$$

Conceptual dependency offers a number of important benefits. By proving a formal theory of natural language semantics, it reduces problems of ambiguity. Second, the representation itself directly captures much of natural language semantics, by attempting to provide a canonical form for the meaning of sentences.

For example: "John prevented Mary from giving book to bill".

The CD representation for this sentence is shown in Figure 6.11.

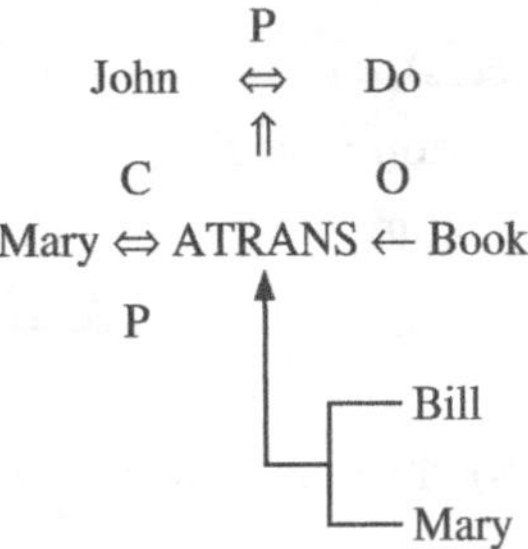

Figure 6.11 CD representation.

Advantages of CD

- Usage of these primitives involves fewer inference rules.

- CD structures already have many of the inference rules.
- The holes in the initial structure help to establish points which are not yet focussed.

Disadvantages of CD

- Knowledge needs to be decomposed to low-level primitives.
- It is always a tedious task to find the correct set of primitives.
- Many inferences are required.
- Even the representation of a simple action can become complex.

6.5 Scripts

A script is a structured representation describing a stereotyped sequence of events in a particular context. The script was originally designed by Schank in 1977. The knowledgebase representation in terms of the situations that the system is supposed to understand. They are extremely beneficial because of the reasons as follows:

- The events always tend to occur in known patterns or runs.
- There is always a relationship between the existing events.
- There exists an entry condition(s) which allows event to take place.
- There are prerequisites for the events that are taking place. For example, when a student progresses through a degree scheme or when a customer buys a product.

The components of scripts are as follows:

- **Entry conditions** include open conditions, satisfied before the events described.
- **Result** or 'fact' is true once the script has terminated.
- **Props** or the 'things' that support the content of a script. The set of props supports reasonable default assumptions about a situation.
- **Roles** are actions that an individual participant performs.
- **Tracks** are the specific variation on a more general pattern that is represented by a particular script.
- **Scenes** are the actual sequence of events that occur.

The events described in a script form a giant causal chain. A set of entry conditions are enabled in beginning of the chain which enables the first event of the script to occur. The end of the chain is the set of results which may enable later events or event sequences to occur. Within the chain, events are connected to earlier events that make them possible, and to later events that they enable.

Example: A railway ticket reservation script

Script : Railway station

Track : Ticket booking

Props : Counters

D = Display board with information
F = Ticket
Credit card
Money

Roles:

S = Traveller
W = Person on the ticket counter (Current tickets)
C = Person issues Tatkal reservation tickets
M = Ticket examiner
O = Railway management

Every condition:

S required a ticket
S has money

Result:

S has less money
O has more money
S needs not reservation
S is pleased (optional)

Scene 1: Entering
S PTRANS S into station
S ATTEND eyes to counter
S MBUILD where to take
S PTRANS S to counter
S MOVE S to ticket counter
Scene 2: Reserving

(D on his hand) (W points out D) (S asked for D)

S PTRANS Menu to S	S MTRANS signal to W
	W PTRANS W to counter
↓	S MTRANS need D to W
	W PTRANS W to D
	↙ W PTRANS W to counter

*S MBUILD Choice of F
S MTRANS signal to W
W PTRANS W to counter
S MTRANS 'I want F' to W
W MTRANS 'Tatkal is available' to S
S MTRANS 'I want Tatkal F' to W
W PTRANS W to C

W MTRANS (A TRANS F) to C

↙ ↘

C MTRANS 'Print F' to W — C DO (prepare F script)
W PRANS W to S — to scene 3
W MTRANS 'Print F' to S
(go back to *)
Scene 3: Checking
C ATRANS F to W
W ATRANS F to S
S INGEST F
(Option: returns to scene 2 to more tickets, otherwise go to scene 4)

↙ ↘

Scene 4: Leaving — S MTRANS to W
(W ATRANS money to S) — W Move (give credit card)
W PTRANS W to S
W ATRANS credit card to S
S PTRANS S to M
S ATRANS money to M
S PTRANS S moves out of the station

Some additional points to note on scripts:

- For applying a script, it needs to be activated, and the activation depends on the significance of the particular script.
- If a subject is noted in passing then a pointer to that script could be held.
- For important topics the script should be opened.
- It is not desirable in many active scripts that might be like having too many windows open on the display or too many recursive calls in a program.
- For events following known trail, we can use active scripts to represent the actions involved and use them to answer detailed questions.
- Different trails may be allowed for different outcomes of scripts (For example, the restaurant script).

Advantages

- Ability to predict events.
- A single coherent interpretation may be built up from a collection of observations.

Disadvantages

- It is not as general as frames.
- It is not suitable to represent all kinds of knowledge.

6.6 Semantic Network

A semantic network or net is a graphic notation for representing knowledge in patterns of interconnected nodes and arcs. The primary version of semantic networks was used in areas of philosophy, psychology and linguistics but now computer implementations of semantic networks are developed for artificial intelligence and machine translation. It was Collins and Quillian who developed semantic networks during their research on human information storage and response times.

A semantic network representation is similar to network model of database management system. That is, entities are connected by relationship. It is understood that semantic network representation gives us clustering of several units namely entities through relationship. Network representation is necessary because many of the English sentences cannot be exactly represented by predicate calculus. A few of the relationships used in semantic nets are:

isa
a kind of - AKO
Instance-of F
Member-of
Has-parts

An example is given in Figure 6.12. In this example, Tweety is a yellow bird which has wings and can fly. This statement gives a network architecture. A semantic network to represent birds is shown in Figure 6.13.

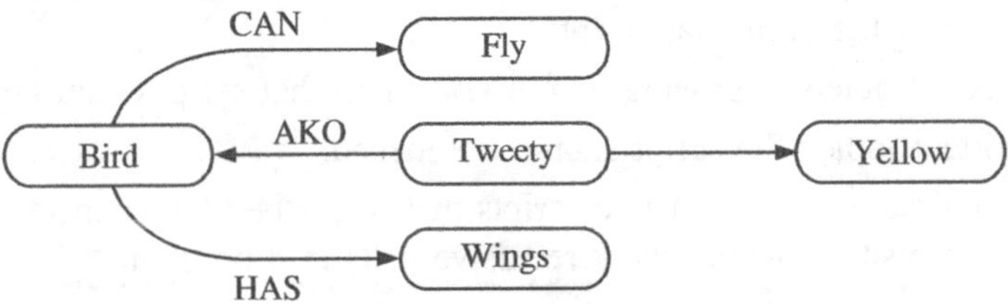

Figure 6.12 A simple semantic network.

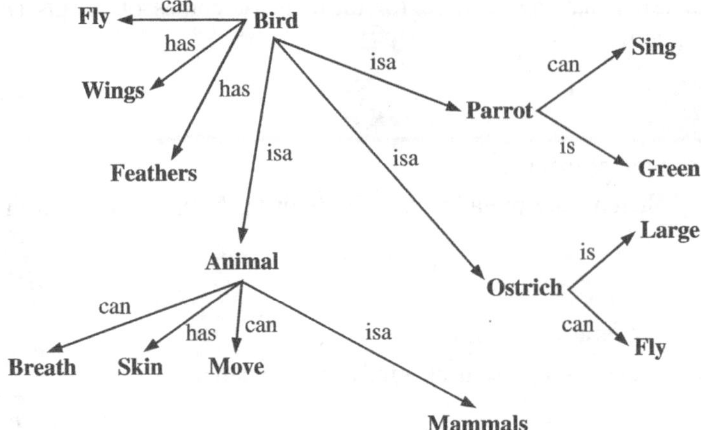

Figure 6.13 A semantic network—birds.

For all semantic networks there is a declarative graphic representation which is used for representing knowledge or to support automated systems for reasoning about knowledge. Many versions are highly informal, but there are versions which are formally defined by systems of logic. The below mentioned are the most common types of semantic networks.

1. **Definitional networks** highlight the subtype or 'isa' relation between a concept type and a newly defined subtype. This network is called a generalisation network hierarchy. There are rules of inheritance that give support for copying properties defined for a super-type to all of its subtypes. The assumed information in these networks is true by its definition and operates in semantic architecture.
2. From the asset propositions, we can design **assertional networks**. It assumed contingently true information in an assertional network compared to definitional networks. In addition, assertional networks are explicitly marked with a modal operator. The conceptual structures underlying natural language semantics are proposed as models.
3. If a network uses implication as the primary relation for connecting nodes, it is named as **implication network**. The representation patterns use beliefs, causality, or inferences for building the network.
4. Mechanisms like the transient or attached procedure used in **executable networks** that helps to create inferences, pass messages, or search for patterns.
5. **Learning networks** acquire knowledge from the real-time examples. Addition or deletion of arcs or nodes from previous network gives new knowledge. If this knowledge is using numeric values as its weight then it can exchange the weight associated with each arc or node.
6. A combination of two or more techniques can incorporate to form a new architecture called **hybrid networks**. This network is a single or a separate network that needs to be a closely interacting network.

Examples of semantic networks are given in Figure 6.14 and Figure 6.15. By traversing network (Figure 6.14) it is found that, Hari has a head (by inheritance) and that certain concepts are related in some ways (For example, apples and elephants). It is not well-defined the meaning of semantic networks, always. Some times confusions arises while asking questions from the semantic networks. A set of questions such as:

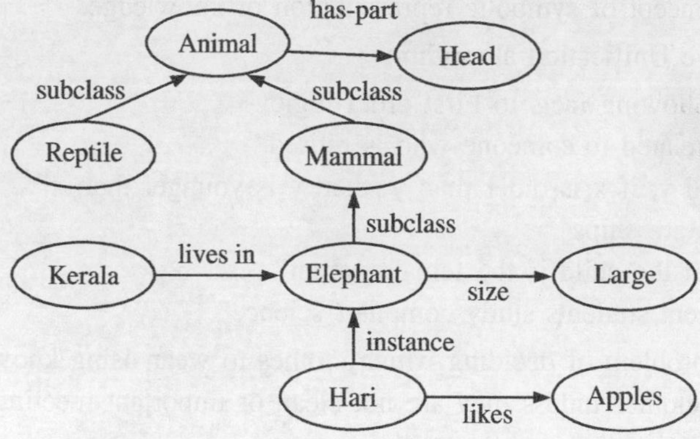

Figure 6.14 A semantic network—animals.

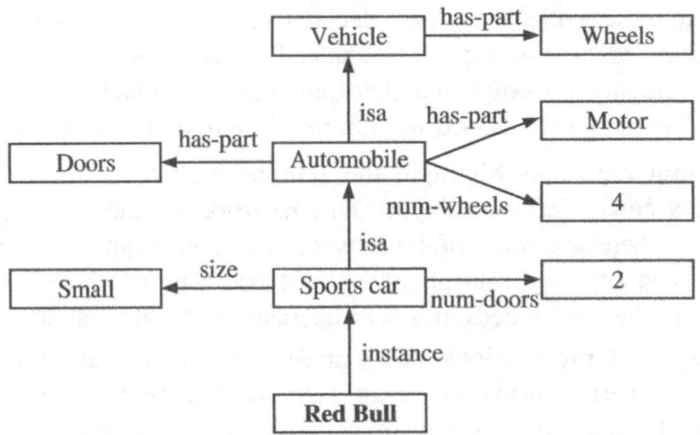

Figure 6.15 A semantic network—vehicle.

Are all elephants big, or just typical elephants?
Do all elephants live in the same Kerala?
Do all animals have the same legs?

These things are defined by machine processing. Modern trend is to have particular knowledge representation languages that look a bit like frames to users. However, which use logic to define a relation mean and do not provide the full power of predicate logic, but a subset that allows efficient inference.

There are some problems with semantic networks like they are computationally costly at run-time as we need to traverse the network to answer some question. We need to traverse the entire network in the worst case, because it is a graph structure. The search may be failed if the knowledge is not existing in the network node. Since the model is like human associative memory, i.e. stores information using associations. It is not practical to build a large semantic network because that is logically inadequate.

6.7 Exercises

1. What do you mean by knowledge representation?
2. Explain the concept of symbolic representation of knowledge.
3. State and prove Unification algorithm.
4. Convert the following facts to First order logic:
   "Everyone is related to someone who is retired."
   "For any x and y, if x is older than y, then y is younger than x".
   "All freighters are ships".
   "The mother of the child is the female parent".
   "Some intelligent students study computer science"
5. Consider the problem of deciding which clothes to wear using knowledge such as:
   "Wear casual clothes unless they are not clean or important meeting occurs today".

"Wear a sweater if it is cold".
"The winter is usually cold".
"Wear shorts if it is warm".
"The summer is usually warm".

6. Consider the following facts:
   - John likes all kinds of pets.
   - Dogs are pets.
   - Cats are pets.
   - Any animal anyone owns and is not killed is a pet.
   - Reji owns a goat and is still alive.
   - Vinod owns everything Reji owns.

   (a) Translate the facts into formulae in predicate logic.
   (b) Convert the formulae into clausal form.
   (c) Prove that Jack likes goats using resolution.

7. Given the following a knowledgebase:
   - If x is on top of y, y supports x.
   - If x is above y and they are touching each other, x is on top of y.
   - A cup is above a book.
   - A cup is touching a book.

   (a) Represent statement in predicate logic.
   (b) Translate statements 1 through 4 into clausal form.
   (c) Show that the predicate support (book, cup) is true using resolution.

8. Formalise the following sentences in first-order logic, using the following predicates: S(x): x studied for the test; W (x): x did well in the test; B(x, y) : x did better than y in the test.
   - Some students who did well did not study for the test.
   - Every student who studied did better than every student who did not study.
   - Some students did better than everyone else.
   - The student who did the best in test studied for it.

9. Consider the following facts:
   - Team India
   - Team England
   - Final match between India and England
   - India scored 350 runs, England scored 350 runs, India lost 5 wickets and Australia lost 7 wickets
   - The team which scored the maximum runs wins
   - If the scores are same then the team which lost minimum wickets wins the match.

   Represent the facts in predicate, convert to clause form and prove by resolution "India wins the match".

10. Consider the following axioms:

- Every child loves every candy.
- Anyone who loves some candy is not a nutrition fanatic.
- Anyone who eats any pumpkin is a nutrition fanatic.
- Anyone who buys any pumpkin either carves it or eats it.
- John buys a pumpkin.
- Lifesavers is a candy.

Conclusion: If John is a child, then John carves some pumpkin. Prove by resolution.

11. Consider the following axioms:

- Every child loves Anand.
- Everyone who loves Anand loves any reindeer.
- Rudolph is a reindeer, and Rudolph has a red nose.
- Anything which has a red nose is weird or is a clown.
- No reindeer is a clown.
- Scrooge does not love anything which is weird.

Conclusion: Scrooge is not a child.

Represent these axioms in predicate calculus; skolemise as necessary and convert each formula to clausal form. (Note: 'has a red nose' can be a single predicate. Remember to negate the conclusion). Prove the unsatisfiability of the set of clauses by resolution.

12. Consider the following axioms:

- Anyone who loves any lottery is a gambler.
- Everyone who favours the lottery proposition loves some lottery.
- Everyone favours the lottery proposition or opposes the lottery proposition.
- If every Baptist votes and opposes the lottery proposition, then the lottery proposition does not win.
- Every Baptist who is faithful is not a gambler.

Conclusion: If every Baptist votes and the lottery proposition wins, then some Baptist is not faithful. Prove it by resolution.

13. Consider the following axioms:

- Anand is a town in Gujarat state.
- Gujarat is a state in India.
- Amul is the milk product from Anand.
- White revolution was a great success in India.
- Amul is the proof of white revolution success.

Conclusion: White revolution was a great success in Anand. Prove it by resolution.

14. Consider the following axioms:

- Everyone who feels warm is either drunk, or every costume they have is warm.
- Every costume that is warm is furry.
- Every artificial intelligence student is a CS student.

- Every artificial intelligence student has some robot costume.
- No robot costume is furry.

Conclusion: If every CS student feels warm, then every artificial intelligence student is drunk. Prove it by resolution.

15. Consider the following axioms:

- Every bird sleeps in some tree.
- Every loon is a bird, and every loon is aquatic.
- Every tree in which any aquatic bird sleeps is beside some lake.
- Anything that sleeps in anything that is beside any lake eats fish.

Conclusion: Every loon eats fish. Prove it by resolution.

16. Consider the following axioms:

- Anything that is played by any student is tennis, cricket, or chess.
- Anything that is chess is not vigorous.
- Anyone who is healthy plays something that is vigorous.
- Anyone who plays any chess does not play any cricket.

Conclusion: If every student is healthy, then every student who plays any chess plays some tennis. Prove it by resolution.

17. Consider the following axioms:

- Anyone whom John loves is a cricket star.
- Any student who does not pass does not play.
- Vinod is a student.
- Any student who does not study does not pass.
- Anyone who does not play is not a cricket star.

Conclusion: If Vinod does not study, then John does not love Vinod. Prove it by resolution.

18. Express the following expressions as semantic nets:

1. in long form
2. and in short form

- John eats more slowly than Vinod.
- For a member of the CS group, John eats slowly, but not for a member of the EC group.
- It is hardly to be assumed that Vinod will ever eat as slowly as John.

19. The following statements are given:

- A house is a (kind of) building.
- A house has at least one storey (the number of storeys of a house is one or more).
- A house is used for living in.
- A single-storey dwelling is a (kind of) house.
- A single-storey dwelling has one storey.
- My house is an instance of a single-storey dwelling.

- My house has its roof colour red.
- My house has its walls made of brick.

Express these statements as semantic net in short form.

20. Express the following information:
   (a) in an appropriate logic
   (b) as a semantic net (short form)
   (c) and as frames

   A is an action of type "give" with agent Prometheus, object fire, and patient mankind. Prometheus is a subject of Zeus.

   A is an action of type that "disallows" with agent Zeus. The object of A is any action of giving fire to mankind. If the ruler disallows an action, which will be performed by one of his subjects, another action takes place, in which the ruler punishes the subject.

21. Construct semantic net representations of the following:
   (a) John is welsh, John is a professor.
   (b) Hari lent his new data structures CD to his best friend.

22. Represent the following in partitioned semantic networks:
   (a) Every player kicked a ball.
   (b) All players like the referee.
   (c) John believes that there is a fish with lungs.

23. Devise algorithms that enable reasoning with frames. Discuss how.
   (a) Inference through inheritance can be achieved.
   (b) Matching can be achieved.

24. Construct CD representation of the following:
   (a) John begged Mary for a pencil.
   (b) Vinod stirred his coffee with a spoon.
   (c) Reji took the book off Jim.
   (d) On my way to home, I stopped to fill my car with petrol.
   (e) I heard strange music in the woods.
   (f) Drinking beer makes you drunk.
   (g) John killed Mary by strangling her.

25. Write a script to allow for both outcome of the Bank robbery: Getaway and going wrong and getting caught.

26. Write a script for enrolling as a student for an examination.

27. How do we precisely define the semantics of a frame system or semantic network?

28. How can knowledge be represented using predicate logic?

29. Can a frame inherit attributes from more than one parent?

CHAPTER 7

Reasoning

The mental ability found in humans, which is able to generate conclusions from assumptions or premises can be termed as reasoning. In other words, it is amongst other things and means by which rational beings propose specific reasons, or explanations of cause and effect. In contrast to reason as an abstract noun, a reason is a consideration which explains or justifies. When we need a system to perform a task which has not been told, then the system needs to reason it. It must figure out what to do, from the knowledge it already has. We have already come across such examples of reasoning in previous chapters. For example, if we know Parrots is a bird and all birds have wings. Then we may ask: do Parrots have wings? This is a simple example for reasoning; there are many situations where we need to answer the questions to get an inference.

7.1 Types of Reasoning

There are many types of reasoning methodology, which generate couple of knowledge in artificial intelligence. The major types of reasoning are explained with the help of natural sentences as follows:

7.1.1 Abductive Reasoning

Abduction is the process of creating explanatory hypothesis. This is the process by which we try to explain things which were observed or felt in some manner but we have no existing knowledge or proof to substantiate the phenomenon. It thus needs some hypothesis which needs to be further worked on to verify whether it was true or not.

For example, when a doctor encounters a patient who has the same symptoms of someone whom he had treated before, may wonder whether the present condition is some what similar to that.

Abduction is a reasoning process that tries to form reasonable explanations for abnormal observations and which is distinctly different from induction and deduction. Uncertainty is an important issue in abductive reasoning and has formalisms in representation of reasoning about its uncertainty. Mycins (an expert system for medical diagnosis) certainty factors, truth maintenance systems, Bayesian belief networks, fuzzy logic and non-monotonic reasoning are some abduction systems. For example, consider a medical diagnosis:

Facts: symptoms, lab test results, and other observed findings.

Knowledgebase: causal associations between diseases and indicators.

Reasoning: casually explain the occurrence of the given indicators through one or more diseases whose presence is in the facts.

Natural language processing, image understanding and criminal investigation are some abductive reasoning systems. In abductive reasoning, conclusions are hypothesis not theorems. For example, rules A → B and C → B, and fact B, both A and C are plausible hypotheses. These hypotheses can be ranked based on their credibility.

7.1.2 Deductive Reasoning

It starts with a general rule and moves to a specific direction. This means the process starts with a general case and deduces itself to specific instances. 'Hypothetical deduction' starts with an assumed hypothesis or theory. The argument it is not questioned because the assumption may be well-accepted or it may be rather shakier. Scientists use deduction, who take a general scientific law and apply it to a certain case. The scientists are assuming that the law is true. Deduction can be used to test an induction by applying it. Although in the test cases, the initial theory is assumed to be true only temporarily.

For example

'You have tickets of cricket match. You agree to meet John and Maya at the corner of special box or at the seats'.

If you see Maya on the corner of special box, you expect to see John as well.
If you do not see either of them at the corner, you expect to see them at the seats when you get to the stadium.
Logical arguments
(John AND Maya) will be located at corner OR
(John AND Maya) will be located at seats
NOT (John AND Maya) located at corner
(John AND Maya) located at seats

7.1.3 Inductive Reasoning

Starting from specifics and deriving a general rule is the key for induction. Inductive reasoning, or induction, is reasoning technique that forms from a specific case or cases and reaches a general rule. From observations, it draws inferences in order to make generalisations.

Inference can be done in the following four stages:

1. **Observation:** Without a bias, collect facts.

2. **Analysis:** Identifying patterns of regularity by classifying the facts.
3. **Inference:** These patterns are used to infer generalisations about the relations between the facts.
4. **Confirmation:** For further observation testing the inference is done.

In an argument, we might do the following steps:

1. Derive a general rule in an accepted area and then apply the rule in the area where we want the person to behave.
2. Give those lots of detail, and then explain what it all means.
3. Think about the benefits of all the parts and only get to the overall benefits later.
4. Take situations that have already happened and provide explanations on why it happened.

Inductive arguments can include:

- **Part-to-whole:** Where the whole is assumed to be like individual parts (only bigger).
- **Extrapolations:** Even the unstudied area is considered and treated as those which was already learned.
- **Predictions:** Where the future is assumed to be like the past.

Difference between deduction, abduction and induction can be explained in a simple manner. Reasons from causes to effects is deduction but reasons from effects to causes is abduction. Induction is reasons from specific cases to general rules. Figure 7.1 gives differences between three reasoning rules. We can explain these three reasoning through natural sentences.

(a)	(b)	(c)
$A \Rightarrow B$	$A \Rightarrow B$	Whenever A then B
A	B	------------------------
---------	------------	Possibly
B	Possibly A	$A \Rightarrow B$

Figure 7.1 (a) Deductive reasoning (b) Abductive reasoning (c) Inductive reasoning.

Deduction:

Major premise: All jerseys in the bag are blue.
Minor premise: These jerseys are from the bag.
Conclusion: These jerseys are blue.

Abduction:

Rule: All jerseys in the bag are blue.
Observation: These jerseys are blue.
Explanation: These jerseys are from the bag.

Induction:

Case: These jerseys are from the bag.
Observation: These jerseys are blue.
Hypothesised rule: All jerseys in the box are blue.

7.1.4 Analogical Reasoning

It is relating things to novel other situations. In this method of reasoning for every situation an analogy of the similar one is taken into consideration. It then tries to create a mapping on the attributes of the analogical situation to the present problem.
To use an analogy, the following steps are involved:

- Fix and start with the target problem (domain) for which you need to create new understanding.
- Find a similar domain so that we get some attributes similar to our target.
- Find exact items from the matching domain.
- Find associated items in the target domain.
- Transfer attributes from the matching domain to the target domain. A mapping must be created from the corresponding matching and target domain.

For example,

'This organisation is like an express train. It runs fast and reaches the destination in-time, and now it needs cleaning and rest for a while'.
'Today is neither sunny day nor rainy day. We do not need an umbrella'.

7.1.5 Cause-and-effect Reasoning

This reasoning shows causes and resulting effect. When we deal with a problem, it will show a set of causes and effects that are involved in it. It thus gives an understanding why a particular event happened or how this will happen if we move forward. This also draws an inevitable link between what happens first and what happens next. If we look over the correlation that may show coincidence to give certain evidence of causality. If you cannot have a causal linkage, then you may be successful just by asserting it. This is because of few people who will challenge a cause-and-effect assertion.

For example,

'If I transfer the money to your bank account, you will be more rich. I will transfer the money'.

'Cause' is the reason why something happens and 'effect' is the result of something happening. For example, consider the following sentences:
'I did not bring my homework, because my dog ate it'.
'I am happy, because today I got 100 marks in mathematics'.

The first sentence is an example for a cause in reasoning. That is, 'failed to bring the homework' is the action that was caused because of damage by dog. The second sentence gives an example of effect in reasoning. The happiness is effected due to 100 marks in mathematics subject.

7.1.6 Comparative Reasoning

It may compare one thing against another. Comparative reasoning helps to find the importance of some attributes by comparing it with other. The gap size between the things compared indicates importance. If we compare against a high grade attribute, then we feel ours as unimportant and while comparing against a low grade will make our attribute look good. When a size gap between the objects are compared, it is always important to create logical arguments that we need to fix the validity of comparison benchmark. For less logic, we may merely assume the benchmark. There are various ways to compare, for example:

- Comparison of people what they have got (or not got) against what others have.
- Past is compared with future.
- Comparison of what is actual with what is ideal.
- Words and actions are compared against values.

For example,
'You are better than John but not up to the mark of Salim'.
'Present salary is not enough for living, so think about doubling it'.

7.1.7 Conditional Reasoning

Conditional reasoning is based on *if-then* statements. For example, if we generate a reasoning based on *if* A *then* B it means B will work only when the condition A is true. There is more scope as it opens door for situation in which what will happen if A is false. Because it is not explained whether B will be true or false when A is false. This reasoning system is highly useful when we design a rule based system for a problem. These rules are forwarded from an expert's viewpoint. For example,

If tank is full
then
switch-off the pump

7.1.8 Criteria Reasoning

It is the process of comparison against an already established criterion. It begins with defining the criteria from which the outcome of a particular decision is judged. The outcomes are then evaluated and the best criteria are selected. In logical arguments, it takes more time to establish a criterion but in less logic situation, we assume the criteria and start working on it. Criteria which appeal to common values are likely to be easily accepted.

For example,

'For getting an admission for this institution you must have 80 percent marks in the qualifying examination'.

'I am looking a fair girl rather than a dark girl'.

7.1.9 De-compositional Reasoning

As the name suggests, it is the reasoning by which you break down the whole problem into small understandable parts. Then these small parts are analysed and the conclusion is drawn on how they fit together. Finally the conclusion on the whole matter is drawn.

For example,
'To know how a Rubik cube operates. I will pull it apart to see its hidden workings. By slowly reassembling it, I am able to explain its apparently magical organisation as a whole in terms of geometry'.

7.1.10 Exemplar Reasoning

The use of examples in argument is the main use of exemplar reasoning. This reasoning may give a story or may be a short comparator as an example. The duplication is possible in a situation or may be a relatively distant metaphor for the reasoning. A known person, recognised situation or something not directly known to the other person is useful in exemplar reasoning.

For example,
'You should go out more frequently. I have a pal who used to stay in and was never really joyful'.

7.1.11 Modal Logic and Traditional Logic

It is the argument about necessity and possibility. It describes the situations based on necessity and possibility. Also explore how they intertwine. The process is not carried out by stating the truth but by its likelihood. Thus, it uses words like 'can', 'may', 'should', 'ought', 'must', 'have to'. How true or necessary something is gives you more potential in arguments. If you have an analogue continuity of alternatives, rather than the black-and-white binary decision of simply whether something is true or false, necessary or unnecessary are major concerns of this logic.

Traditional logic was originated by Aristotle. It obeys formal rules and facts. It starts with premises that are assumed to be true then uses only logical rationale to derive a conclusion. Always check if it is applied correctly. Keep emotion well out of it.

For example,
'Some applies are green. Some apples are red. I do not know if there are any green and red apples'.

Based on extension, traditional logic works and based on intention, modal logic works. We have to find supporting statements with many situations that it is difficult to determine exact truth.

7.1.12 Pros-vs-cons Reasoning

Pros are argument of a case and cons are arguments against the case (cons). Pros-vs-cons reasoning seeks to weigh up the arguments for a case against the arguments against the case.

A conclusion is arrived from the arguments that have checked whether the pros or cons were stronger. Thus, precipitating a 'reasonable' conclusion is formed. Things that will make a 'pro' stronger (and vice-versa) include:

- By including logical arguments.
- By providing evidence that could be well displayed. This includes actions and perceptions of other people.
- Greater emphasis being put on keywords.
- Stick with the subject and provide more arguments for the case.

We may start discussions either by supporting or standing against the situation. While starting with the favoured side it allows you to fill the other person's mind with the key points, such that the second list becomes less easy to absorb. But if you are starting with the disfavoured side it allows you to make it sound reasonable, then knock down each of the disfavoured arguments with stronger arguments for the contrary case. Any how the final conclusion will always be drawn by checking which was stronger, i.e. pros or cons.

For example,

'Vinod likes it, John likes it, Raji likes it, Hari likes it. Only Anand and Maya do not like it. Most of them like it'.

7.1.13 Set-based Reasoning

Set-based reasoning is normally used in set theory. Its arguments range around whether things are members of named groups or not. For instance, a 'cat' is an animal but not a 'vegetable'. These members can be indifferent groups. A basic assumption of set-based reasoning is that an item can be categorised into a given group or set. This also assumes that both the item and the set exist in the first place. The following arguments then may include consideration of the go beyond sets and the implications of this. Set reasoning thus often includes statements along the lines of:

A is a B
If A is a B then...
A is not a B, but it is a C
A is both C and D, therefore...

For example,
"He works for Defence. Defence people are intelligent. Therefore he is intelligent".
"If this is an international standard product then it will use ISO standard encryption coding".

7.1.14 Systemic Reasoning

This is a most generic approach, we can try to understand a case by considering the whole system. The various attributes are checked and analysed. The relation between them is also validated. You can use de-compositional scheme of reasoning to identify case which does not belong to your consideration.

For example,

'I am planning to purchase a villa in the middle of the city by considering landscaping, swimming pool and tall building around it. I am also considering pleasant movement of the people during the day time especially on Sundays'.

7.1.15 Syllogistic Reasoning

During conversation we make many sentences and try to imply logical connection between them. But many a times the truth and logic that we assume will not be there. The following statement and conclusion will help in understanding more about it. Consider the following statements and conclusions:

Statement 1: "All human beings are animals".
Statement 2: "Some animals are aggressive".
Conclusion: "Some people are aggressive".

This seems to be a reasonable conclusion, but then consider the following:

Statement 1: "All human beings are animals"
Statement 2: "Some animals are male".
Conclusion: "Some human beings are male".

Reasoning is exactly the same as in this example, conclusion appears to be ridiculous and false. See the first example, it arrived a false conclusion. The animal who are aggressive are not necessarily human being. We know that it is true logic for statement in certain situations. When speaking about assumptions and beliefs to less acceptable ends, we can use Venn diagram to visualise reasoning by set theory.

7.1.16 Bayesian Reasoning

The principle of uncertainty is distributed over input, output and knowledge. The uncertainty of inputs is due to missing and/or noisy data. The uncertainty in knowledge is due to several causes leading to multiple effects, incomplete record of conditions, or effects, and probabilistic effects. The output uncertainty is due to inherently uncertainty in abduction and induction. It may also be due to incomplete deductive inference.

Bayesian reasoning based on a Bayesian network can overcome such problems using the probability principles. Bayesian networks are solid representation of probability distribution over a set of propositional arbitrary variables taking advantage of independence relationships. The probability of model *m* is obtained by Bayes rule as follows:

$$P(m|E) = \frac{P(E|m) \times P(m)}{P(E)} \tag{7.1}$$

$P(E|m)$, is the probability of the model that would have produced the dataset. When the model is good to fit to the data, then the value is high and the value is low when the model would have predicted different data. Bayes theorem is the base of the Naive Bayesian classifier.

Inside the predictors there are independent assumptions. Due to power of probability and statistics, Naive bayesian classifier outperforms in the classification and is widely used as a machine learning technique.

From the Bayes theorem, the posterior probability $P(c|x)$ is calculated from $P(c)$, $P(x)$ and $P(x|c)$. Figure 7.2 gives the basic equation of Bayes theorem. It is assumed that, for a given class c the effect of value of a predictor x is independent of the values of other predictors.

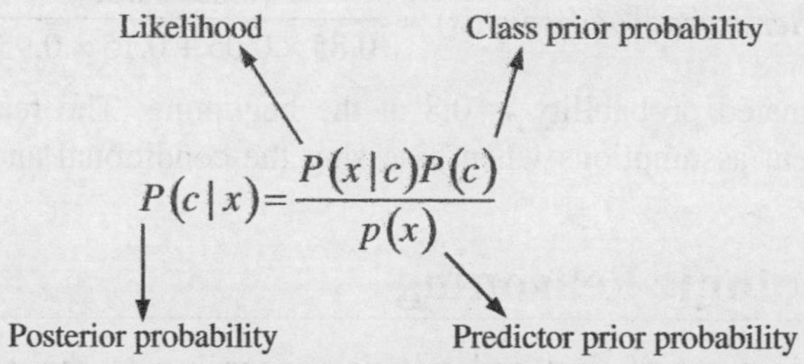

$$P(c\mid x)=\frac{P(x\mid c)P(c)}{p(x)}$$

$$P(c \mid x) = P(x_1 \mid c) \times P(x_2 \mid c) \times \cdots \times P(x_n \mid c) \times P(c)$$

Figure 7.2 Bayes formula.

Where $P(c|x)$ is the posterior probability of target class for a given predictor, $P(c)$ is the prior probability of class, $P(x|c)$ is the likelihood which is the probability of predictor given class, and $P(x)$ is the prior probability of predictor.

If all rules in the knowledgebase are represented like,

IF E is true THEN H is true {with probability P}

This rule implies that if event E occurs, then the probability that event H will occur is P. When we make an expert system, H usually represents a hypothesis and E denotes evidence to support this hypothesis. The probability values are required as the primary input of the framework of Bayesian reasoning. A human judgement is involved in the assessment of these values. Normally such a human intervention is not possible with the Bayesian rules. The probabilities given by the expert are normally taken because they are consistent.

For example, a car that does not start and makes odd sounds when you press the starter. The conditional probability of the starter being faulty if the car makes odd sounds may be expressed as:

IF symptom is 'odd sounds' THEN the starter is bad {with probability 0.8}

The car that does not start and makes odd sounds when you press the starter. The conditional probability of the starter being faulty if the car makes odd sounds may be expressed as:

P(starter is not bad | odd sounds) = P(starter is good | odd sounds) = 1 – 0.8 = 0.2

We can obtain a companion rule that states as follows:

IF the symptom is 'odd sounds' THEN the starter is good {with probability 0.2}

It is not possible to deal with conditional probabilities by the domain experts and often deny the very existence of the hidden implicit probability. Consider the following rules:

IF the starter is bad THEN the symptom is 'odd sounds' {probability 0.85}

IF the starter is bad THEN the symptom is not 'odd sounds' probability 0.15}

We need prior probability for using Bayesian rule, the probability that the starter is bad if the car does not start. Suppose, the expert supplies us the value of 5 percent.
Now, we can apply the Bayesian rule to obtain:

$$P(starter\ is\ bad|odd\ sounds) = \frac{P(0.85 \times 0.05)}{0.85 \times 0.05 + 0.15 \times 0.95} \tag{7.2}$$

The expert estimated probability is 0.8 at the beginning. The reason for inconsistency is expert made different assumptions when assessing the conditional and prior probabilities.

7.2 Non-monotonic Reasoning

The knowledgebase is a set of first order logics according to the rules of inferences. In monotonic reasoning, all assumptions are true facts. But in non-monotonic reasoning facts are true assumptions and hypotheses that have an assumption are presumed to be true. The non-monotonic reasoning is good because we can update the knowledgebase by adding or deleting a belief as and when required. Also, we can use the existing knowledge to solve conflicts if there are contradictory facts. Conventional reasoning using predicate logic is based on the following three important assumptions:

1. The predicate descriptions must be sufficient with respect to our application domain. That is, all information necessary to solve the problem must be represented.
2. The information base must be consistent, that is, pieces of knowledge cannot contradict each other.
3. Through the use of inference rules, the known information grows monotonically.

If any of these is not satisfied, the conventional logic-based approach will not work. Non-monotonic systems address each of these three issues.

1. Reasoning systems are often faced with a lack of knowledge about domain.
2. An assumption required to traditional logic-based systems is that the knowledge supporting reasoning must be consistent.
3. If we wish to use logic to address the problem of how a knowledgebase is updated.

Motivation of the non-monotonic reasoning is to define a strong and flexible system to deal with situations of incomplete knowledge by considering syntax and semantic rules. So, it has a practical implementation. We can simply define a monotonic property as "Consider a theory T and its complement T', then $T \vdash A - T \rightarrow T' \vdash A$". Here non-monotonic system violates this property.

Non-monotonic reasoning are of two types—auto-epistemic reasoning and default reasoning. In auto-epistemic reasoning incomplete representation of complete knowledge is given depending on context. It is like 'P is valid because otherwise I would have information that leads to $\neg P$'. There are two types of auto-epistemic reasoning—explicit auto-epistemic reasoning and subjective auto-epistemic reasoning. The first one deals with using an explicit convention to draw conclusions. For example, 'My next class will be on Monday'. The convention used is 'Classes are every Monday unless we are told otherwise'. In subjective auto-epistemic reasoning, there is no convention given but we must believe that *P* must be true or we will have the evidence to the divergence. For example, "The college is still on strike". The briefing is "If the strike was over someone would have told me".

Default reasoning has a rational conclusion from partial information. It has a new evidence that can invalidate the conclusion. There are four types of default reasoning such as prototypical, no-risk, best-guess and probabilistic. In prototypical default reasoning typically *P* is true so it is rational to assume it is true now. For example, Bird (Tweety), typically birds can fly, therefore Fly (Tweety). In no-risk default reasoning, the cost of making a mistake that $\neg P$ outweigh the cost of making a mistake that *P* then assume *P*. For example, "The course may be be cancelled but just in case I will do the exercise anyway because if I am wrong I will fail in the course". It is assumed that course will be cancelled and is wrong that the cost of assuming the course will not be cancelled and may be wrong. In best-guess reasoning no evidence supports *P* or $\neg P$ choosing whichever is more suitable. For example, 'John likes to enrol the college but he does not know which gate is open to enter that is closest'. In probabilistic reasoning if *P* has a high probability then assume *P*. For example, 'The probability that John will expect the marks 60 for Data structure course is 0.8, therefore he can assume that he will get at least 60 marks'.

7.2.1 Non-monotonic Inference Methods

These methods are not valid forms of inferencing. But they are important since they are being used in everyday activities. These are as follows:

1. **Abductive inference:** This is based on the use of known knowledge to explain or justify a valid conclusion, given the truth of propositions P and Q and the implication, $P \rightarrow Q$, conclude P.

   For example: People who are drunk stagger when they walk. It is not reasonable to conclude that a person who is staggering is drunk.

2. **Inductive inference:** This is based on the assumption that a recurring pattern observed for some event or entity implies that the pattern is true for all entities in their class.

   For example: given $p(a_1)$, $p(a_2)$, ... Then conclude $p(x)$

3. **Analogical inference:** This is a form of inference obtained with experience. When we find that A is related in certain way to B and A′ is similar in some context to A′, then we conclude that B′ has some relation to A′ in same context.

   Example: If a 2 × 2 system of linear equations can be solved in some manner, then the same method can be adopted for a 3 × 3 system also.

Truth Maintenance Systems (TMS) have been developed as a variety of means implementing non-monotonic reasoning systems. The truth maintenance system is a constraint system among objects in first order predicate logic that keeps a consistent subset of theorems, according to the constraints. It executes inferences of the form 'If P is consistent then Q'.

7.2.2 Truth Maintenance Systems

The idea of a Truth Maintenance System (TMS) grew out of dependency-directed backtracking. A TMS may be employed to protect logical integrity of the conclusions of an inferring system. A TMS allows assertions to be connected via a spreadsheet like network of dependencies. In a Justification-based TMS (JTMS), the TMS itself does not know anything about the structure of the assertions themselves. The TMS's only role is not to serve as a book keeper for a separate problem-solving system, which in turn provides it with both assertions and dependencies among assertions.

Applications:

- Solve search problems, example, schedule meetings so that everyone can attend.
- Determine what derived conclusions are still true when some facts change.
- Determine what part failures could have caused a machine to fail.

ATMS

Assumption-based Truth Maintenance System or ATMS is a powerful tool for different applications in the domain of uncertain reasoning. ATMS keeps for each proposition a set of sets of assumptions under which the proposition is true. A proposition is believed at a given time if one of the assumption sets supporting it are satisfied. A traditional ATMS is based on propositional logic, but it is limited to Horn clauses and to simple queries. Horn clause is a clause (a disjunction of literals) with at most one positive literal. An ATMS requires a subset of propositions declared as assumptions. Assumptions are needed to express uncertainty. The fundamental ATMS problem is to identify combinations of assumptions (labels) such that a given query holds. The weak point of the traditional ATMS concept is the restriction to Horn clauses. In many cases, also non-Horn clauses are needed to express the knowledge. The advantage of ATMS lies in its efficient computations.

JTMS

Justification-based Truth Maintenance System (JTMS) keeps for each derived proposition a list of sets of propositions from which it was derived, and for each proposition, a list of propositions it supports. When a proposition is retracted, the system checks each proposition that depends on it, and if the proposition no longer has support, retracts it. There are three main operations that are performed by the Justification-based TMS.

1. The JTMS inspects the network of justifications. This inspection can be triggered by queries from the problem solver such as:
   "Should I believe proposition P?",
   "Why should I believe proposition P?" and
   "What assumptions underlie proposition P?"
2. The second operation of JTMS is to modify the dependency network, where modifications are driven by information supplied by the problem solver. Modifications

include adding new propositions, adding or removing premises, adding contradictions and justifying the belief in a proposition.

3. The final operation of the JTMS is to update the network. This operation is executed whenever a change is made in a dependency network. The update operation recomputes the labels of all propositions in a manner that is consistent with existing justifications.

Consider the following example:

$\forall x$ good-student (x) $\wedge$ M study-hard (x) $\rightarrow$ study-hard (x) $\forall y$ party-person (y) $\rightarrow$ not (study-hard (y)) $\wedge$ good-student (David) We can make this set of proposition into a justification network given in Figure 7.3.

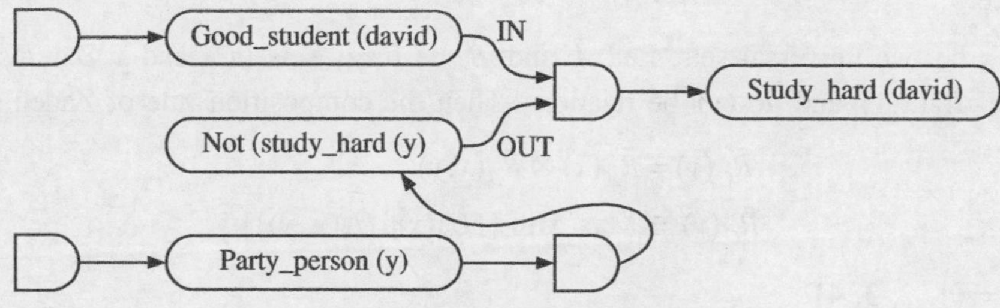

Figure 7.3 JTMS network.

7.3 Reasoning with Fuzzy Logic

Fuzzy logic is a different approach to representing uncertainty in problem-solving. Fuzzy logic focusses on describing events ambiguities rather than the uncertainty about the occurrence of an event. Modification of the definitions of set theory and logic results in the fuzzy set theory. Traditional set theory defines set memberships as a Boolean predicate.

7.3.1 Fuzzy Sets

In the usual set theory, an object is a member of a set or is not a member of the set. Traditional logics are based on the notion that *P*(*a*) is true if a is member of the set, otherwise *P*(*a*) is false. There is no parallel containment. This amounts to the use of a characteristics function for set *A*.

$$\text{If } f_A(x) = 1; \quad \text{if } A \in \text{A}$$
$$= 0; \quad \text{otherwise}$$

We may generalise the notion of *x* by allowing the characteristic functions to have values other than 0 or 1. We define the notion of a fuzzy set with characteristic function '*U*', which maps the universal set *U* to the values in the interval [0, 1]

That is, $u: U \rightarrow [0, 1]$

Let *U* be a set and x' be an element of *U*. A fuzzy set *A* is a set of ordered pair $(x, U_A(x))$, $\forall x \in U$ where U_A is known as membership function whose value lies in the interval [0, 1].

Example: Let TALL(x) → height of a person. We can define the fuzzy set A with the membership function values as follows:

$U_A(0) = 0 = U_A(10) = U_A(20) = U_A(30) = U_A(40)$

$U_A(50) = 0.2$ $U_A(60) = 0.4$ $U_A(70) = 0.6$ $U_A(100) = 1$

All the properties and theorems holding good for the set theory will also hold good for fuzzy sets.

7.3.2 Fuzzy Reasoning

There are different methods of fuzzy inference. We can consider Zadeh's compositional rules of inference.

Let x and y be two universal sets. Let $\tilde{A}$ and $\tilde{B}$ be fuzzy sets in x and $x \otimes y$ respectively. Let $\tilde{R}_A(x)$, $\tilde{R}_B(x,y)$ and $\tilde{R}_C(y)$ be relations. Then the composition rule of Zadeh's is

$$\tilde{R}_C(y) = \tilde{R}_A(x) \otimes \tilde{R}_B(x,y)$$

$$\tilde{R}_C(y) = \text{Max Min } [\{U_A(x), U_B(x, y)\}]$$

Let $x = y = \{1, 2, 3, 4\}$

A = Little = {(1,1), (2,0.6), (3, 0.2), (4, 0)}

Let R denotes approximately equal to a fuzzy set (Table 7.1).

TABLE 7.1 Values in fuzzy table

x/y	1	2	3	4
1	1	0.5	0	0
2	0.5	1	0.5	0
3	0	0.5	1	0.5
4	0	0	0.5	1

$$\begin{aligned}\tilde{R}_C(y) &= \text{Max Min } [\{U_A(x), U_B(x, y)\}] \\ &= \text{Max Min } [\{(1,1), (0.6, 0.5), (0.2, 0), (0,0)\} \\ &\quad \{(1, 0.5), (0.6, 1), (0.2, 0.5), (0, 0)\}, \{(1, 0), (0.6, 0.5), (0.2, 1), (0, 0.5)\} \\ &\quad \{(1, 0), (0.6, 0), (0.2, 0.5), (0, 1)\}] \\ &= \text{Max } [(1, 0.5, 0, 0), (0.5, 0.6, 0.2, 0), (0, 0.5, 0.2, 0), (0, 0, 0.2, 0)] \\ \tilde{R}_C(y) &= [1, 0.6, 0.5, 0.2]\end{aligned}$$

We might inference

Premise: x is little

Implication: x and y are approximately equal.

Conclusion: y is more or less little.

7.4 Rule Based Reasoning

Resolution by verification requires set of *wffs* be converted to clause forms but a clause form may not exactly resolving with other clause forms may not give an answer which is required by the problem. So, one can deal with the *wffs* themselves and adopt methods of a given system called a rule based system. In rule base system, we have a set of fact and rules and a goal sentence. A set of rules forms a rule base and each rule can be used to produce a new fact (that is, a conclusion) from one or more established facts. But the process can work in two different ways.

Advantages of rules

1. Knowledge comes in meaningful chunks.
2. New knowledge can be added incrementally.
3. Rules can make conclusions based on different kinds of data, depending on what is available.
4. Rule conclusions provide 'islands' that give multiplicative power.
5. Rules can be used to provide explanations, control problem-solving process, check new rules for errors.

In its place of representing knowledge in a relatively declarative, static way as a bunch of things that are true. The rule based system represents knowledge in terms of a group of rules give a different situation. A rule based system that consists of IF-THEN rules, a group of facts and some interpreter controlling the application of the rules, given the facts.

There are two ways a rule based reasoning systems works: forward reasoning (forward chaining) systems, and backward reasoning (backward chaining) systems. In a forward reasoning system you start with the initial facts, and keep using the rules to draw new conclusions (or take certain actions) given those facts. In a backward reasoning system we start with some hypothesis (or goal). To prove and keep looking for rules that would allow you to conclude that hypothesis, perhaps setting new subgoals to prove as we go. Forward reasoning systems are primarily data-driven, while backward reasoning systems are goal-driven. Both methods are useful for reasoning purpose.

In general, the reasoning in problem-solving is mainly in the forward direction. However, when the forward reasoning cannot solve the problem, an alternative way is to use the backward reasoning technique. Sometimes the forward and backward reasoning are used together for solving problem.

7.4.1 Backward Reasoning

We can adopt either forward reasoning (bottom-up resolution) or backward chaining (top-down resolution). In top-down resolution we obtain a sequence of derives as inferences and two sentences are resolved. This means that in backward chaining we start with the goal sentence, and the base rules are arrived at the given facts where as in forward chaining we start with the facts given in rule based system to obtain goal sentence. Backward-chaining architecture means that no rules are fired upon assertion of new knowledge. If we find an unknown predicate about a known piece of knowledge detected in an operator's condition list, all rules appropriate to the knowledge in question are fired until the question is answered or until quiescence. The backward chaining systems generally work from a goal state back to the start state. The backward procedure is given in Algorithm 15.

Algorithm 15 Backward chaining procedure

Inputs: A finite list of atomic sentences, $q_1, q_2, \ldots, q_n$
Outputs: Yes or no depending on whether a given knowledgebase entails all of the q_i

```
SOLVE[q1, q2, ..., qn] =
  if n = 0 then return yes
  for each clause c ∈ knowledgebase F, do
    if c = [q1, ¬p1, ..., ¬pm] and
    SOLVE [p1, p2, ..., pm, q1, q2, ..., qn]
     then return yes
end for
return no
```

Example 7.1: Consider the facts (F_1, F_2, F_3 and F_4) and rules (R_1, R_2, R_3, R_4 and R_5).

F_1 : Dog (Fido)
F_2 : ¬ Barks (Fido)
F_3 : Wags-Tail (Fido)
F_4 : Mews (Myrtle)

Rules

R_1 : Wags–Tail $(x_1) \wedge$ Dog $(x_1) \rightarrow$ Friendly (x_1)
R_2 : Friendly $(x_2) \wedge \neg$ Barks $(x_2) \rightarrow \neg$ Afraid (y_2, x_2)
R_3 : Dog $(x_3) \rightarrow$ Animal (x_3)
R_4 : Cat $(x_4) \rightarrow$ Animal (x_4)
R_5 : Mews $(x_5) \rightarrow$ Cat (x_5)

'If there a cat and a dog the cat is not afraid the dog'. Solve the problem by backward reasoning.

First we convert the goal into *wff*.

$\exists x : \exists y$: Cat $(x) \wedge$ Dog $(y) \wedge$ Afraid (x, y)

i.e., Cat $(x) \wedge$ Dog $(y) \wedge \neg$ Afraid (x, y)

By backward chaining we start from this goal state. Figure 7.4 shows the backward chaining procedure.

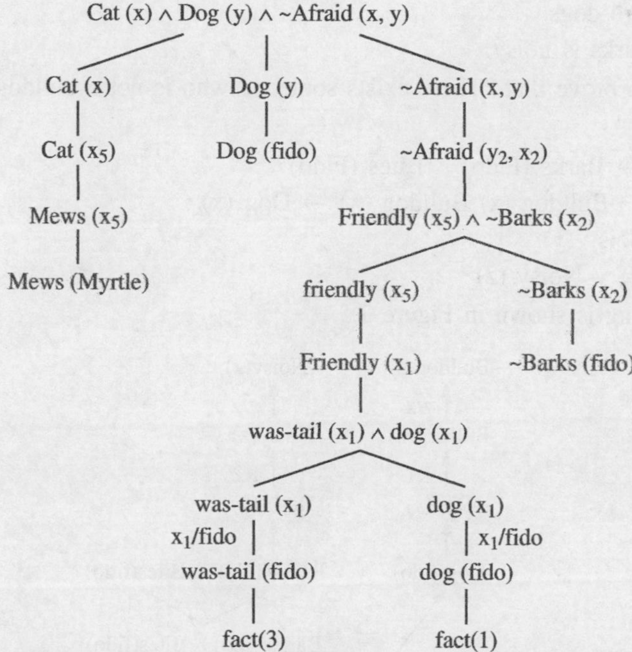

Figure 7.4 Solution using backward reasoning.

7.4.2 Forward Reasoning

In forward reasoning we obtain a sequence of assertions as inferences and two sentences are resolved. Forward-chaining implies that upon assertion of new knowledge, all relevant inductive and deductive rules are fired exhaustively, effectively making all knowledge about the current state explicit within the state. Forward chaining may be regarded as progress from a known state (the original knowledge) towards a goal state(s). The forward procedure is given in Algorithm 16.

Algorithm 16 Forward reasoning procedure

Inputs: a finite list of atomic sentences, $q_1, q_2, ..., q_n$
Outputs: yes or no depending on whether a given KB entails all of the q_i
  if all of the goals q_i are marked as solved, then **return** yes
  Check if there is a clause $[q_1, \neg p_1, ..., \neg p_n]$ in the knowledgebase,
  such that all of its negative atoms $\neg p_1, ..., \neg p_n$ are marked as solved,
  and such that the positive atom p is not marked as solved
  if there is such a clause, mark p as solved and go to step 1
  otherwise, **return** no

Consider the fact(F_1) and rules (R_1,R_2). Solve the following problem by forward chaining:

F_1: Fido barks and bites or Fido is not a dog.

R_1: All bull dogs are dogs.

R_2: Anyone who barks is noisy.

The goal is to prove that "There exists someone who is not a bulldog or who is noisy". Equivalent *wff* s are:

F_1: $\neg$ Dog (Fido) $\rightarrow$ Barks (Fido) $\wedge$ Bites (Fido)

R_1: $\neg$ Dog (x) $\rightarrow$ $\neg$ Bulldog (x) Bulldog (x) $\rightarrow$ Dog (x)

R_2: Barks (y) $\rightarrow$ Noisy (y)

Goal: $\neg$ Bulldog (z) $\vee$ Noisy (z)

The forward chaining is shown in Figure 7.5.

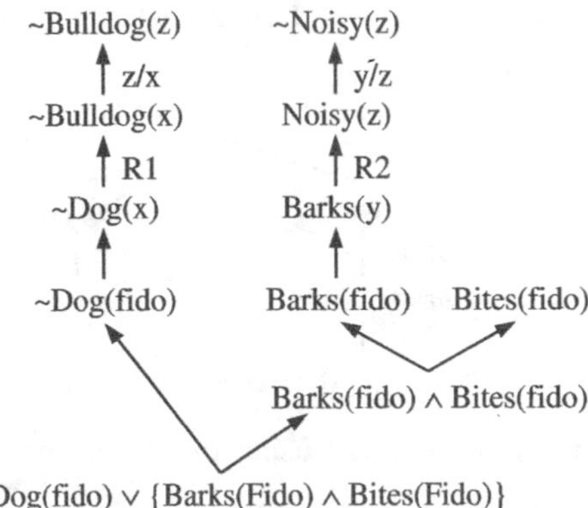

Figure 7.5 Solution using forward reasoning.

7.4.3 A Comparison

The choice of reasoning strategy depends on the properties of rule set. If there is a single goal (conclusion), backward chaining will normally be more efficient, as there is no wasteful generation of irrelevant conclusions. But if there are many different ways of demonstrating any particular fact, backwards chaining may be wasteful. Forward chaining is likely to be more efficient if there are many conclusions to be drawn or where we have a small set of initial facts. It may also be preferable if conclusions tend to have many rules. Backward chaining is likely to be more efficient where there is a single conclusion to be drawn or where the initial set of facts is large.

The approach of backward reasoning is goal driven while backward reasoning is data driven. Backward reasoning begins with the results while forward reasoning starts with new data and facts. The objectives of backward reasoning are facts to support the conclusions but forward reasoning aims to limit the result followed by some sequences. Backward reasoning is conservative approach because a specific goal can only have restricted data. The forward reasoning approach is opportunistic because it could produce different results. Backward reasoning works in reverse order and forward reasoning flow is from the antecedent to consequent.

In summary, forward reasoning is the process of searching for a solution path connecting initial fact(s) with desired conclusion(s). States are combinations of facts and each rule is a method for generating a single successor (that is, it defines a single transition). Backward reasoning is the process of searching for a solution path connecting some final conclusion with one or more initial facts. States are combinations of required conclusions and the transitions are defined by the rules. Each rule is a method for generating further 'required conclusions' from existing required conclusions.

7.5 Case Based Reasoning

Case Based Reasoning (CBR), as the name implies, is a reasoning, based on Cases. We may define CBR as the process of problem-solving where the results are based on the solution of previous similar problems. For finding solution to a problem, solutions that were opted for similar cases in the past are retrieved. It is then used as a baseline for solving the present problem. The solution derived will be tested and retained. CBR is well accepted in many cases because, we can create solution from some background knowledge rather than from scratch.

Case based reasoning is a particular technique used for the development of a knowledgebased system. CBR compares a current situation (or case) with situations that have been encountered in the past to see if one or more of the earlier situations can provide a model for how to act in the current situation. Knowledge representation is based on different rules with a different reasoning paradigm that closely reflects how experts solve problems.

For example,

'How do you find your way home?'

'Remember the way and generate new path'

Another example,

'What if it is a new location?'

'Remember a close location then adapt it'

What is 12×13?

You know 12×12 = 144

Remember this and adapt it by adding 12 giving 156.

7.5.1 How CBR Works?

Aamodt and Plaza described case based reasoning as a cyclic process comprising of four areas (Figure 7.6), namely Retrieve, Reuse, Revise and Retain(4R).

Whenever a new problem arrives, the database is checked for similar cases. Whenever it finds a match the particular case is retrieved. The retrieved problem will already have a result; it is then reused and tested for validity. If the retrieved case is not similar to the current problem, result needs to be revisited and the result thus obtained is retained for future references.

There are various CBR tools which are used for the auto-retrieval of cases. But it always requires an expert to revisit the cases and to produce valid results.

The simple CBR cycle can be further divided to a hierarchy of CBR tasks. Figure 7.7 gives a general idea on the various subtasks that are involved.

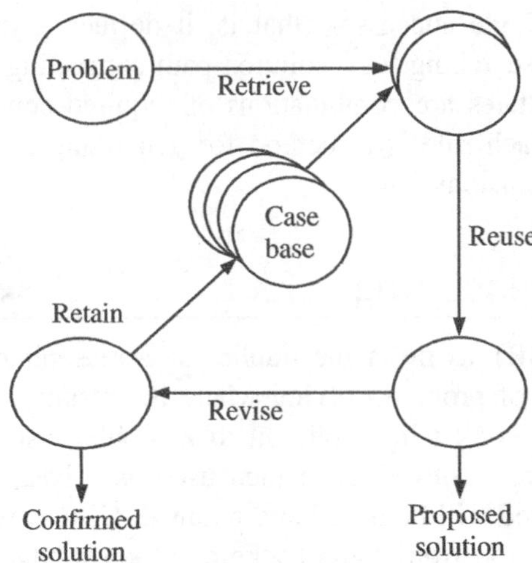

Figure 7.6 CBR steps.

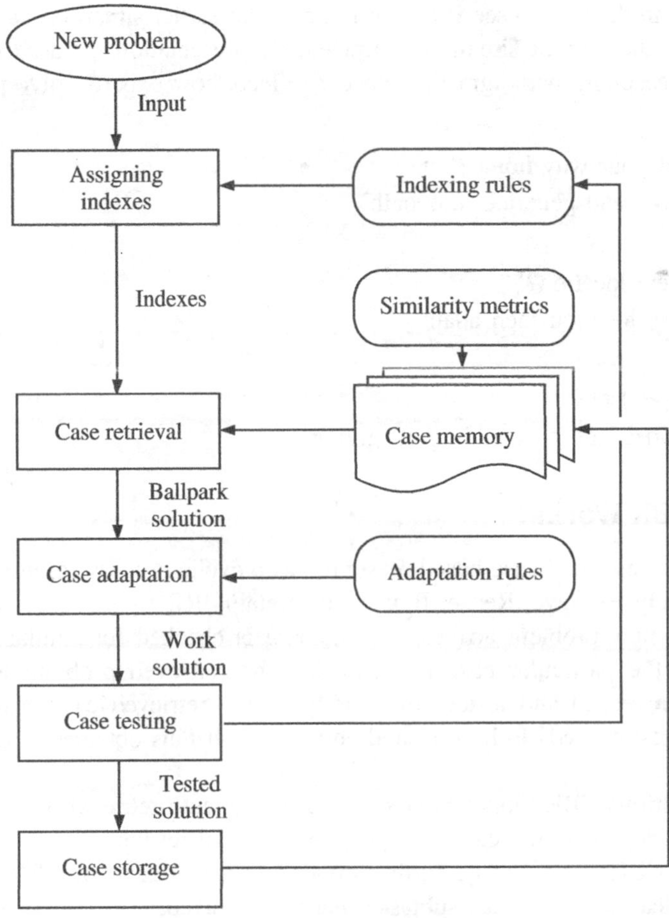

Figure 7.7 Tasks included in the CBR.

7.5.2 Case Base and Case Representation

The case base can be easily termed as a database of cases. A case is a contextualised piece of knowledge representing an experience. A case contains the past lesson that is the content of the case and context in which the lesson can be used. Typically a case comprises of the following:

- Problem description: It details the state of world when the case occurred and, if appropriate, what problem was needed to be solved at that time.
- Solution: It details solution to the problem specified in the problem description or the reaction to the given situation.
- Outcome: The resulting state of the world when the solution was applied.

However, the concepts of problem and solution have no general definitions and vary from application to application. There should be some method in which the cases need to be stored for easy retrieval. For this the most commonly used method is indexing. The indexing always needs to be predictive and must be concrete enough to be recognised in future. The cases are normally stored in a dynamic memory model, which helps to find the cases easily.

Reuse the Case(s) to attempt to solve the problem

When the match is found out, the system tries to use the same solution that is retrieved for this current problem too. However, the current problem will not be similar to the one which is so retrieved; the use of copy of the retrieved solution will not give a proper solution. Hence, the solutions need to be adjusted and modified depending on the parameters and situation where the problem is being used. This is termed as adaptation of the retrieved solution to fit the current problem. For the process of adaptation the reviewer must have good in-depth domain knowledge and must be able to adjust the parameters and configuration according to the rules and constraints. The adaptation required depends on the type of application.

Revise the proposed solution

The major task in CBR is to refine the result. As case based reasoning suggests only the solution of previous matches, it is a hard task in validating the result. The correctness can be found out by applying the revisited solution in a real environment or in some simulation. If the result is found successful, the solution is retained and stored for future references. When the generated result is found to be failure, then arises the opportunity of learning from failure. The case is revisited by an expert and he changes the user inputs and tries to create a solution specific to the needed domain. It involves not only repairing the current solution but also gives explanation for similar problems. It can also be retained for future references.

Retain new experiences as a new case in the case base

Retaining is the process of storing or incorporating valuable knowledge that we have framed for future reuse. This is the final step in the CBR cycle. It can be considered as the feedback session where the system stores data according to the degree of success. Learning takes place from the given feedback to the overall system in order to improve a certain performance using some experience or instructions. Learning can appear in many ways, like through inductive inference or through decision tress. Induction helps to refine the solution based on the examples available with their solution. Wearning methods can be used in order to improve the knowledge container of a CBR system (adding, creating, deleting cases) of the similarity measure (adjusting weights) and of the solution transformation (new adaptation rules), as well as, techniques from statistics and information theory.

7.5.3 Case Based Reasoning Systems

Case based reasoning is one of the particular techniques used for the development of a knowledgebase system. Case based reasoning compares a current situation (or case) with situations that have been encountered in the past to see if one or more of the earlier situations can provide a model for how to act in the current situation. Case based reasoning can trace its roots to the field of psychology and theories about how human memory works. The case based reasoning is shown in Figure 7.8. Episodic Memory provides a method for storing and recalling large chunks of related information such as events, scenes, occurrences, and stories schema reasoning, the process of applying chunks of information to new situations. Case ultimately refers to an experience that everything is a case!, we can just define.

'A case is contextualised piece of knowledge representing an experience that teaches a lesson fundamental to achieving the goals of the reasoner'.

The major advantages are as follows:

- Increased availability and re-usability of expertise at reduced cost.
- Fast, consistent response.
- Increased safety.
- Learning component enables adaptation to similar situations.
- Works well in conjunction with a human operator (system can make suggestions in unusual situations).

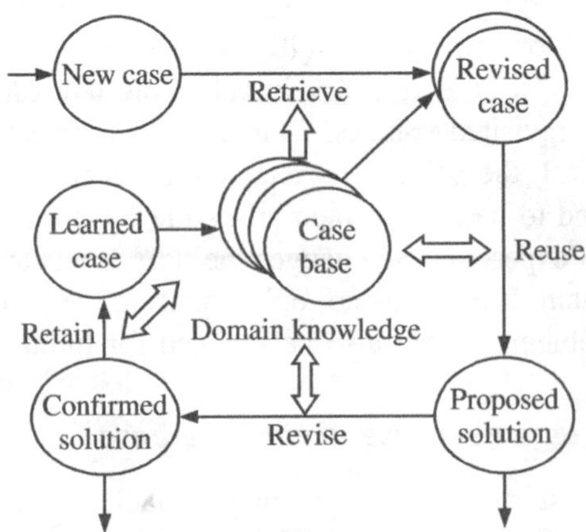

Figure 7.8 Case based reasoning system.

Major challenges with practical issues of case based reasoning are as follows:

- Domain knowledge acquisition.
- Indexing and retrieving case information.
- Completeness of case base.
- Maintenance of case base.
- How many cases are needed.
- How to remove overlapping cases.
- How to efficiently search.
- What features to use for indexing.
- How to weight the features.
- How to acquire adaptation knowledge.

7.5.4 Where to use CBR?

In which all situations can we employ the CBR? The most commonly used situations are mentioned as follows:

- When we have a previous record of solved problem.
- When the expert finds difficult to find a solution from scratch.
- When we have a large domain knowledge obtained through examples.
- When we have experience of handling similar situation.

While using CBR, for categorising the cases we do not really need a big knowledge acquisition phase. Case based reasoning makes the cases to grow incrementally and will maintain such cases in the CBR libraries. The major users of CBR are as follows:

Lawyers: Argue legal cases based on precedence. It is need to find a similar case and show why it is relevant to the current case.

Estate agent: Value of new house based on cost of similar properties in the area.

Doctors: Feel a case like this before and this is what previously did.

7.5.5 Applications of CBR

There is a wide range of applications where CBR is being employed. In the case of medical field CBR is most commonly used for diagnosis. For diagnosis, the system will try to retrieve a past case with the same symptom and nature. Thus, providing faster service. In customer service also CBR plays an important role. There are situations where same compliant or same help is needed for a lot of customers, so if we already have a scheme implemented by checking past details. A perfect help desk can be made up. In financial market also CBR is used to determine the values of variables by comparing with something similar. Assessment tasks are quite common and they make use of CBR. Even in applications of decision making and for creating new designs in architecture and industrial set-up, we always check for similar cases. Thus, in those fields also, CBR is used.

7.5.6 CBR Issues

We have stated the various advantages of CBR and the areas where it is used. But we need to keep in mind the various issues CBR can get through if a proper scheme is not being employed. The various sensitive problems that the CBR faces include:

- The way in which the cases are presented.
- The indexing which is used for easy retrieval.
- The problems while handling larger databases and multimedia data.
- A general adaptation scheme to make the retrieved solution to fit the current situation.

Hence, there is still a wide area open for exploring the case based reasoning.

7.6 Model Based Reasoning Systems

This type of reasoning refers to an inference method used in expert systems. Basically, model based reasoning uses a model of the physical world. The main focus of application development is creating a model with observed data to derive conclusions such as a diagnosis or a prediction. The model based reasoning schematic is shown in Figure 7.9.

The model based reasoning system is an engineering model that forms the for diagnosis which interrogation of fault propagation graphs is very efficient. A hybrid approach is used to combine these techniques and the system give more flexibility. Major challenges include model building and validation which is not easy and the scalability is an issue for model based reasoning systems.

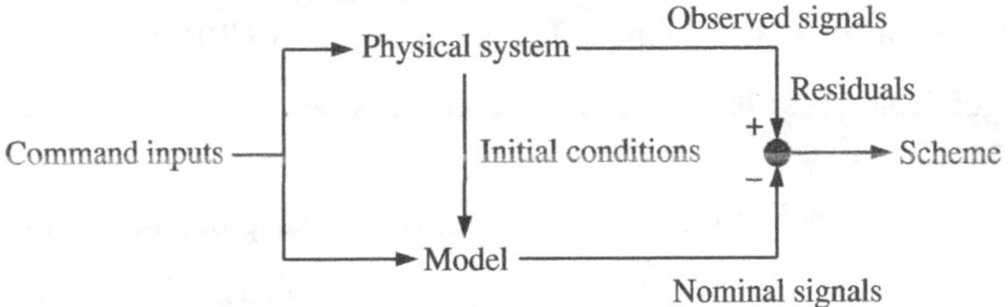

Figure 7.9 Model based reasoning.

7.7 Exercises

1. Consider the problem of deciding which clothes to wear using knowledge such as:

   "Wear casual clothes unless they are not clean or important meeting occurs today".

   "Wear a sweater if it is cold".

   "The winter is usually cold".

   "Wear shorts if it is warm".

   "The summer is usually warm".

(a) Construct a JTMS network to represent these facts.
(b) Try to solve the problem 'In winter, do I wear shorts?'
(c) Answer the question What shall I wear today? (You may assume that the system knows the time of year).

2. Construct a JTMS and ATMS to represent the following:
(a) If you have spots and a fever you have measles.
(b) If you have a runny nose then unless it is hay fever season you have a cold.

3. Use an ATMS to solve the following car diagnostic problem: You may assume the following:
"If the battery is OK and the headlight bulbs are OK then the headlights will work".
"If the battery is OK and the starter is OK then the engine will start".
Find explanations as to why:
(a) The engine will not start but the headlights work.
(b) The headlights will not work but the car starts.

4. In this fault-diagnosis rule set, each line is a rule constructed from a conclusion (the word after 'implies') and one or more conditions (words before 'implies').
unexpected Behaviour and data Corruption ⇒ diskOverflow
unexpected Behaviour and network Exposure ⇒ virusInfection
broadband ⇒ networkExposure
broadband ⇒ intenseNetworkUsage
attachments Opened ⇒ networkExposure
games Downloads ⇒ networkExposure
ADSLConnection ⇒ broadband
cable Connection ⇒ broadband
network Exposure ⇒ firewallNeeded
network Exposure ⇒ emailCapture
crashing ⇒ unexpectedBehaviour
freezing ⇒ unexpectedBehaviour
(a) Estimate the branching factor for the search space for backward-reasoning with these rules.
(b) Identify a minimum alteration of the rules which will enable the conclusion 'disk Overflow'.
(c) Show all the conclusions which can be established using forward reasoning from the fact 'ADSLConnection'.
(d) Draw out these conclusions in the form of a tree of deductions with 'ADSLConnection' at the root.
(e) Using backward reasoning, show whether 'virus Infection' may be concluded on the basis of the facts 'ADSLConnection', 'games Downloads' and 'freezing'.

5. Explain the difference between forward chaining and backward chaining and the situations where each one is more appropriate.
6. Discuss the main advantages of applying fuzzy logic in knowledgebase systems?
7. Why is backward chaining used for diagnostic problems?
8. State for each of the following frame properties, an axiom schema that defines it. Prove that the schema is valid in each frame that has the property. Furthermore, state an appropriate frame that does not have the property and in which the axiom is not valid.
   (a) vRu and vRw implies uRw.
   (b) For each v there exists at most one w with vRw.
   (c) uRv and uRw implies that there exists an x with vRx and wRx.
9. Translate into first-order default logic and check whether the given conclusions are credulous and/or skeptical.
   By default, students are not lazy. But computer science students are typically intelligent, and intelligent students are usually lazy. Salim and Maya study humanities, Anand and John study computer science. Conclusions: Anand and John are lazy; Maya and Salim are not.
10. Suppose one writes a logic program that carries out a resolution inference step. That is, let *Resolve* (*c*1, *c*2, *c*) succeed if c is the result of resolving c1 and c2. Normally, Resolve would be used as part of a theorem prover by calling it with c1 and c2 instantiated to particular clauses, thereby generating the resolvent c. Now suppose instead that we call it with c instantiated and c1 and c2 un-instantiated. Will this succeed in generating the approximate results of an inverse resolution step? Would you need any special modification to the logic programming system for this to work?
11. Describe the type of problems where rule-based systems are more appropriate than CBR systems.
12. How physical systems help a model building?

CHAPTER

8

Learning

Learning is the process by which system improves its performance from experience or environment. Learning may also be defined as the art of acquiring new or modifying existing knowledge and skills. In short, we can define learning as a change in behaviour. Learning has many roles in artificial intelligence applications. For example, in robotics or expert system development, learning plays a separate module in the system. Machine learning is a computerised process which gives an improved performance in the application development with human computer interaction.

What is learning?

If computers are able to learn new things, and adapt new situations, then we say that the computer has been equipped with 'learning programs'. Learning is an area of artificial intelligence that focusses on processes of self-improvement. Information processes that improve their performance or enlarge their knowledgebases. Computers do the learning process through the following steps:

1. One way of learning is data acquisition of new knowledge. Any program that does such data acquisition is a learning program.
2. The problem-solving is another component of learning.

Learning is hard because

1. Intelligence implies that an organism or machine must be able to adapt to new situations.
2. It must be able to learn and must do new things.
3. This requires knowledge acquisition, inference, updating/refinement of knowledgebase, acquisition of heuristics, applying faster searches, etc.

How do computers learn?

Many approaches have been attempted to provide a machine with learning capabilities.

This is because learning tasks cover a wide range of phenomena.
Listed below are a few examples of how one may learn.

Skill refinement: One can learn by practice, for example, playing the guitar.

Knowledge acquisition: Learning through experience and storing it in a knowledgebase is knowledge acquisition. Rote learning can be cited as an example of knowledge acquisition.

Taking advice: It is similar to rote learning, but the input knowledge may need to transform effectively by an order.

Problem-solving: If we solve a problem systematically, we will get experience to solve similar problems. For the next time we can solve similar problems more efficiently. Usually, this does not involve new knowledge gathering but may involve reorganisation of data or remembering how to achieve solution.

Induction: One can learn from examples. Humans often classify things in the world without knowing explicit rules. Usually a teacher or trainer is involved to aid the classification.

Discovery: Here one learns knowledge without the aid of a teacher or a supervisor. New knowledge elements are inferred from the knowledgebase obtained from a supervisor.

Analogy: It may be able to transfer little knowledge for getting a better solution of the task in hand a system can use. This system finds and stores the similarities in the information.

8.1 Types of Learning

Learning is acquiring new or modifying existing knowledge, behaviours, skills, values, or preferences and may involve synthesising different types of information. The ability to learn is possessed by humans, animals and some machines.

8.1.1 Rote Learning

The simplest kind of machine learning is the recording of data. Data storage programs can be used for making the machines learn. Such type of learning is known as rote learning.

In the game of chess, checkers or Tic-Tac-Toe, one can find the backup value of a configuration (board states) by Minimax procedure. If the backed up value is stored along with storage of the board position, this can be used in the next game. When the same configuration arises once again, this backed up values are consumed without using Minimax procedure once again. Here the machine has learned the best move for a given board position. A few more capabilities for rote learning can be included in the following:

1. **Organised storage:** There is an organised storage of information in order to use the stored value, the board position should be organised in such a way that the process of retrieval will be fastest. This is done by indexing the board position by the number of pieces on the board.
2. **Generalisation:** To keep the number of stored objects at a manageable level, some kind of generalisation is necessary.

3. **Direction:** The program should try to find required stored value intelligently for a given board position. So, we have to focus the attention to a single processing direction.

8.1.2 Learning by Parameter Adjustment

Game play determines the best moves by using static evaluation function or by Minimax procedure. Pattern classification does this by assigning weights to the various features that a particular pattern contains in chess game. This is done by considering a polynomial of the form $C_1t + C_2t^2 + C_3t^3 + \cdots + C_{16}t^{16}$ where one can use 16 features and determine 16 weights. The adjustment of $C_1, C_2, ..., C_{16}$ depending on the configuration is known as parameter adjustment. The value will be set with the best values obtained by experience. The computer plays the game with another copy of itself which was configured earlier. If the new sets of parameter values are better in the sense of winning the game. They are changed to the new values and a copy of it is made. This process is repeated for several times. This is the way by which the parameters are adjusted for a game. This learning process is applicable for initial game knowledgebase building.

8.1.3 Learning by General Problem-Solving

Learning can be thought as a form of problem-solving methods. General Problem Solver (GPS) provides a design of a learning program. To use this in a particular table, it is necessary to have a table that were operate relevant to reducing the important differences between a given state and the goal state. Suppose we decide that rather than providing such table to GPS, we want the program to teach the table itself from its own experience. Then we say that GPS is another type of learning program.

GPS was an attempt to construct a general problem-solving mechanism that could solve problems in a new area given domain-specific knowledge about that area.

GPS is a generalised state-space search mechanism. It has the following components:

1. A set of states and operators that change states. Specification of start and goal states.
2. A procedure for identifying differences between states.
3. A table of connections that connects observed differences with operators that may be relevant for reducing those differences.

GPS works by determining the difference between the current state and the goal state and selecting operators relevant to reducing that difference. Hopefully this will give direction to the search and make it more efficient than a blind state-space search.

To apply GPS, the problem of discovery of good set of differences used by another copy of GPS working in a given problem domain is necessary to do the following:

1. Input a set of operators that will be available in the application domain.
2. Define a goal for the learning task. This goal is to provide a good set of difference for the application task.
3. Define a set of difference used by the learning task.
4. Define a set of operators for the learning task.

8.1.4 Concept Learning

Classification is the process of assigning to a particular input, the name of class to which it belongs. Classification is an important component of problem-solving task. The idea of producing a classification program that can evolve its own definition and constructing these class definitions is called as concept learning. The technique used depends on the way the classes are to be desired.

Suppose **if**:
the current goal is to go from place A to place B and there is a wall separating the two places **then**:
look for a doorway in the wall and go through it.

To use this rule successfully, the systems matching outlived must be able to identify an object as a 'wall'. Without recognising the object to be the 'wall' the rule can never be invoked.

8.1.5 Learning by Analogy

Analogy involves a complicated mapping between what might appear to be two dissimilar concepts. Analogy is a powerful inference-rule that allows similarities between object to be stated. Analogical learning system uses frame structure.

For example, Suppose we wish to say that, "John is like a turbo-jet".

This means, a person 'John' who's movement or speech is very fast. Here the turbo-jet is an object that maps to the person 'John'.

Analogical problems are giving learning process which maps in two different ways. These two methods of analogical problem-solving that have been studied in artificial intelligence are transformational analogy and derivational analogy.

Transformational analogy

Look for a similar solution and copy it to the new situation making suitable substitutions wherever appropriate (Example, geometry). In 1983, Carbonell described this method. The idea is to transform a solution to previous solved problem into a solution for the current problem (Figure 8.1). Transformational analogy does not look at how the problem was solved,

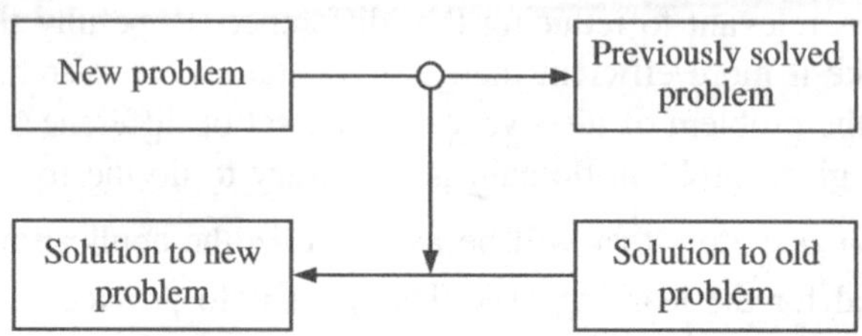

Figure 8.1 Transformational analogy.

it only looks at the final solution. For example, we know how to solve problem related to isosceles triangle. The same methodology can be adapted to solve a problem related to equilateral triangle.

Derivation analogy

The history of the problem solution, the steps involved, are often relevant. In 1986, Carbonell showed that derivation analogy is a necessary component in the transfer of skills in complex domains (Figure 8.2). For example, we know how to solve the area of a triangle and a square. Derivation analogy will help to solve the volume of a pyramid from this knowledge.

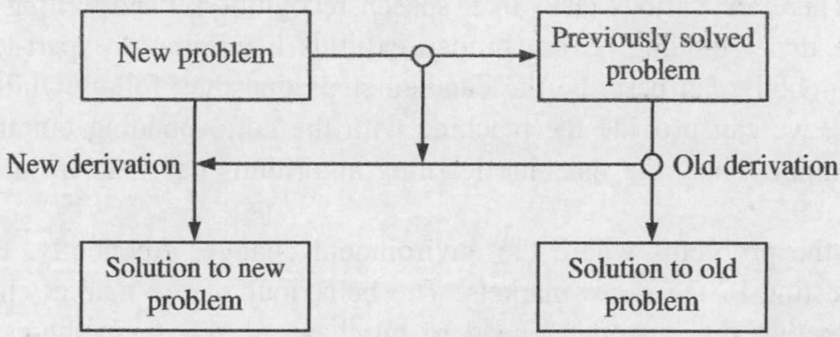

Figure 8.2 Derivational analogy.

8.2 Machine Learning

Machine learning is the branch of artificial intelligence concerned with the design and development of algorithms. It allows the computer to behave in a way that is based on the empirical data, such as from electronic sensors or databases. A major focus of machine learning research is to automatically train the system for recognising complex patterns and to create intelligent decisions based on the provided inputs. The real set of possible data will be too large when compared to the inputs that we supply. Hence, it is always difficult to produce an effective output. All machine learning techniques are falls in any of one of the category—supervised, unsupervised and reinforcement learning strategies (Figure 8.3).

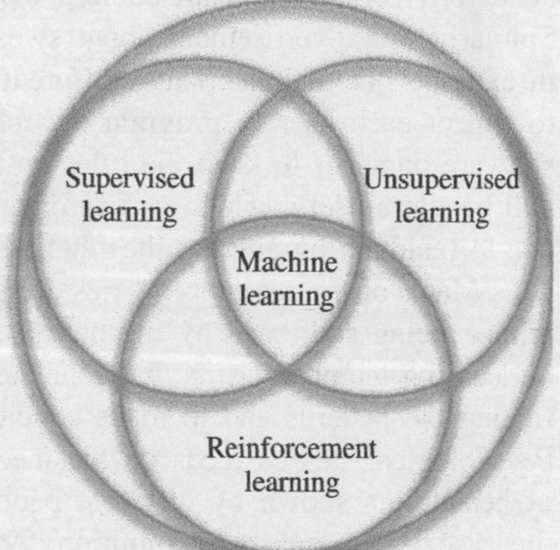

Figure 8.3 Types of machine learning techniques.

8.2.1 Why Machine Learning?

Machine learning is the field of study that gives computers the ability to learn without being explicitly programmed. The main aim of machine learning is to develop systems that are too difficult or expensive to construct manually with ease. The system must automatically adapt and customise themselves to individual users like personalising news or mail filtering.

Computers are used in various fields of science and for various tasks. Normally the programmers design and implement software that they need for particular functions. However, there are many task for which the implementation is difficult or impossible. We can generally classify such problems into four categories as follows:

First, for the problems where no human experts exist, for example, in modern automated manufacturing facilities, we need to predict the failure of machines before it happens by analysing the sensor details. As the machines are new and no experts exit for such task, a machine learning system can be employed to study the sensor data and to predict subsequent machine failures.

Second, for the problems where human experts exist, but they will be unable to explain their expertise. There are various tasks like, speech recognition, hand-writing recognition, and natural language understanding. Every human exhibits a minimum expert-level abilities for these tasks, but nobody can describe the detailed steps that they follow while performing the task. Fortunately, we can provide the machine with the corresponding output that we expect for a particular input. Thus, the machine learning algorithms can help in mapping the inputs to the outputs.

Third, for the problems where the environment changes frequently. For example, we have people investing in the share markets. The behaviour of the market changes frequently and according to that the customers need to purchase or sell their shares. To predict the future behaviour of the stock market even an expert programmer needs to rewrite the codes of his program frequently. A learning program can relieve the programmer from the burden of constantly modifying and tuning a set of learned prediction rules.

Fourth, for applications that need to be customised for individual users or for a group of users. For example, a program to filter unwanted electronic mail messages. The advertisement about refrigerators and air-coolers will always be a spam message for people in Alaska. Similarly the advertisement about sweaters and jackets will be of no use for people living in extreme hot regions. Thus, different users will need different filters. It is not reasonable to expect each user to program his rules, and it is infeasible to provide every user with a software engineer to keep the rules up-to-date. In such a situation a machine learning system will be more useful which can automatically learn which mail messages to reject and which one to maintain by altering the filtering rules.

Many of the research queries are same as in the fields of statistics, psychology and data mining being addressed by machine learning but with a different outlook. Statistics focusses on learning the phenomena that generate the data. Data mining studies are always interested in finding the patterns and to make conclusions which will be easily understandable by people. Psychological studies focus on the mechanism that underlies for the reason of various types of behaviours shown by different people. We can illustrate this with an example; consider the case about speech recognition. A machine learning approach focuses on building an accurate and efficient speech recognition system. But the statistician might collaborate with a psychologist to find hypotheses about the mechanisms underlying speech recognition. A data mining technique will look for patterns in speech data and a person will be interested if he can apply it to some group of people according to gender or age.

Now, let us identify the types of problem in which machine learning is being used.

8.2.2 History of Machine Learning

1950's can be termed as an exploratory period where many general techniques were born. It was during this time, Samuel Checkers-playing program which is considered as the world's first self-learning program was developed. It demonstrated the fundamental concept of artificial intelligence. Selfridge's Pandemonium model proposed by Selfridge in 1959 also noted its success in modelling human pattern recognition.

During 1960's the concepts of neural networks were developed. Rosenblatt developed the 'perceptron', which was modelled after neurons. It was the precursor to latter work in neural networks. In 1969, Minsky and Papert wrote a paper proving that the perceptron was inherently limited, which discouraged research in this area for almost 20 years.

1970's were characterised by the development of more practical algorithms, often using symbolic techniques. Winston's important work on learning (Winston's arch learner) in blocks-world domain happened in 1970. The idea of macro-operators was developed by Fikes during 1972. Mathematical discovery with AM by Lenat in 1977 and Scientific discovery with BACON by Langley in 1978 were the major discoveries.

Explosion of different techniques and increased emphasis on evaluation happened during the 1980's. The notion of 'version spaces' was developed by Mitchell in 1982. Quinlan in 1983 created the ID3 decision tree algorithm. It was Valiant in 1984 who defined the idea of Probably Approximately Correct (PAC) learning, which is still the dominant theoretical paradigm. Connectionist research became active again. The backpropagation algorithm (Rumelhart in 1986) overcame many of the limitations of perceptron. The COBWEB clustering system was developed by Fisher in 1987. Explanation-based learning, speed-up learning, and case-based learning became very popular during this era.

The rapid changes in 1990's were triggered by the data mining applications. Data mining emerged as an important tool for many business activities. Reinforcement learning and inductive logic programming were also developed during this period. Baye's net learning also grabbed pace during the second half of 1990's.

It was during early 2000's when studies on Support Vector Machines (SVM) and Kernel methods became prominent. Compilers, debugging tools and graphic interfaces were introduced in the process of learning. Applications to robotics, adaptive software agents, bagging, boosting, and stacking also caught their pace during this period.

8.2.3 Timeline of Machine Learning

Machine learning is the youngest child of statistics. The major machine learning activities like prediction, classification, clustering, automation, etc. start from the basic components of probability and statistics. In this section, a detailed discussion on the evolution of machine learning is included.

Bayes' Era: The major contribution of every machine learning algorithm will have a trace back to the Bayes' theorem put forward by Thomas Bayes during 1763. The strong backgrounds of classification used in major application are derived from this. An expanded work on Bayes' theorem was put forwarded by Pierre Simon Laplace in the year 1812.

Markov and Turning Era: Markov proposed a power statistical model during 1900s. The applications of this model were realised only in 1970s, followed by a series of applications implemented by Markov models. The first signs on making a machine to learn things started when Alan Turing proposed genetic algorithms during 1950s. This was followed by the introduction of the first neural network machine SNARC by Marvin Minsky and Dean Edmonds. The checkers game was introduced in 1952 when IBM launched the first ever machine learning game programs. Arthur Samiel led the development initiative of this project.

Era of Neural Networks: Neural nets are one of the backbones of machine learning. The discovery of perceptron generated a great deal of excitement among the scientists. The self-learning structure which triggered the learning activities was discovered by Frank Rosenblatt during 1957. Recurrent neural networks were developed by John Hopfield during 1982 which were used as content addressable memory systems. Backpropagation algorithm was developed during 1986, which facilitated machines to do multi-level learning.

1989: Genetic algorithm packages became commercial, Axcelis Inc, released Evolver during this period. This was the first such package that was made available in personal computer. Machine learning started becoming commercial.

1992: Gerald Tesauro developed the boardgame TD-Gammon, which uses an artificial neural network based temporal difference training. The game was able to rival but not consistently surpass. Still the learning system was in place, which always paved way for system improvement.

1995: Support Vector Machine (SVM) was introduced by Corinna Cortes and Vladimir Vapnik, which is a discriminative classifier formally defined by a separating hyperplane. One of the most commonly acceptable classifier is SVM. Random decision forests were also discovered during this year.

1997: The epic episode of machine learning, IBM's Deep Blue beated the world chess champion Kasparov. That is when real artificial intelligence began. The recurrent neural network was given a prominent add-on by giving LSTM (Long Short Term Memory) which significantly increased the efficiency. It was Sepp and Jurgen who put forward the LSTM phase.

1998: For most applications in machine learning, lack of datasets are always an issue. In 1998 the MNIST dataset was released by a team led by Yann LeCun. This is considered as a basic benchmark for evaluating handwriting recognition.

2002: There was initial difficulties in implanting the machine learning applications due to the less software support. During 2002, Tourch—a software library for machine learning was released. This drew a large crowd of people to focus their attention to the machine learning world.

2006: As the world changed its focus to predictions and analysis, every place had an application of machine learning. Netflix launched a competition where a recommendation system was used to predict the rating that a user gives to a new film by considering its previous rating. This was a major area where the public actually got to know the real strength of machine learning applications.

2009: ImageNet was created during this period. It is a large visual database developed from the stanford university by Fei Fei Li. His inference on ML was "no algorithm is going to provide you accurate results/predictions unless we train them with real-time dataset". For many, ImageNet was the real catalyst for the artificial intelligence boom.

2010: Application based competitions started in artificial intelligence platform. Kaggle was the first website launched for moderating various machine learning competitions.

2011: Using a combination of machine learning, natural language processing and information retrieval techniques, IBM's Watson beated two human champions in a jeopardy competition.

2012: The Google Brain team, led by Andrew Ng and Jeff Dean, created a neural network that learned to recognise cats by watching unlabelled images taken from frames of YouTube videos.

2014: Face recognition and speech identification was yet another application which paved way for millions of applications starting with phone lock to missing person identification. Deepface was a system developed by Facebook researchers which makes use of neural networks for face recognition. During 2014, Google developed a proprietary platform for massively parallel machine learning algorithms to make predictions and recommendations.

2016: Google's AlphaGo program becomes the first Computer Go program to beat an unhandicapped professional human player using a combination of machine learning and tree search techniques. Later improved as AlphaGo Zero and then in 2017 generalised to Chess and more two-player games with AlphaZero.

2018: Captured the voice and turned into commands. Like the voice assistance Siri and Go Google, a new prototype where the system always listens our speech and decodes it using NLP techniques and gives the apt results on the Go. This is the future of prediction.

2019: Works on introducing AR-VR in an *n*-dimension space which has always been the dream of computer scientists. Initial work on making this a reality has started as the primary step from Google.

The world have seen the exponential growth of machine learning which the world still considers as a preliminary stage. Magic of machine learning is yet to come.

8.2.4 Types of Problems in Machine Learning

Machine learning is all about algorithms and designing of algorithms for computer to train them to improve its performance. It is not just the search of consciousness in machines but also coming close to the traditional goals defined by artificial intelligence. There are various machine learning algorithms available but selecting an apt algorithm for the kind of problem that we need to solve is the thumb rule in the process of learning. Learning is just the matter of finding statistical regularities or other patterns in the data. In a broad sense we can classify the machine learning problems as follows:

- Classification problems
- Decision problems
- Learning problems

The types of learning algorithms fall along several classifications. One classification is the type of result expected from the algorithm. For example, some problems are classification problems and other may be used for solving the problem. Let us take an example of a classification problem. Handwritten recognition (Figure 8.4) and digit recognition are problems where the computer needs to learn how to classify the given inputs by checking the patterns which have been already fed to it. There are various hand written recognition softwares which give extremely good results. The neural network community works on the digit recognition problem. For the same problem, support vector machines offer a better solution classifier. The deep networks play new role in prediction of handwritten characteristics with promising results.

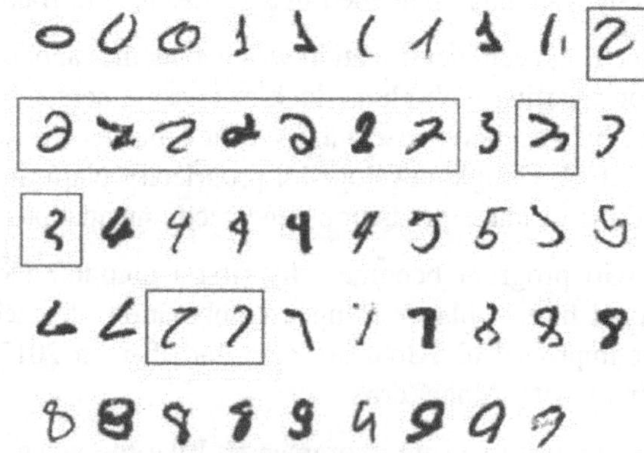

Figure 8.4 Handwritten digits—digit 2 is marked.

We can have machine learning for problem-solving too. It is mainly used in medical diagnoses and in computational biology—for instance, if the computer is trained to distinguish between cancer cells and normal tissues, the machine learning technique is trained by giving a set of attributes and the system will be able to distinguish between likely cancer cells and normal tissues. Neural networks or support vector machines can sort out the learning process of the problem. Another learning technique that can solve the problem is decision trees.

Decision trees are a relatively simple classification technique that follows the concept of following a tree of questions and answers. If the answer to a question is yes, then the algorithm proceeds down one branch of the tree and if the answer to a question is no, the algorithm takes the other branch. Finally the algorithm will reach a leaf node with the final classification. Learning decision trees is straight-forward and requires less work than neural nets. So, you need to be aware that even the simplest learning algorithms could accomplish great deals. Thus, a decision tree learning algorithm can be used in any scenario where you can collect a reasonable number of attributes and a classification system that can expect effective results.

Let us look at an example of classification learning, i.e. speech recognition. Usually, the computer will be given a set of training instances which consist of sounds and what word that particular sound corresponds to. This type of learning can be well carried out with neural networks, though it is hard to imagine that the problem is simple enough for decision trees. One of the machine learning techniques by name hidden Markov model, is designed specifically for working with time series data like this, which has shown good results in speech processing.

There is a common type of learning which is not designed to create classifications of inputs, but to make decisions. Such type of learning is called decision problems. They normally require various sorts of assumptions about the present state of the system to predict the rules. Decision problems sometimes need to make a single decision or multiple decisions. When there are multiple decisions to be made, the problem becomes trickier because it needs to take into account all the possibilities and all the consequences of actions that the system performs.

For understanding both classification and decision problems, we make use of the concept of utility function. The basic idea is that the computer will gain a particular amount of value while performing some actions. The agent needs to learn what actions resulted in positive utility pay-off and what actions resulted in negative utility. For example, if you are making a robot to explore a jungle, sometimes you will never know where exactly the robot is located. We are not aware of what will happen to the robot when he steps forward or turns left. Thus, we are either not sure of the state of world or not sure whether its actions will work as expected.

The computer (agent) needs to learn the various aspects of the model which is not provided. In our example, either the location or the action needed will be unknown for a particular instant. The system needs to track that and learn how the world looks like or what effect a particular action can cause. Once these different functions are learned, the correct action to take is simply a matter of deciding which action maximises the 'expected utility' of the agent. Expected utility can be calculated by multiplying probability of each pay off by pay off and taking the total sum. In cases of making multiple decisions and when reasonably small groups of possible states are available, an algorithm known as a Markov decision process is commonly used.

In some cases, the actual calculation of the utilities can be avoided as some learning algorithms incorporate the feature of predicting future from the information they learned. Reinforcement learning is a common method for these kinds of situation.

Supervised learning is the type of learning where the training is done by providing correct results. Thus, a feedback system will be equipped in order to get the correct result and to check the progress of learning. In unsupervised learning, the goal is harder because there is no pre-determined output and sample patterns. The system needs to undergo lot of training to get results.

Developing a system that is too difficult/expensive to construct manually because they require specific detailed skills or knowledge tuned to a specific task is one of the knowledge engineering bottlenecks. Develop a system that can automatically adapt and customise themselves to individual users. Data mining techniques are used to discover new knowledge from large databases. The training may have some experience like direct experience or indirect experience. Direct experience is the given sample input-output pairs for a useful target function, and the indirect experience is the given feedback that is not direct input-output pairs for a useful target function.

8.3 Aspects of Machine Learning

One useful aspect on machine learning is that it involves searching a very large space of possible hypotheses to determine one that best fits the observed data and any prior knowledge held by the learner. For example, consider the space of hypotheses that could in principle be outputted by the above checkers learner. This hypothesis space consists of all evaluation functions that can be represented by some choice of values for the weights w_0 through w_6. The learner's task is thus to search through this vast space to locate the hypothesis that is most consistent with the available training examples. The Least Mean Square algorithm for fitting weights achieves this goal by iteratively tuning the weights, adding a correction to each weight each time the hypothesised evaluation function predicts a value that differs from the training value. This algorithm works well when the hypothesis representation considered by the learner defines a continuously parameterised space of potential hypotheses.

Many of the chapters in this book present algorithms that search a hypothesis space defined by some underlying representation (For example, linear functions, logical descriptions, decision trees, artificial neural networks, etc). These different hypothesis representations are appropriate for learning different kinds of target functions. For each of these hypothesis representations, the corresponding learning algorithm takes advantage of a different underlying structure to organise the search through hypothesis space.

Throughout this book we will return to this perspective of learning as a search problem in order to characterise learning methods by their search strategies and by the underlying structure of the search spaces they explore. We will also find this viewpoint useful in formally analysing the relationship between the size of the hypothesis space to be searched, the number of training examples available and confidence of hypothesis consistency.

8.3.1 Inputs to Training

The learning model requires two inputs to the training process—training set and existing knowledge. Most of the learning algorithms are heavily example based that have pure deductive learning that uses no examples and only makes use of existing knowledge. The major classifications are supervised vs. unsupervised training (Class information on training example is available or not), Incremental vs. non-incremental training (Training examples are given one time or consequentially) and homogeneous vs. heterogeneous (Described by only one type of data like numeric, symbolic, etc. or several types of data.) These aspects are required to consider while designing a machine learning model.

Supervised vs. Unsupervised

As we have seen, supervised learning is employed when we have a sample data. So, it is most commonly used in classification problems because the goal is to get the agent learn a classification system that is created. The digit recognition problem can be considered as an example for supervised learning. In the case of neural networks and decision trees we have pre-determined classification and hence supervised learning is commonly used. In the case of neural networks, the classification is mainly used for determining the error of the network and is used to minimise it. In case of decision trees, the classification is used for determining the apt attributes which provide more information that can be used for solving the puzzles. In short, supervised learning will always have some 'supervision' for problem-solving. To minimise the errors and to adjust the input parameters hidden Markov and Bayesian network relies a lot on supervised learning. Thus, we may say that in classification problems, goal of the learning algorithm is to minimize the error with respect to given inputs. These inputs are called the training set, which is an example set. The training set will have both positive and negative inputs. Then only we can understand the behaviour of system in both situations.

Unsupervised learning is normally treated as difficult because goal of this process is to have the computer learn something that we do not tell it how to do. Two approaches are mainly employed for unsupervised learning.

First one is to teach the system not by giving sample categories but by providing some positive flag as a reward to denote success. This type of training is mainly used for decision making problems and not for classification because the systems always try to get maximum rewards. This approach is similar to the real-world scenario where some positive actions are rewarded and the others are punished. In unsupervised learning the system always tracks the reward for its previous actions and learns what needs to be done for a similar problem. This type of learning is not extremely beneficial because the learning is very time consuming and the process goes through trial and error methods.

The second type of unsupervised learning is clustering. Here the system tries to find similar data and groups them into a single cluster. The goal is not to maximise a utility function for this type of learning. The similarity in the training data is identified for further training. For example, clustering individuals based on genders will result in a cluster of the males in one group and females in another. This data driven approach works well only if there is sufficient data. This principle is used by websites like *amazon.com* to recommend certain books for a particular group of users. Algorithms which are used for these purposes are called information filtering algorithms. Information about the members in a cluster always helps to produce meaningful results. Unsupervised learning also has a problem of overfitting the training data.

Incremental vs. Non-incremental

The dataset can be either incremental or non-incremental depending on the growing nature of the dataset. For example, when we design a model for prediction of faults of the machines in a process plant, the data from the input sensors are considered as incremental. Because at every second, the sensors are updating the values in the database for prediction of the faults. We have to predict the output from the current sensor values updated by the sensors. Another example is when we require to predict the class of animal, from a set of picture dataset, the data is non-incremental. The number of data in this set is fixed and these pictures are obtained from a particular geographical location.

The terms incremental and non-incremental are opposite in statistics. In machine learning when we have to consider the inputs, dataset may vary with respect to the updation. If machine learning model is ready to adapt training examples that are given one time or consequentially it is categorised as non-incremental or incremental dataset.

Homogeneous vs. Heterogeneous

Homogeneity and heterogeneity are properties of dataset in statistics and these two have opposite meaning. A homogeneous data are things that are similar to each other. For example, a set of student's examination results of computer science subject.

Before applying a statistical technique, it is important to fix a set of data if homogeneous. The homogeneous data is derived from a single population. The changes in the data potentially affect statistical properties of the observation which leads to a poor machine leaner. In real-time, the homologous data are difficult to pool into machine learning technique. For example, data of large scale of climatic change is homogeneous. The natural variations in the climatic changes cause substantial errors in the dataset while updating. The interpretation of conclusions from the results of a learner depends on the time scales considered by an expert.

Heterogeneous data are with high variability of formats and data types. Due to high data redundancy, missing values and dishonesty, the heterogeneous data are ambiguous. The heterogeneous dataset is difficult to integrate with a machine learning technique because data quality is poor. For example, the dataset obtained from Internet of Things (IoT) is considered as heterogeneous data. There are many acquisition devices, connected geographically in different locations that collect sensor values recorded into a datastore. These data are in large scale and is not in unique nature. If the machine learning technique requires historical data for the training, concurrent data is used for the prediction of results. The historical data should be a time frame based data having a strong correlation between space and time. Since, the location of these devices are geographically different, there is a time stamp of the data generated by these devices. A small amount of these data may be required for the prediction but collection of the entire data is essential for storing in the history database.

The heterogeneity in a dataset occurs in different ways, i.e. syntactic, conceptual, terminological and semiotic. It occurs when two data sources are not expressed in the same language. For example, if we consider a set of words in English and Hindi, a dataset used for training a machine learning model consists a mixture of these two language words. Conceptual (semantic or logical) heterogeneity occurs different model design for a single domain. For example, design of two separate language classifiers from a frequent set of words, terminological differences in names of same entities from different data sources, semiotic (pragmatic) heterogeneity created by people for different interpretation of entities.

8.3.2 Output of Training

To a large extent the outputs of learning are determined by the application to develop. Machine learning systems spread over as classifiers or predictors. The output of the machine learning system is to be designed for the purpose of an input dataset. These dataset contain positive and negative examples. Figure 8.5 shows different groups of birds and non-birds. Examples of birds are considered as positive dataset and non-birds as negative dataset. The purpose of learning of the model is to determine the common features. For example, features of birds are extracted as positive data and features of non-birds are extracted as negative data for the training of the model. There is a classification rule as follows:

if example satisfies condition
then assign it to class X
In our example X is either birds or non-birds

Figure 8.5 Examples of birds and non-birds.

Conjunctions of simple tests on attributes 'attribute1 = value1 and attribute2 = value2, etc.' More complex forms of output may arise in robotics, theorem provers, planning, etc.

8.3.3 Learning Systems

For decision making, classification and prediction problems, we can use machine learning approaches. How can we evaluate performance of the developed model? How will we quantify the performance of these models? Many standard methods are suitable for measuring performance of a classification system.

It was Arthur Samuels who developed the first machine learning system at IBM in 1959. Typical learning system is given in Figure 8.6. There are two approaches to develop a learning system, i.e. experimental and theoretical. Experimental approach conducts controlled experiments to compare various methods on a variety of benchmark datasets and gathers data on their performance. For example, test accuracy, training-time and testing-time. This method should analyse differences for statistical significance in the results.

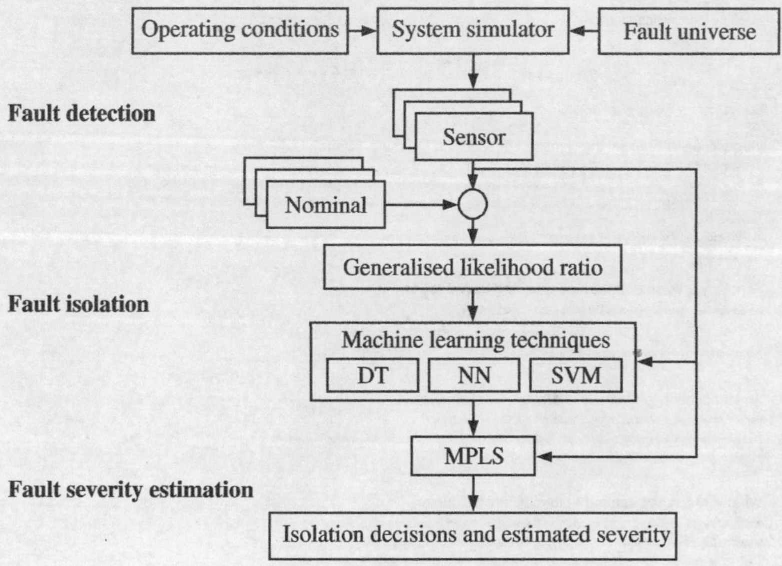

Figure 8.6 Learning systems.

Algorithms are analysed mathematically using theoretical approaches. Prove theorems in terms of computational complexity and calculate the ability to fit the training data and sample complexity. This needs a number of training examples needed to learn an accurate function. The major advantage is data-driven approach that is able to transform high-dimensional noisy data into lower dimensional information. This provides monitoring capability and facilitates model-building via identification of dynamic relationships among data elements. High dependence on quantity and quality of system operational data is one of the major challenges.

8.4 Machine Learning Applications and Examples

How does search engine return the result? Once a query is submitted, it checks out the server for the most relevant results. Server returns the page which has a maximum relevance. This entire process comes under the concept of webpage ranking. Figure 8.7 shows the example of the query result processing.

Search engine, irrespective of the built platform, always returns a sorted list of webpages for each query submitted. Sorting depends on a matrix linked to each webpage. Whenever a query runs, if the user selects a page then its relevance matrix value is increased, thus the page with highest matrix value appears on top of the results. Relevance of the site is measured using the frequency of hits, webpage structure, information links, etc. Machine learning tools are used to find out the sites using background information and filtering. This is a clear replacement for guesswork with clever engineering. The data mined through various websites are also used for generating knowledge. While ordering a book through *amazon.com* or any other prominent commercial site, they capture our data and advertise us back with book suggestions. Similar data from social media is also used for sentimental and behavioural analysis.

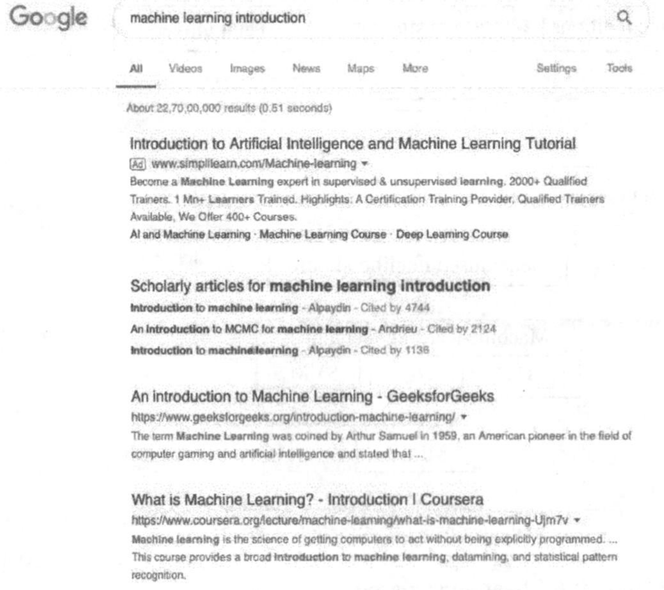

Figure 8.7 Query result processing.

Whenever we write some reviews about a movie, some machine learning techniques evaluate it and the analytical engine suggests which genre the movie belongs to, according to user reviews. Final decision or prediction is the combined inputs of similar users. Figure 8.8 shows an example of how a recommendation system for a book search 'Fundamentals of Statistics'.

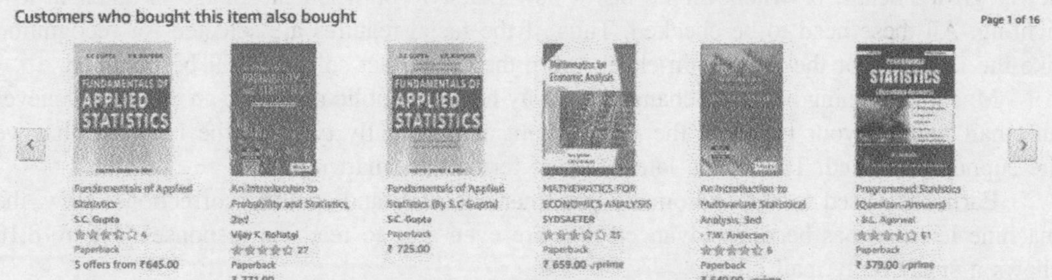

Figure 8.8 Recommended systems.

These types of recommendation algorithms always work on strong mathematical background and hence there is no room for a guess work. Yet another application is in case of natural language processing. If we have a particular sentence in English and if we need to translate it to French, the normal algorithms directly map the words in a language to their equivalent. But many times, grammatical issues and sentence construction issues are not taken care of. This problem can be overcome by building an intelligent translator on top of a machine learning platform. Various applications which run natural language processing these days use heavy machine learning platform for such tasks. Various other applications like face recognition for providing access, number plate detection, road topography detection, handwriting detection, fire or any other natural calamity detection works with less than 1% error rate while deployed in machine learning. An improvement in face detection was actually suggested where motion from a video footage can be easily detected. Hence, we do not need any security personal who checks the CCTV footage. Whenever there is a motion, the system automatically detects and informs the system admin. For any sort of machine learning activities we need training and testing datasets. A commonly used face detection sample dataset is shown in Figure 8.9.

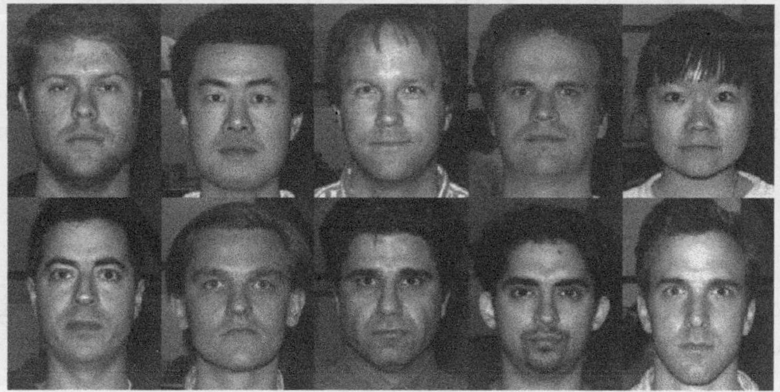

Figure 8.9 Face recognition dataset.

While running the algorithm, all images need to be classified into the corresponding name or it should be classified as an unnamed tag if the image is a new one. There are various criteria to be checked, even if we have the photograph of a person, there are situations where the algorithm needs to identify him even if he wears a new pair of glasses, or if he has a newly grown beard, or whenever he has a new hairstyle or when the image is taken in low lighting. All these need to be checked. Thus, if the facial features are selected for recognition like the iris data, or the cross chin-cheek ratio, then a perfect solution will be obtained.

Machine learning has even changed the way how the public replies to an email. Whenever an email pops to your mailbox, the email client automatically types in the response that we are supposed to send. This is the latest feature termed as smart reply.

Earlier we used to have automatic grammar checking and spelling corrections. Now, the machine learning has become advanced it is are even able to read the responses. Figure 8.10 shows a smart reply mail client.

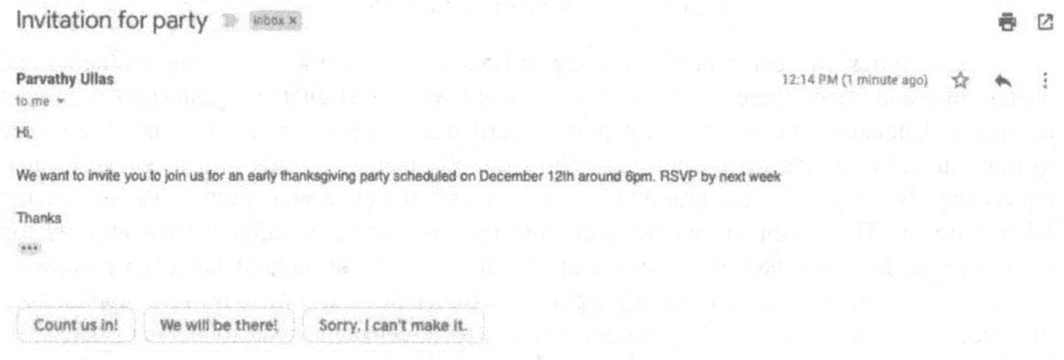

Figure 8.10 Smart reply.

Machine learning also has significant applications in medical fields. Various types of predictions can be of very useful for the doctors. It is used in situations where an application is used to detect and classify cancerous and non-cancerous cells. There are applications where survival prediction after a surgery is predicted using machine learning process. It is also used in cases of finding the best match during organ transplants.

Classification is also yet another area where machine learning is prominently used. Consider the case of spam mail filtering, how does this process happen automatically? There are in-built learning databases which is the basic backbone of this process. Whenever a new mail enters our mailbox, it undergoes a filtering process and finally a learning scheme is built over it and a recommendation system generates the output as whether it is spam or not. Figure 8.11 shows email spam filtering process.

Everything mentioned in the previous section is just the tip of an ice-burg. The real story and application of machine learning is still being discovered and used in plenty. Starting from the basic prediction, classification, recommendation and association, the usage of machine learning platforms is growing at an exponential rate.

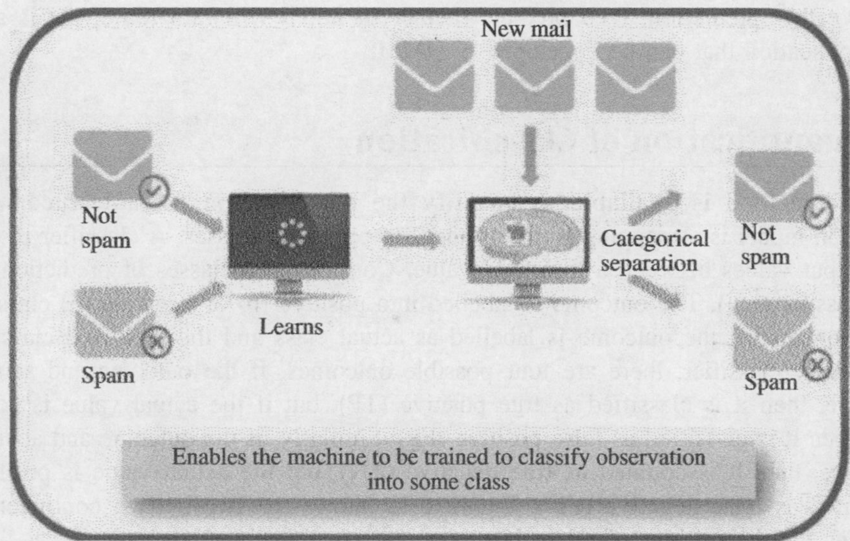

Figure 8.11 Spam filtering process.

Major applications of machine learning can be summarised as follows:

- Optical character recognition: Which maps the images of handwritten characters into corresponding letters or digits. ANN and HMM are widely used for optical character recognition.
- Face detection: To find faces in images or to indicate if a face is present. HMM is suitable for developing such application.
- Spam filtering: To identify e-mail messages as spam or non-spam from a bunch of e-mails. Any of the machine learning techniques like SVM, ANN or HMM can be used to develop such applications.
- Topic spotting: In order to categorise news articles (say) as to whether they are about politics, sports, entertainment, etc. HMM or SVM can make an influence in this application.
- Spoken language understanding: Within the context of a limited domain, determine the meaning of something uttered by a speaker to the extent that it can be classified into a fixed set of categories. For language classification, HMM is highly useful.
- Medical diagnosis: To diagnose a patient as a sufferer or non-sufferer of some disease, any rule based system can be useful.
- Customer segmentation: To predict, which customers will respond to a particular promotion in a domain task. This prediction problem can be incorporated with any of the prediction techniques like SVM or ANN.
- Fraud detection: To identify a credit card transaction and to identify whether a fraud happens. This prediction system used a supervised machine learning technology that can be applied to make a solution.

- Weather prediction: To predict whether or not it will rain tomorrow is a type of application that can be developed in HMM.

8.5 Quantification of Classification

A standard method is available to quantify the classification for a learned model. A classification model is a mapping from instances to predicted classes. A classifier results a set of real output values between a threshold value. Consider two classes of prediction problem (binary classification). The outcome is labelled into positive (p) or negative (n) class. For the classification model, the outcome is labelled as actual class and the predicted class {Y, N}. For the binary classifier, there are four possible outcomes. If the outcome and actual value are positive, then it is classified as true positive (TP), but if the actual value is counted as negative then it is classified as false positive (FP). Similarly, if the outcome and actual values are negative, then it is counted as true negative (TN), but the actual value is positive, then it is counted as false negative (FN). Figure 8.12 shows the two-by-two contingency table representing the dispositions of set of instances.

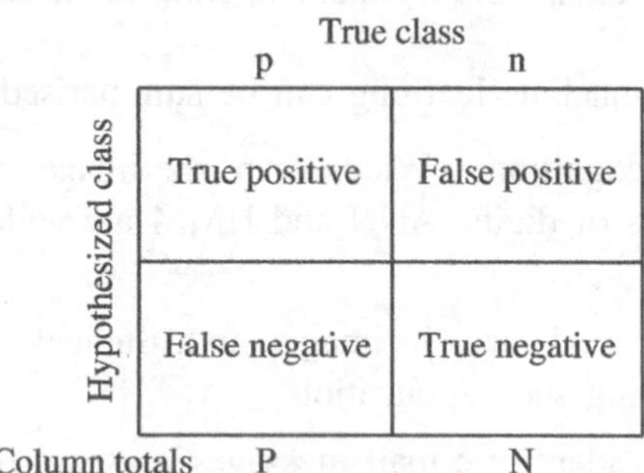

Figure 8.12 Confusion matrix.

The hit rate or True Positive Rate (TPR) of a classifier is estimated as:

$$\text{TPR} \approx \frac{\text{Correctly classified negatives}}{\text{Total positives}} = \frac{\text{TP}}{\text{TP} + \text{FN}} \tag{8.1}$$

True positive rate is also called Sensitivity (Se).

$$\text{FPR} \approx \frac{\text{Incorrectly classified negatives}}{\text{Total positives}} = \frac{\text{FP}}{\text{TN} + \text{FP}} \tag{8.2}$$

False Positive rate is also called 1 – Specificity (Sp) i.e.,

$$\text{Specificity (Sp)} = \frac{\text{TN}}{\text{TN} + \text{FP}} \tag{8.3}$$

The accuracy can be estimated as

$$\text{Accuracy (Ac)} = \frac{\text{TP} + \text{TN}}{\text{TP} + \text{TN} + \text{FP} + \text{FN}} \tag{8.4}$$

Classification error rate or Misclassification = 1 – Accuracy

Classification precision is the proportion of correctly predicted positive cases and is estimated as:

$$\text{Precision} = \frac{\text{TP}}{\text{TP} + \text{FP}} \tag{8.5}$$

F-measure gives a single measure that combines recall score and precision score to analyse the classification performance. It is the harmonic mean of precision and recall that helps in information retrieval. F-measure can be estimated as:

$$\text{F-measure} = 2 \times \frac{(\text{Precision.Recall})}{(\text{Precision} + \text{Recall})} \tag{8.6}$$

8.5.1 Threshold Fixing

Consider a dataset that contains data of people from a hospital. The dataset consists of cancer diseased people data and non-cancer diseased people data. The dataset is mixture of positive (suggesting presence of disease) and negative data (non-diseased). The individuals can test either positive or negative for the disease. Figure 8.13 shows the test hypothesis. A false positive occurs if an individual is not diseased but our system predicts that he is diseased. This is a Type I error. A false negative occurs when the system fails to predict an individual as a diseased person but he is. This is called Type II error.

Disease \ Test	Not rejected	Rejected
No disease (D = 0)	☺ Specificity	X Type I error (False +ve)
Disease (D = 1)	X Type II error (False –ve)	☺ Sensitivity

Figure 8.13 True disease state vs. test result.

We have a detailed analysis on this by an example. There are many people (individuals) undergone our system. The system predicts either diseased person or not a diseased person. Hence, this is considered as a multiple sequence test. Figure 8.14(a) shows a bunch of test results from a test dataset. The results are plotted on two graphs—Patients without the disease and Patients with disease. That is, patients without the disease are called as 'negative' and patients with disease are called as 'positive'. Figure 8.14(b) shows these two class criteria.

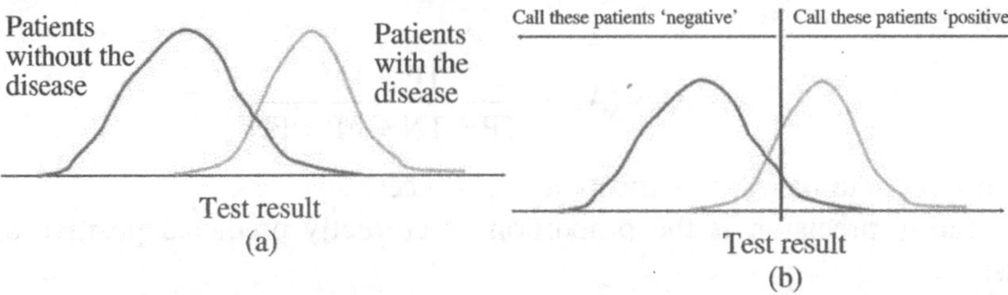

Figure 8.14 (a) Specific example (b) Threshold fixing.

Now, we want to evaluate our system with more test procedures. A multiple test is conducted that points out the graph into positive and negative points plotted in the given two graphs. The plotting is either in a correct place (positive prediction in the positive region or negative prediction in negative region) or in a wrong place (positive prediction in negative region or negative prediction in positive region). The positive prediction in positive region is considered as true positive (Figure 8.15(a)) and negative prediction in negative region is considered as true negative (Figure 8.16(a)). Similarly, group of the test results those who have no disease but predict as positive (suggesting presence of disease) is considered as false positive (Figure 8.15(b)). Also, the group of test results under the category of disease which are indeed under the negative class (non-diseased) is considered as false negative (Figure 8.16(b)). These two error predictions create vulnerability in a prediction/classification system.

To improve the performance of the prediction system, consider a pointer called 'threshold', which is used to maximise to reduce false negative and false positive rates. The threshold is pointed in negative and positive prediction in multiple tests. Figure 8.17(a) shows the threshold fixing at the diseased side (patients 'negative') and Figure 8.17(b) shows the fixing a threshold at the non-diseased side (patients 'positive'). A combination of these two thresholds gives a good prediction system.

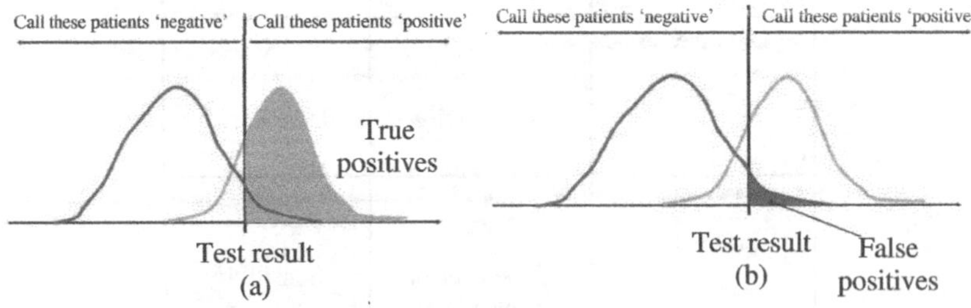

Figure 8.15 (a) True positive (b) False positive.

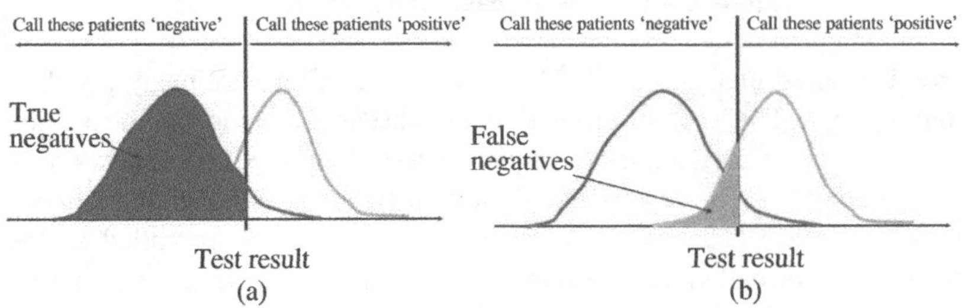

Figure 8.16 (a) True negative (b) False negative.

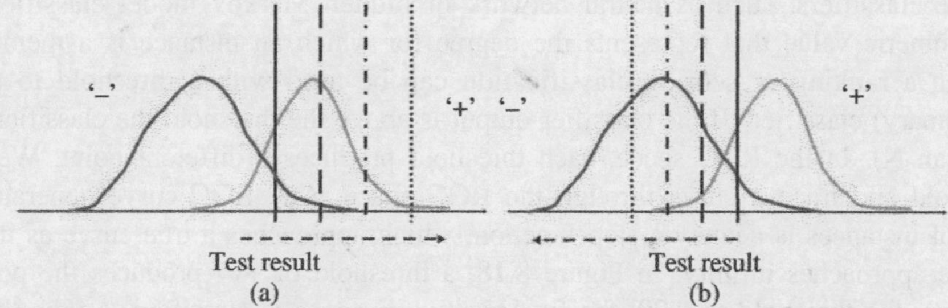

Figure 8.17 (a) Moving the threshold: right (b) Moving the threshold: left.

8.5.2 ROC Graphics

ROC graphics was used in electronic signal detection theory (1940s–1950s) and became very popular in biomedical applications, particularly radiology and imaging. This technique is highly used in machine learning applications to assess classifiers and can be used to compare tests/procedures.

Receiver Operator Characteristic (ROC) graph is a technique for visualising, organising and selecting classifiers based on their performance. The ROC analysis is widely used to analyse the machine learning classifiers and generally useful as performance graphing method. ROC is a graph with true positive rates (sensitivity) on X-axis and false positive rate (1-specificity) on Y-axis. Figure 8.18 shows a basic ROC graph. The lower left point (0, 0), represents a no false positive error and also gains no true positives. The point (1, 1) shows the opposite strategy of unconditionally issuing positive classification. The point (0, 1) represents perfect classification.

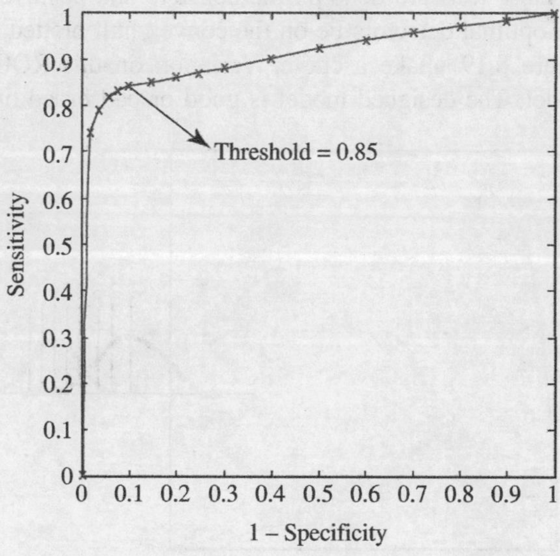

Figure 8.18 ROC graph created by threshold of a test set. Sensitivity (true positive rate) on X-axis and 1-specificity (false positive rate) on Y-axis. There are 17 threshold scores assigned to the classifier.

Some classifiers, such as neural network or hidden Markov model classifiers yield a score, a numeric value that represents the degree for which an instance is a member of the class. Such a ranking or scoring classification can be used with a threshold to produce a discrete (binary) classifier (If the classifier output is above the threshold the classifier produces a Y, else an N). In the ROC space, each threshold produces a different point. We can vary the threshold and trace a curve through the ROC space. Any ROC curve generated from a finite set of instances is acutely a step function, which approaches a true curve as the number of instance approaches infinity. In Figure 8.18, a threshold of $+\infty$ produces the point (0, 0). As we lower the threshold to 0.99, the first positive instance is classified. As the threshold is further reduced, the curve climbs up to the right and ends up at (1, 1) with a threshold 0.01. It seems that, the ROC point at (0.1, 0.8) produces its high accuracy (84%). At this point, the classifier shows a better accuracy by identifying likely positives than at identifying likely negatives. So the best accuracy is found to be at a threshold 0.85.

Classifiers appearing on the left-hand side of an ROC graph, near to X-axis show positive classifications only with strong evidence so they make a few positive errors, but with low true positive rates. Classifiers on the upper right-hand side of an ROC graph may produce positive classifications with weak evidence so they classify nearly all positives correctly, but they often have high false positive rates.

8.5.3 ROC Formulation

Consider our previous example of cancer diseased individuals. Suppose we have eight datasets which contains data of individuals. Each set contains equal or unequal number of data (individuals). We have a test procedure for the designed model. Each test gives a point that is marked across X-Y axis of the ROC graph. Figure 8.19 shows a set of fixed points in the ROC graph. Note that, these tests are done on the negative and positive curves by varying the threshold value. All the optimal detectors lie on the convex hull plotted across the line through these eight points (Figure 8.19) make a curve. Variation on this ROC curve is resulted the performance of the model. The designed model is good or bad based on the ROC test results.

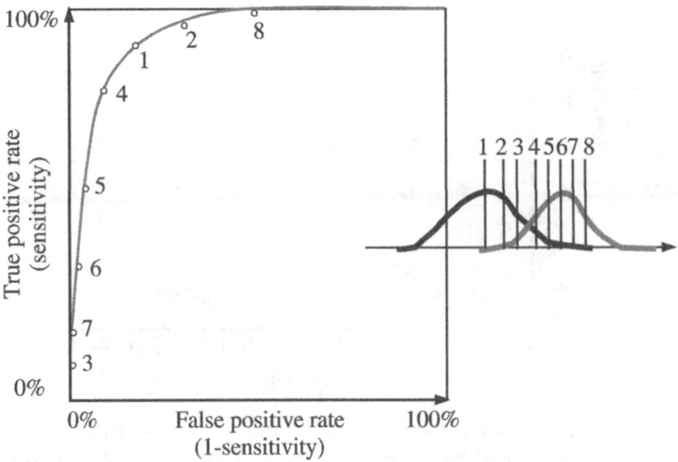

Figure 8.19 ROC graph formulation.

Figure 8.20 and Figure 8.21 give the various ROC results of different models. These graphs show whether a designed model is good or not.

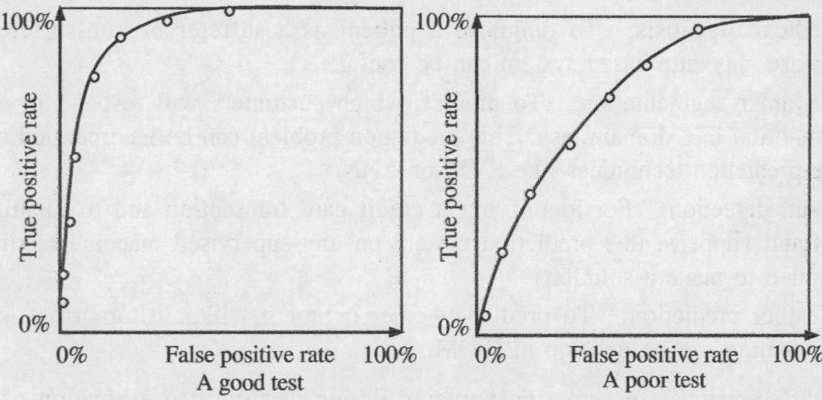

Figure 8.20 ROC graph for a good test and a bad test.

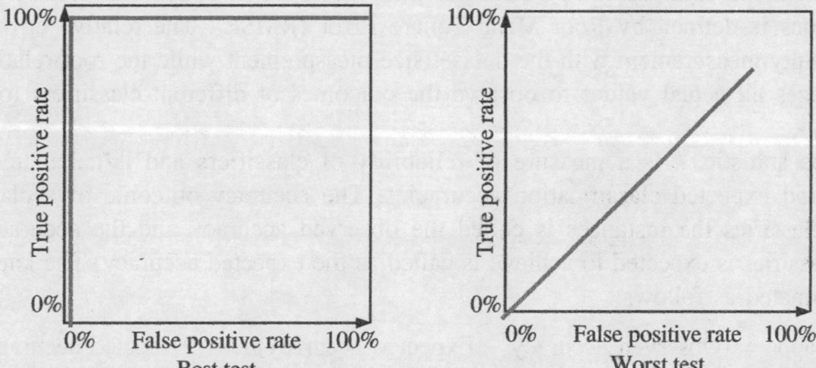

Figure 8.21 ROC graph for a best test and a worst test.

Major applications of machine learning that use ROC graphics are summarised as Follows:

- Optical character recognition: To categorise images of handwritten characters by the letters represented in a text. ANN and HMM are widely used for optical character recognition.
- Face detection: To find faces in images or indicate if a face is present. HMM is suitable for developing such application.
- Spam filtering: To identify e-mail messages as spam or non-spam from a bunch of e-mails. Any of the machine learning techniques like SVM, ANN or HMM can be used to develop such applications.
- Topic spotting: In order to categorise news articles (say) as to whether they are about politics, sports, entertainment, etc. HMM or SVM can make an influence in this application.

- Spoken language understanding: Within the context of a limited domain, to determine the meaning of something uttered by a speaker to the extent that it can be classified into one of a fixed set of categories. For language classification, HMM is highly useful.
- Medical diagnosis: To diagnose a patient as a sufferer or non-sufferer of some disease, any rule based system can be useful.
- Customer segmentation: To predict, which customers will respond to a particular promotion in a domain task. This prediction problem can be incorporated with any of the prediction techniques like SVM or ANN.
- Fraud detection: For identifying a credit card transaction and to identify whether a fraud happens, this prediction system on any supervised machine learning can be applied to make a solution.
- Weather prediction: To predict whether or not it will rain tomorrow is a type of application can be develop in HMM.

The different types of errors encountered during performance evaluation of classifiers are quantified in terms of various error rates. Mean Absolute Error (MAE) measures the closeness of actual and predicted outcomes. The difference in actual and predicted values in classifications is defined by Root Mean Square Error (RMSE). The relative error compares the uncertainty measurement with the dataset size measurement while the root relative squared error averages all actual values to observe the outcomes of different classifiers for the same problem.

Kappa statistic, k is a measure of reliability of classifiers and is calculated from the observed and expected classification accuracies. The accuracy outcome of a classifier that correctly classifies the instances is called the observed accuracy and the accuracy that any random classifier is expected to achieve is called as the expected accuracy. The kappa statistic can be estimated as follows:

$$\text{Kappa} = (\text{Observed accuracy} - \text{Expected accuracy})/(1 - \text{Expected accuracy}) \quad (8.7)$$

In machine learning, kappa statistic is used to reliably compare classifiers constructed on multi-class datasets. As kappa statistic scales both observed and expected accuracies in a common platform built, it can reflect the performance of the classifier for all instances more accurately.

8.6 Case Studies

8.6.1 Learning Engine for Process Power Plants

Today, supervision of power plant is more and more complex to watch not only because of the number of parameters to monitor but also because of numerous interactions existing between them. This is due to the processes in a power plant which are highly automated. Whenever there emerge new problems in a power plant, it is very difficult to diagnos manually and fix the problem. It may be time consuming to manage such problems. Most of the critical systems such as power plants, water purification plants, and railway control systems and transportation,

Supervisory Control And Data Acquisition (SCADA) system play a key role. SCADA system, distributed control system and other supervisory control systems are seen in industrial sectors and critical infrastructures. These systems are commonly called industrial control system. Power plant is one of the industrial control systems, which generate electricity. A control system is a device or set of devices to manage, command, direct or regulate the behaviours of other devices or systems. Most of the power plants are controlled with computer systems and SCADA systems.

Objective of the work is to develop a learning engine that will enhance the working of real-time expert system. With the new electronic and software technologies, SCADA systems are widely used in thermal power plant automation. A typical thermal power generating electricity using dry water steam to drag the alternator in rotation is analysed. This steam is generated, transforming the energy of naphtha fuel.

Most of the thermal power plants' operational controls are automatic and controlled by computers. However, at times, manual intervention may be required. Thus, the plant is provided with monitors and alarm systems that alert the plant operators when certain operating parameters are seriously deviating from their normal range. Thermal power plant generates power from heat energy as its driving force. With reference to the power plant considered (Figure 8.22), heat energy is efficiently used by two gas turbines, i.e. GT_1 and GT_2, and a steam turbine ST. In gas turbines, the dynamic pressure from flowing gases (air and combustion products) directly operates the turbine. Natural gas fuelled and oil fuelled combustion turbine plants can start rapidly and so are used to supply energy during periods of high demand of power loads, though at higher cost than base-loaded plants. The output heat energy of the gas turbine (greater than 500 degrees Celsius) is not directly exhausted into the air, but used to generate steam for the steam turbine. Then the steam is condensed in a condenser and is reused. After the condensation, water is transmitted from the tanks to pumps of extraction in the station of base pressure to be warmed progressively before being sent back to the heater through the intermediary of the pumps. This warms progressive of water for goal to increase the output of heater and to avoid all thermal constraints on its partitions and this station of water is composed of a certain number of intersections that is nourished in steam of three bodies of the turbine. Finally, the cycle reproduces indefinitely since steam and water circulate in a closed circuit.

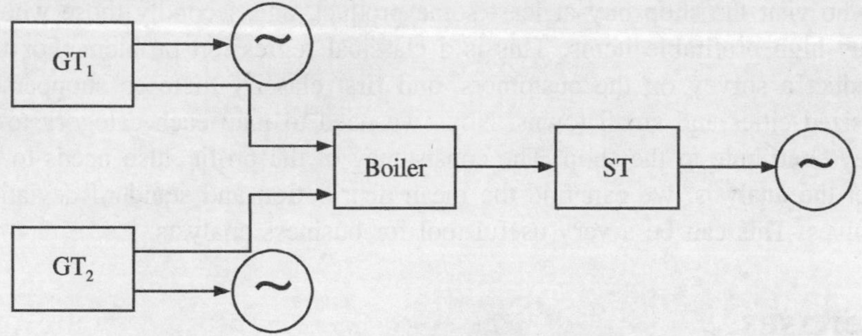

Figure 8.22 Functional schematic of a thermal power plant.

A rule based expert system can fix problems related to thermal power plants by diagnosing the root causes. The rules are incorporated with tags, values and flags. The forward as well as backward chaining can be used while developing the expert system shell. This expert system shell component can be developed in a machine learning platform. The learning process is implemented by inductive learning process from the historical SCADA database of the plant by giving rule templates. A method for forecasting possible faults occurrence of the power plant after learning the rules from the historical SCADA database.

Learning engine for real-time expert system in process power plant is an extension of the expert system shell where new rules have to be learned by the shell for process plants. The plant should be continuously available for power generation for a grid. In case a problem happens, the cost and time to reinstate the power plant is a crucial issue. So, better option to maintain the power plant is accommodating a facility for preventive maintenance. Knowledgebase of the real-time expert system should be updated with new rules by analysing the historical data. The learning engine for real-time expert system proves to be a good addition to real-time expert system. In this case study, we have to create a knowledgebase with a set of rules created by the use of inductive learning. This knowledgebase is useful for making solutions to future problems.

8.6.2 Elephant's Identification

Let us take the case of a research to study the elephant's details from its footprints. Samples of leg lengths and skull size from a population of elephants are collected. The two variables selected in the study are 'leg length' and 'skull size' that is associated in some way. The elephants with short legs may have big heads. So, an association may be found in these variables. Regression is an appropriate bag to describe a relationship between head size and leg length. Is it correct? If skull size increases, the length of leg increases? Is size a cause for small skull? Answers to these questions can be concluded by the regression analysis of two variables.

8.6.3 Retail Analysis

A manager of the retail shop basically has two objectives. Firstly, to make sure that the customers who visit the shop buy at least some product, and secondly those who are buying products, buy high profitable items. This is a classical regression problem. For the analysis, we can conduct a survey on the customers, and first classify them as shoppers from large cities, mid sized cities and small towns. Now, we need to map each category to the amount of profit they contribute to the shop. The consistency of the profit also needs to be checked. Once we get the analysis, we can find the mean distribution and standard deviation to know the exact values. This can be a very useful tool for business analysts.

8.7 Exercises

1. Is the learner guaranteed to converge to the correct hypothesis in the limit as the number of training examples increases indefinitely?

2. How many training examples are needed for a learner to construct (with high probability) a highly accurate concept?
3. How will machine learning be helpful in job scheduling?
4. What are robotic agents? How do they reduce human efforts?
5. What are the advantages that one gets when machine performs intelligently?
6. Cite areas where machine cannot excel human being?
7. How will machines make decision in various problems?
8. What is knowledge-based learning? Describe explanation-based generalisation. In what sense is this learning? What is the role of the data in driving explanation-based generalisation?
9. Suppose that you have learned a classifier for some Boolean property. You run your classifier over a randomly selected test set of 1024 instances, and find that it gives the correct answer three-quarters of the time. Give the formula for the 95% confidence interval for the true accuracy of the classifier. What is the meaning of 95% confidence interval?
10. In the above problem, if you increase the size of the test set by a factor of 16 (that is, use a test set of 16,384 instances), what happens to the confidence interval?
11. In tuning the parameters of a learning algorithm to a given application, why is it necessary to use a validation set that is separate from the test set?
12. What is stratification? What is cross-validation? What is 'leave-one-out' cross-validation? Why is cross-validation useful and what are its drawbacks?
13. What is the theory of minimum description length learning? How is this theory used with imperfect classifiers? Explain, using an example, how it attempts to address the problem of overfitting.
14. A learning theory may be divided into the following parts: A hypothesis space; A representation for hypothesis; A preference criterion over hypothesis, independent of the data; A measure of how well a given hypothesis fits given data, and A search strategy to find a good hypothesis for a given dataset. For the machine learning method of your choice explain what each of these are?
15. How are true positive rate and recall related?
16. What do you understand by Type I vs Type II error?
17. When is Ridge regression favourable over Lasso regression?
18. You are given a dataset. The dataset has missing values which spread along one standard deviation from the median. What percentage of data would remain unaffected? Why?
19. You run a classifier on a test set with the following results:

Item	Labelled sports	Labelled non-sports
Predicted sports:	10	20
Predicted non-sports:	5	2000

What is the precision of your classifier? What is the recall of your classifier? What is the accuracy of your classifier? Is this a useful way to evaluate your classifier here? Why or why not?

20. Suppose we use a decision-tree induction algorithm and bagging to learn a classifier that is an ensemble of 20 decision trees, for a two-class problem.

(a) Describe a procedure for computing a Precision-Recall (PR) curve for this classifier, given a held-aside set of labelled examples.

(b) Describe a procedure for computing an ROC curve for this classifier, given a held-aside set of labelled examples.

(c) For both types of curves, describe how you would informally determine if the decision-tree ensemble has any predictive value.

(d) Suppose we modify our task to be a three-class problem. Could we still use PR and ROC curves? If so, how could we compute them?

21. How can one find the cut-off for test with discrete rating values using ROC curve?

22. How to calculate best cut points for surveys using ROC curve?

23. What is the difference in using of ROC curve and sigmoid logistic regression?

24. How to generate a REC-Curve for multilevel regression models?

CHAPTER 9

Supervised and Unsupervised Learning

Suppose we have two datasets A and B containing some medical image data which transmit the diseases information of an organ. We know the labels of all data in the dataset A, that is, whether one image represents diseased or not. But the images in dataset B are not labelled, that is, we do not know whether the organ is diseased organ or not. How do we classify the images in second dataset B? In this case, we can build a classification model using dataset A (since all images are labelled as diseased or not). After fine tuning the model we can use it to classify the data in dataset B (since, the data are not labelled before). This classification model is built by supervised learning techniques. Here dataset A is called training dataset and dataset B is called testing dataset.

Suppose we do not know the labels of both datasets A and B. How do we group the data in A and B in diseased class? Using unsupervised techniques, we can group the data in set A and B. Usually supervised techniques are used for classification or prediction and regression problems. Unsupervised techniques are used in dimensionality reduction and clustering problems.

Learning is the process of acquiring knowledge. As mentioned in earlier chapters, we have basically three types of learning, i.e. supervised, unsupervised and reinforcement learning. In addition to these three techniques, evolutionary learning techniques have also evolved in the machine learning. But this chapter considers only supervised and unsupervised learning paradigms. Chapter 10 deals with reinforced learning paradigm. The evolutionary learning techniques are incorporated in second part of the book (from chapter 'Models of Life and Intelligence'). Figure 9.1 gives the three main machine learning paradigms, discussed in this chapter.

In simple words, supervised learning can be defined as the process of learning with the help of a supervisor or a teacher. In supervised classification, the classes will be predefined and even the test data will be classified to these classes. In short, the process will already have a defined schema. Let us try to explain the concept with a simple example. Consider the situation where 'John' wants to travel from 'Trivandrum' to 'Agra'. 'John' is totally unaware of the route and has no prior experience of travelling to 'Agra'. Supervised learning is the process where a supervisor gives 'John' the route map and the kind of knowledge for travelling from Trivandrum to Agra. Thus, he is fully equipped for the process of travelling.

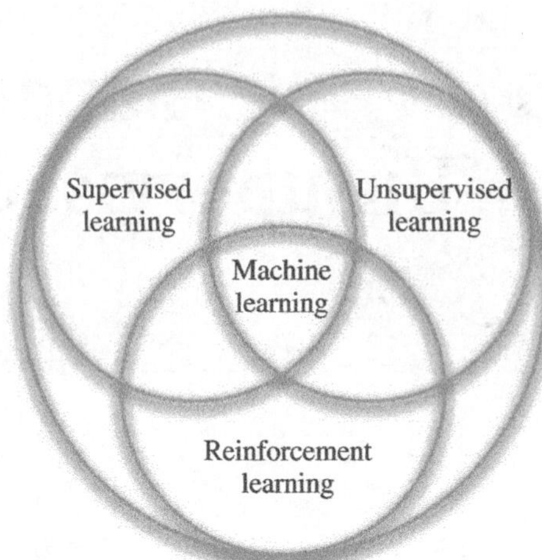

Figure 9.1 Learning paradigms of machine learning.

The most important part of finding good solutions to problems using machine learning algorithms primarily depends on choosing relevant data and learning methods. The algorithm should produce sensible outputs (responses) for the set of input data that were not encountered during learning. At the same time these algorithms should also deal with small inaccuracies or errors in the data called noise that are inherent in measuring any real-world process. Learning is also from the real-world experience or from nature. Considering these scenarios, machine learning is categorised into supervised learning, unsupervised learning, reinforcement learning and evolutionary learning.

Supervised learning

A training sample set with correct responses are provided. Based on this training set, the algorithm generalises to respond correctly to all possible inputs.

Unsupervised learning

Correct responses are not provided; instead the algorithm tries to identify similarities between the inputs so that inputs that have something in common are categorised together.

Reinforcement learning

This type of learning falls between supervised learning and unsupervised learning. The algorithm gets intimated when there is incorrect answer, but fails to get intimated how to correct it. It has to explore and try out different possibilities to get the correct response. That is, the learning process is outcome based.

Evolutionary learning

The principle of biological evolution by 'survival of the fittest theory' is modelled as a learning process in this type of learning. Based on the fitness and survival rates of data, the algorithm responds with a score that measures the goodness of current solution.

9.1 Supervised vs. Unsupervised Learning

Supervised learning or classification can be normally employed in situations where the training data are fully available. In cases where the part of data is hidden or missing, this type of learning will fail. Throughout this book we are focussing on the various types of supervised classifiers namely, support vector machine, inductive logic programming, case based reasoning, ensemble classifier, nearest neighbourhood classifier and fuzzy networks. Major difference between supervised and unsupervised learning is illustrated in Figure 9.2(a) and Figure 9.2(b) respectively. Later, we can see how problems are solved by supervised learning techniques.

Figure 9.2 (a) Supervised learning (b) Unsupervised learning.

In supervised learning, the classes for prediction will be predetermined. It can be viewed as a finite set, which is already defined by humans. Hence, the main task of machine learning is to search for patterns and to create models according to the predetermined formats. Decision tree induction, Naive Bayes etc., are the examples of supervised learning algorithms.

Unsupervised learning is a method where we assign values for each action. The major task is to develop the classification automatically. It then checks the data to find the similarity and groups them into a particular cluster. It has been found successful in many applications like creating and training machines capable of driving automobiles. Clustering is useful when we have enough data to form clusters. The data about the members in a cluster can be used to produce more accurate rules or to draw conclusions. Classification learning is used in cases where the decision made by an algorithm is used as input somewhere else. Both techniques give accurate result but the only thing we need to be careful about is that the selection of learning algorithm depends on the circumstances and the kind of problem in which it is applied.

9.1.1 Prediction System

A prediction model is given in Figure 9.3. The model can be built in either supervised or unsupervised terminology. Initial experiments are conducted and fed into the designed model, but this model may not be good enough to predict best results. A re-define process on the parameters and model components improve the performance of designed model. Using further experiments on this model followed by refining the parameters and re-designing the model once again. Using optimisation techniques or cleaning mechanisms on the parameters or on the models gives a good prediction system. These processes are considered as a continuous activity repeated on the processes and machine learning model.

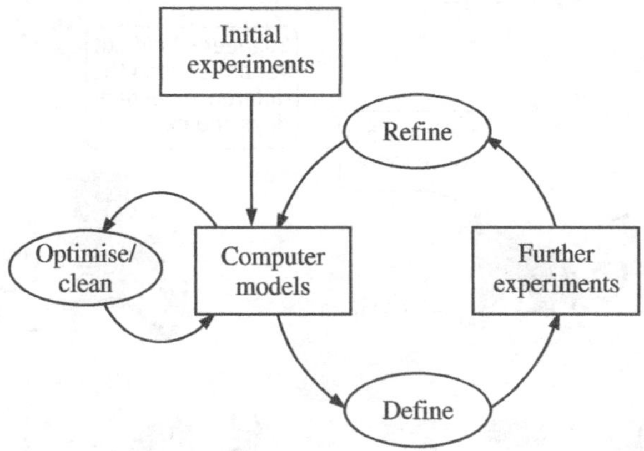

Figure 9.3 Prediction system.

9.1.2 Training, Testing and Validation Datasets

The training dataset is the group of data that is used for training purpose. Formulate a separate sample dataset used for unbiased evaluation of the final model, which is called test dataset. After completely training the model the testing process is initiated. Hence, the test dataset is also prepared at the time of training dataset preparation.

For an unbiased evaluation of a machine learning model a separate dataset is required which is called validation dataset. This will help a model to tune with hyper parameters. A hyper parameter is a parameter whose value is set before the beginning of learning process. Complex machine learning models require hyper parameters but simple models do not require any (for example, least square regression). The other parameters are derived at the training time.

The dataset is incorporated into the model configuration for making a perfect model. Use of same dataset for testing and validation process is not a good practice. The distribution of training, testing and validation dataset distribution is given in Figure 9.4(a). Note that, the validation dataset is used to tune the hyper parameters (the architecture) of a classifier model.

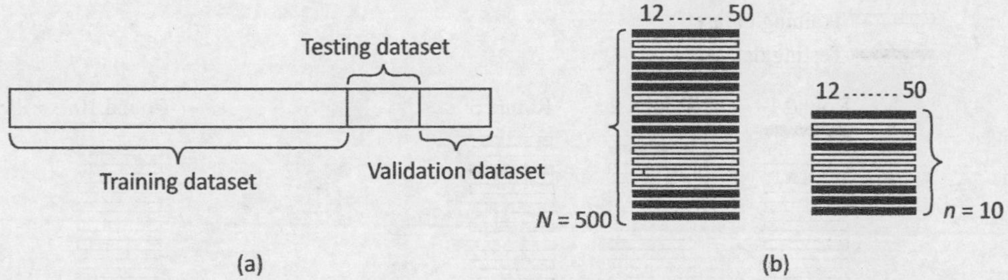

Figure 9.4 (a) Dataset distribution (b) Training and testing data distribution.

How will you make a clear split-up of a dataset for training, testing and validation purpose? The machine learning models with few hyper parameters will be easy to validate and tune. Hence, this dataset size can be reduced. But if the model has a large number of hyper parameters, the size of validation dataset must be increased. The validation dataset can be avoided if the model has no hyper parameters or ones that cannot be easily tuned.

The sample dataset distribution for the training and testing a model is given in Figure 9.4(b). This is a mixture of positive and negative datasets. To formulate test dataset, make sure that it is large enough to yield statistically meaningful results and ensure that the data characteristics of training and testing are the same. In normal cases, the model designers split a dataset into two—training and testing. They are randomly chosen N% of data for the training purpose and the remaining (100 – N)% data for testing purpose. For example, 80% data for the training and 20% of the data for testing can be used from a dataset.

9.1.3 Cross Validation

Cross validation is a rotation estimation or out-of-sample testing technique for a machine learning model with a test process by statistically independent dataset. It is used to estimate how accurately a predictive model performed in practice. Major goal of the cross validation is to test the model's ability to predict data that was not used in estimating. To avoid overfitting or underfitting and perform a better result in independent dataset we can use the cross validation technique. The independent test dataset is an unknown dataset from a real problem.

k-fold cross validation

In the *k*-fold validation, a single parameter called *k* which refers to the number of splits in a dataset that is used for training and testing process. If we choose $k = 10$, then a 10-fold cross validation is applied to the machine learning model. In Figure 9.5, 10-fold cross validation cycles are given. The training and testing process is conducted from round 1 through round 10 by splitting the data into two sets—training and testing. This technique is very useful to estimate the skills of a machine learning model on unseen data. When number of samples (data) are limited, the performance of the model needs to be at par with an expected level. Cross validation is an efficient method to evaluate the performance of a classifier model when limited number of data are available. The procedure for making the dataset is as follows:

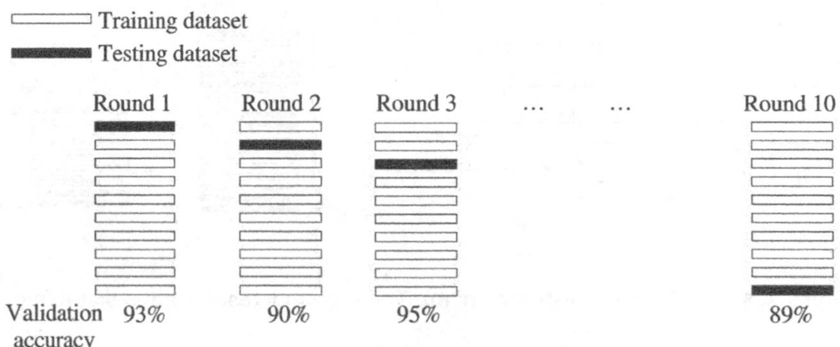

Figure 9.5 A 10-fold cross validation.

1. Shuffle the dataset arbitrarily.
2. Split the dataset into k groups.
3. For each unique group:
   (a) Take the group as a hold out or test dataset.
   (b) Take the remaining groups as a training dataset.
   (c) Fit a model on the training set and assess it on the test set.
   (d) Retain the assessment score and discard the model.
4. Summarise the skill of the model using the sample of model assessment scores.

If there is a minimum number of data available and there is a big difference in quality or different optimal parameters between folds then the cross validation is a good choice. The value of k is shown empirically (by choosing $k = 5$ or $k = 10$) to yield test error estimates that suffer neither from unreasonably high bias nor high variance. A poorly chosen k gives for a model a bad representative idea such as high variance score or a high bias, that is, overestimates the skill of the model. A common solution to this problem is choosing, $k = 10$ found through experimentation which generally fixes the skill of the model. Another idea is based on the size of dataset and the value of k can be fixed, $k = n$, where n is the size of dataset. This approach is called leave-one-out cross-validation.

9.2 Supervised Learning Model

Supervised learning is like training by a teacher. A supervised learning model is depicted in Figure 9.6. Each instance of the training set contains a class information. During the training process, the actual output from the model is compared with desired output and error is fed back to the model. This error signal is used to adjust parameters of the model, and this process repeats until the system converges.

In Figure 9.7, positive and negative data are given and are counted as a datasets of shapes. These data are a mix of positive (select) and negative (not select) items with a set of properties such as colour, shape and size. From this dataset, first we have to identify the parameters from the properties (colour, shape and size) of the given shapes. In the next step, we have to label the item parameters into Blue (B), Red (R), Torus (T), Square (S), Star (R), Arrow (A),

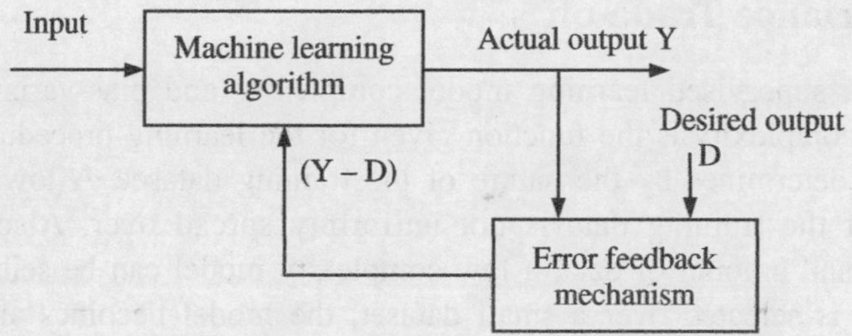

Figure 9.6 Supervised learning model.

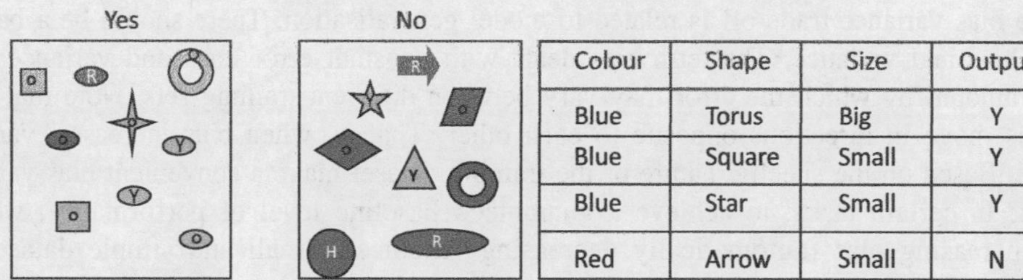

Colour	Shape	Size	Output
Blue	Torus	Big	Y
Blue	Square	Small	Y
Blue	Star	Small	Y
Red	Arrow	Small	N

Figure 9.7 Positive and negative dataset.

Big (B), Small (S). Now, we have three parameter set say, X_1, X_2, X_3. There are two parameters in X_1, four parameters in X_2, and two parameters in X_3. We need to classify whether an item can be selected or not (Y/N). Hence, the output is either in Y or N. In this case, we have to prepare a separate training dataset and test dataset. Figure 9.8 shows the corresponding supervised learning technique designed for this scenario. Here, we have to formulate a hypothesis, obtained from the designed learner. The prediction from test dataset is done by the learner using the hypothesis. Note that, the parameters are reassigned numeric or different letters. To understand the supervised learning terminology for the above case, assign a letter of each parameter (first letter in most of the parameters).

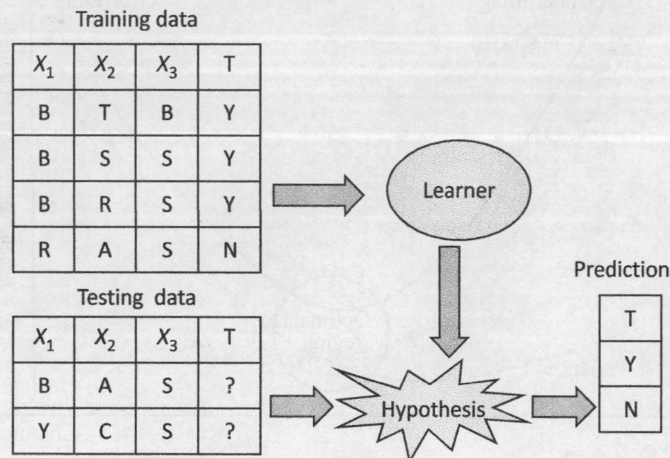

Training data

X_1	X_2	X_3	T
B	T	B	Y
B	S	S	Y
B	R	S	Y
R	A	S	N

Testing data

X_1	X_2	X_3	T
B	A	S	?
Y	C	S	?

Prediction

T
Y
N

Figure 9.8 Supervised learning technique.

9.2.1 Bias-variance Trade-off

While designing a supervised learning model complexity and bias-variance, trade-offs are important. Model complexity is the function given for the learning procedure which is similar to a function that determines by the nature of the training dataset. A low complexity model can be selected if the training data is not uniformly spread over. Also, when a training dataset contains small amount of data, a low complexity model can be selected. When a high complexity model is selected over a small dataset, the model becomes an overfitted model. That is, first learning a function by the training data and not generalising the other data in the dataset.

The bias variance trade-off is related to model generalisation. There should be a balance between bias and variance. The term bias deals with constant error term and variance deals with the amount by which the error may vary between different training sets. Note that these two terms move in directions opposite to each other. That is, when bias increases, variance decreases. Based on the specific nature of the training dataset make a convenient bias-variance spectrum. In certain tasks, to achieve a guaranteed baseline level of performance, we can choose increasing bias (automatically decreasing variance). Small and simple dataset are usually learned by low variance models. The complex or large dataset requires a high variance models to learn the structure of data completely. The trade-off between biased and variance is given in Figure 9.9. An optimal point is reached with a specific bias and variance in the dataset. At this point, there should not be underfit or overfit conditions met. The way of selecting a model in supervised learning can be obtained from these conditions.

The underfitting occurrs due the failure to capture underlying pattern of data. Property of this model is high bias and low variance due to the training dataset. When we build this model as a linear model, we have very less amount of data which is nonlinear in nature.

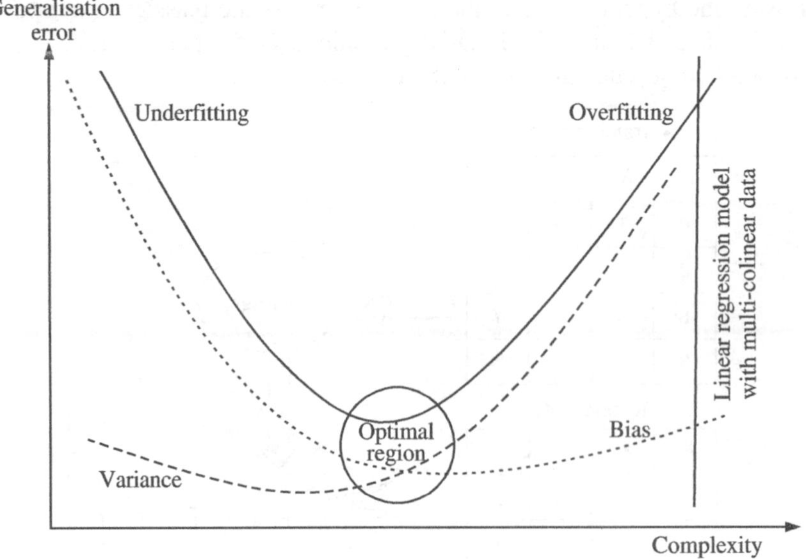

Figure 9.9 Bias-variance trade-off.

The overfitting occurrs due to the noise captured along with the underlying pattern in data. The noisy data is used to train the model to produce low bias and high variance.

Parametric or linear machine learning algorithms often have a low variance but a high bias and non-parametric or nonlinear machine learning algorithms often have a high variance but a low bias. Examples of low bias and high variance machine learning algorithms are support vector machines, decision trees and *k*-nearest neighbours. Examples of high bias and low variance machine learning algorithms are linear discriminant analysis, linear regression and logistic regression.

9.2.2 Classification Problems

Classification is a modelling that works as a mapping function *f* from input variables *X* to discrete output variables *Y*. The mapping function predicts the classes labelled in the output variables. Consider an example of a set of e-mails in an inbox of classified as 'spam' or 'non-spam'. It can be a classification problem with two classes with real valued or discrete input variables. Since, there are only two classes in this problem, it can be treated as a binary classifier. The output probability of the prediction system decides whether an e-mail is a spam or not while arriving the inbox. For example, probabilities of 0.1 as being 'spam' and 0.9 as being 'not spam'. Figure 9.10 shows how a classification is done in a set of data.

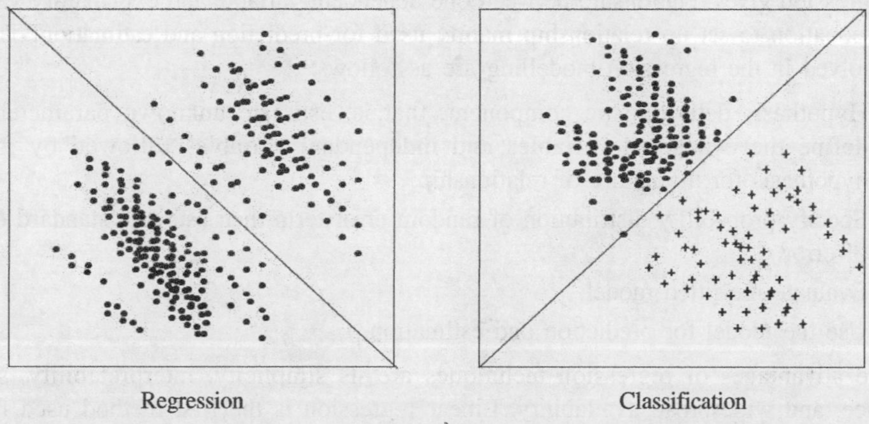

Figure 9.10 Supervised learning techniques in a dataset.

The response category value coming into classes like 'yes' or 'no' is basically included in binary classifier. For example,

- Is the picture of a dog or not?
- Is the mail being spam or not?
- Will it rain today?
- Can I play tennis today?

In multi-class classification, more than one group is required to predict. For example,

- Is the mail coming under category 'social', 'spam' or 'promotion'?

- Is the picture of a cat, dog or a monkey?
- What is the climate today, i.e. rainy day, sunny day or cloudy day?

If we need to predict a continuous response value, we can use the regression techniques. For example,

- What is the value of a house in New Delhi?
- How many moves were played in the chess game?
- What is the value of the present stock exchange?
- What will be the expected Human Development Index (HDI) value of India in 2025?

Regression is a powerful statistical method that can be used in machine learning to predict continuous dependent variable from a number of independent variables. That is, regression creates a linear relationship between two or more variables. Usually variable Y is predicted from X. The variables used in regression can be either continuous or discrete. The simplest form of regression is linear regression that shows a relationship between the variables X and Y formulated as follows:

$$Y = a + bX,$$

where X is the explanatory variable and Y is the dependent variable. Slope of the line is b, and a is the intercept (the value of y when $x = 0$).

Regression gives relationship between one dependent variable and explanatory variable(s) that use equation to set up relationship mainly used for prediction and estimation. The major steps involved in the regression modelling are as follows:

- Hypothesis deterministic component, that is, estimate unknown parameters. Here, define the dependent variables and independent variables followed by making a hypothesis for the nature of relationship.
- Specify probability distribution of random error term then estimate standard deviation of error.
- Evaluate the fitted model.
- Use the model for prediction and estimation.

The advantages of regression techniques are its simplicity, interpretability, scientific acceptance, and widespread availability. Linear regression is the first method used for many machine learning problems. Analysts can use linear regression together with techniques such as variable recoding, transformation, or segmentation. Do not think that regression is always a good solution for every prediction problem. It may fail when false assumptions are made. For example, considering more independent variables than the observations, false assumptions are made in a prediction problem. Regression only looks a linear relationship between dependent and independent variables. When a straight line relationship is between two variables, for example, income and age, there may be a curve which can be fitted which is incorrect because of the income.

Example: Consider a small dataset given in Figure 9.11(a). The three parameters are Name, Rank and Years of service. We have to predict if a person is 'tenured' or not. In this case parameter name does not play any importance, but other two parameters contribute in the prediction process. Figure 9.11(b) shows how the prediction is done for the test dataset and an unseen data. Here we can use a regression function that will work as a prediction function.

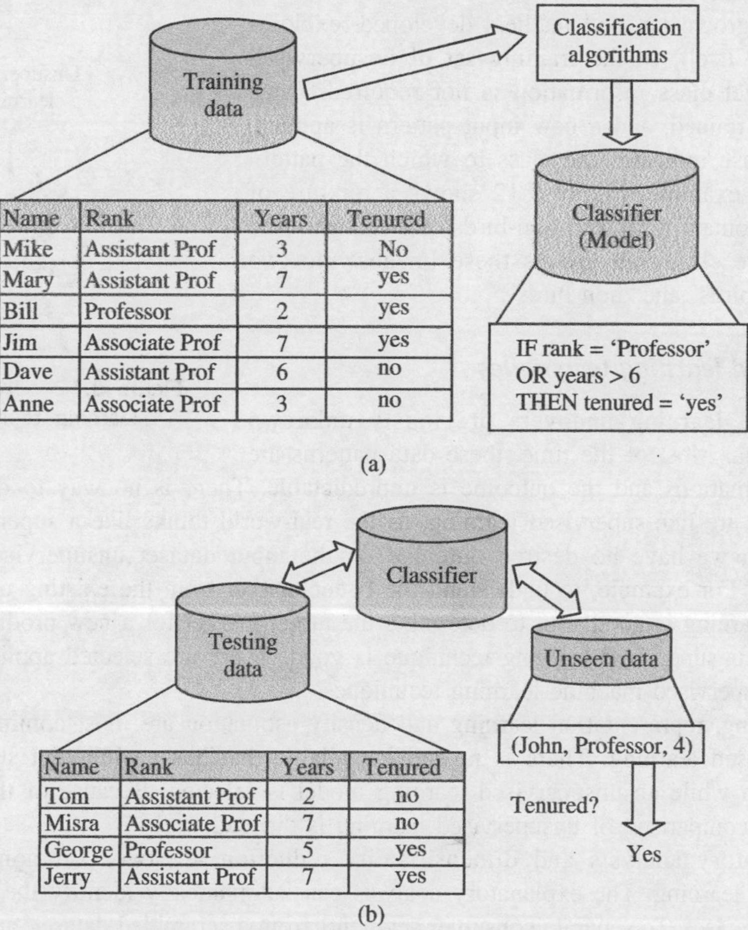

Name	Rank	Years	Tenured
Mike	Assistant Prof	3	No
Mary	Assistant Prof	7	yes
Bill	Professor	2	yes
Jim	Associate Prof	7	yes
Dave	Assistant Prof	6	no
Anne	Associate Prof	3	no

Name	Rank	Years	Tenured
Tom	Assistant Prof	2	no
Misra	Associate Prof	7	no
George	Professor	5	yes
Jerry	Assistant Prof	7	yes

Figure 9.11 (a) Training model—Supervised learning (b) Testing cases and classification for an unseen data.

9.3 Unsupervised Learning Model

Supervised learning is mostly used in classification or prediction problems. We can map the input to the output labels while designing a classification of prediction problem. The supervised technique can also be used in regression problems while mapping the input to continuous output. In both designs (classification models and regression models) main goal is to find specific relationship or structure in the input data that allows us to effectively produce precise output data. The correct output is determined by the quality of the training data. So, data in the training dataset should be labelled with real-world situations. If data's incorrect or noisy, labels are in the training dataset that reduce the performance of model.

The unsupervised learning process is independent, where training instances organise themselves and form clusters. This is analogous to the case of a baby bird which when grown up, and feathers developed, explores the world by itself. In the training set of unsupervised learning, target class information is not required. Once the model is trained, and a new input pattern is applied, output response indicates the class to which the pattern belongs. For example, Figure 9.12 shows a mixture of images that contain birds and non-birds. An unsupervised technique like clustering groups these images into two classes, i.e. 'birds' and 'non birds'.

Figure 9.12 Unsupervised learning technique.

Unsupervised learning techniques

Unsupervised learning uncovers previously unknown pattern of data. Most of the time, these data patterns are poor approximations and the outcome is unpredictable. There is no way to determine how accurate they are like supervised learning, as the real-world thinks like a supervised learning model. When we have no desired outcome on the input dataset unsupervised learning is very suitable. For example, to understand the better market from the existing consumer base, supervised learning is good. But to determine the target market for a new product (that is not sold before), unsupervised learning technique is good. There are selected applications spread over the unsupervised machine learning techniques.

Clustering, representation learning and density estimation are most common tasks done by unsupervised learning. There is no label available that has an inherent structure in the available data while an unsupervised learning model is selected. Because of this reason, the performance comparison of unsupervised learning is difficult.

Exploratory analysis and dimensionality reduction are two common use case of unsupervised learning. The explanatory analysis can automatically identify the structure of a data from a dataset. To separate consumer segments from a scrambled dataset, an unsupervised model is very suitable. For example, segmentation of cancer images from an image dataset. In this case, unsupervised learning models give an initial insight that can be used to test individual hypotheses.

Dimensionality reduction is another method in the unsupervised learning that represents less columns or features. In this method, a learning procedure is established for the relationship between individual features. The method uses to represent the data that uses latent features that interrelate our initial features. It can further process the data and eliminate redundant features.

9.3.1 Clustering

Clustering is a split of dataset that formulates into groups based on its similarity. The similarity property is based overestimates between groups and not treat the data points as individuals. The different groups are formulated based on some criteria and these criteria can be simple or complex. A few types of different types of clustering are as follows:

1. **k-Means clustering:** The data points are grouped into k number of mutually exclusive clusters. For a larger k, the number of groups become smaller with more granularity. For a lower k, creating more groups creates less granularity. Output of the algorithm creates to assign each data point labelled to one of the k groups. By this way, groups are defined that create a centroid for each group. The centroid is the heart of the cluster that captures the data point and adds to the group which is closest to them. For example, the granularity of items grouped into one as shown in the online website like Amazon. Figure 9.13 gives clustering of data points for $k = 3$.

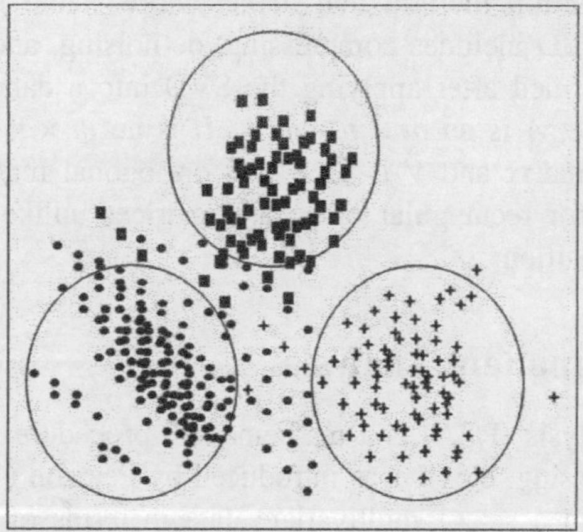

Figure 9.13 Clustering of data points, $k = 3$.

2. **Hierarchical clustering:** It is similar to a regular clustering, process of data points into parent and child clusters. Consider an example, the consumer group is split into younger age older age, and each of those groups into their own individual clusters.
3. **Probabilistic clustering:** Based on a probabilistic value, the clusters are formulated. For example, how many youngsters like drinks—Pepsi or CocaCola, is grouped based on the count (probability). So, clusters are formed based on these probabilities.

The following chapters give a clear outlook of the various clustering techniques with examples.

9.3.2 Data Compression

Data compression is used to make the dataset as small and efficient as possible by using algorithms on dataset with lesser training. This process is called dimensionality reduction that comes in unsupervised learning category. The dimensionality reduction is used when a lot of redundant data is available in the dataset. Reducing the dimensionality of data is an important part of a good machine learning pipeline. The common techniques are principal component analysis and singular value decomposition.

1. **Principal component analysis:** The Principal Components Analysis (PCA) finds the linear combinations that communicate most of the variance in a data. It is a

statistical procedure to label a set of multivariate data. The labelling is done by correlation of variables by relatively few numbers of linearly uncorrelated variables. If the training dataset is explained by multiple characteristics, PCA summarises the characteristics of the dataset and explain the dataset correctly. If a large number of variables in the dataset that are uncorrelated then PCA removes the redundancy or correlation by analyses the data.

2. **Singular-value decomposition:** The Singular Value Decomposition (SVD), factorises a dataset into product of three other smaller matrices. It is a common matrix decomposition method that makes more stable method than other. Major applications of SVD includes compressing, de-noising, and data reduction. There are three matrices formed after applying the SVD into a dataset (in a matrix format)—$A = USV^T$. Where, A is an $m \times n$ matrix, U is an $m \times n$ orthogonal matrix, S is an $n \times n$ diagonal matrix and V is an $n \times n$ orthogonal matrix. As per linear algebra, SVD is defined for rectangular or square matrices unlike the more commonly used spectral decomposition.

9.3.3 Principle Component Analysis

Principle Component Analysis (PCA) is a mathematical procedure widely used in exploratory data analysis, signal processing, etc. It was introduced by Pearson (1901) and Hotelling (1933) to describe the variation in a set of multivariate data in terms of uncorrelated variable sets. It is widely used as a mathematical tool for high dimension data analysis. However, it is often considered as a black box operation whose results and procedures are difficult to understand. PCA has been used in face recognition, motion analysis, clustering, dimension reduction, etc. It provides a guideline for how to reduce a complex dataset into a lower dimensional one to reveal the underlying hidden and simplified structures. There are a lot of mathematical proofs and theories involved in PCA, but here we are not discussing them completly. We are just outlining how the analysis is carried out.

Background knowledge

Variance: In a dataset, the variance is spread over. The variables are transformed that explain variance of the dataset in decreasing manner when we are doing PCA.

Co-variance: To calculate the strength of correlation between two or more datasets, covariance is a measurement. The covariance becomes zero in PCA when the variables are to be transformed. The purpose of PCA is to transform a dataset matrix with diagonal elements zero. The components are explained for some uncorrelated elements. The transformation is done by the eigen vector decomposition of the variance-covariance matrix.

PCA is a non-parametric mathematical technique for mining relevant data from large datasets. PCA can be applied to high dimensional data. It can be calculated efficiently with computer programs. PCA includes getting the covariance matrix, getting the eigen value and eigen vectors of each attribute and selection of main attributes. PCA uses the covariance matrix to measure the average extent of attributes along any axis. To get the covariance matrix, we need to obtain the variance of attributes. With eigen analysis, we could find out the eigen values and eigen vectors of a square symmetric matrix with sum of squares and cross products. The eigen vector related to largest eigen value defines the first principal component. The eigen vector related to second largest eigen value defines the second principal component. The process continues till the final principal component is determined. The sum of the eigen values is equivalent to the sum of diagonal elements in the square matrix. The total number of eigen vectors are equivalent to the number of rows or columns of this matrix. The first attribute extracted in a principal component analysis accounts for a maximal amount of total variance in the observed variables. The second attribute extracted will have two important characteristics. First, this attribute will account for a maximum amount of variance in the dataset that was not accounted for by the first attribute. Again under typical conditions, this means that a second attribute will be correlated with some of the observed attributes that did not display strong correlations with the first attribute. The second characteristic of the second attribute will be uncorrelated with the first attribute. When the analysis is complete, the resulting attributes will display varying degrees of correlation with the observed variables, but are completely uncorrelated with one another. The various steps involved in principle component analysis are given as follows:

Steps involved in principle component analysis

1. **Get the data:** Aim of our study is to analyse the data, so the primary requirement is to get the dataset. We have already mentioned the various methods of data collection schemes. Any of those approaches may be used for collecting the data. In case of result comparisons, already processed datasets will be available in the logs and data bank. Anyhow, acquiring the dataset is the primary step in PCA.
2. **Data adjustment:** For the analysis to carry out smoothly, the data need to be polished. For this process, you have to subtract the mean from each data dimension. That is, all x values will have $\overline{x}$ subtracted and all y values will have $\overline{y}$ subtracted from it respectively (if the selected data is a 2D dataset). This produces a dataset with zero mean.
3. **Covariance matrix is calculated:** Covariance is usually measured between 2 dimensions. But whenever we have more dimensions, more covariant values need to be calculated. In general, for n-dimensional dataset we need to calculate $\frac{n!}{(n-2)! \times 2}$.
   When we are having more dimensions, what we do is, calculate the covariance between two different dimensions. For example, in an 'n' dimension matrix, the entry in row 2, column 3 is the covariance value calculated between the 2nd dimension and 3rd dimension.
4. **Calculate the eigen vectors and eigen values of the covariance matrix:** To do this, we find the values of λ which satisfy the characteristic equation of the particular matrix. If A is the matrix, we find $A - \lambda I$, then find the determinant of $(A - \lambda I)$. Now, solving this obtained equation, the results give the eigen values. Once the eigen values of a matrix A have been found, just find the eigen vectors by Gaussian elimination.

5. **Project and derive the new dataset:** Once we have chosen the components (eigen vectors) that need to be incorporated in our data, we find the new feature by taking the transpose of the vector and multiply it on the left of original dataset. That is,

   $$\text{Final data} = \text{Row feature vector} \times \text{Row data adjust}$$

   Principal component analysis accounts for the total variance of the observed variables. If we wish to see the arrangement of points across many correlated variables, which can use principal component analysis to show the most prominent directions of the high-dimensional data. Using principal component analysis, the dimensionality of a set of data can be reduced. Principal components representation is important in visualising multivariate data by reducing it into two dimensions. Principal components are a way to picture the structure of data as completely as possible by using as few variables as possible.

Example: This is a simple example. When dealing with a large complex problem domain the steps involved are same but the output is different. Consider a dataset in Table 9.1. The steps involved in the PCA are given as follows:

1. **Getting the data:** Table 9.1 shows data and the adjusted data for PCA. A plot can be made using these data as shown in Figure 9.14(a).

TABLE 9.1 Data and data adjust

x	y	x	y
2.5	2.4	0.69	0.49
0.5	0.7	–1.31	–1.21
2.2	2.9	0.39	0.99
1.9	2.2	0.09	0.29
3.1	3.0	1.29	1.09
2.3	2.7	0.49	0.79
2.0	1.6	0.19	–0.31
1.0	1.1	–0.81	–0.81
1.5	1.6	–0.31	–0.31
1.1	0.9	–0.71	–1.01

2. **Data adjustment:** The data is adjusted as shown in Table 9.1. This produces a dataset whose mean is zero.
3. Covariance matrix of values in Table 9.1 is calculated. The covariance is a 2 × 2 matrix and is $\text{Cov} = \begin{pmatrix} 0.616555 & 0.615444 \\ 0.615444 & 0.716555 \end{pmatrix}$

   The non-diagonal elements in this covariance matrix are positive, expect that both the variables increase together.

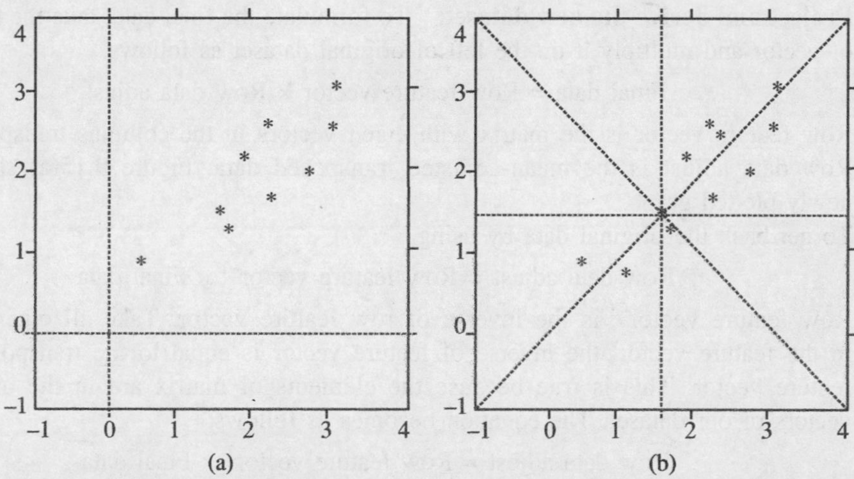

Figure 9.14 (a) PCA example data, original data on the left, data with the mean subtracted on the right, and plot of data, (b) A plot of the normalised data (mean subtracted) with the eigen vectors of the covariance matrix overlaid on top.

4. Calculate the eigen vectors and eigen values of the covariance matrix. The eigen values are calculated by eigen values = $\begin{pmatrix} 0.049083 \\ 1.284027 \end{pmatrix}$

eigen vectors = $\begin{pmatrix} 0.735178 - 0.677873 \\ 0.677873 - 0.735178 \end{pmatrix}$

The plot is shown in Figure 9.14(b). Note that, the eigen values are quite different values, eigen vector with the highest eigen value is the principle component of the dataset. Transformed data with two eigen vectors is given in Table 9.2.

TABLE 9.2 Transformed data

x	*y*
–0.827970	–0.175115
1.777580	0.142857
–0.992197	0.384374
–0.274210	0.130417
–1.675801	–0.209498
–0.912949	0.175282
0.0991094	–0.349824
1.14457	0.0464172
0.438046	0.0177646
1.22382	–0.162675

5. **Project and derive the new dataset:** To formulate the final data, take the transpose of vector and multiply it on the left of original dataset as follows:

Final data = Row feature vector × Row data adjust

Row feature vector is the matrix with eigen vectors in the columns transposed and Row data adjust is the mean-adjusted transposed data. Figure 9.15(a) shows the newly plotted graph.

To get back the original data by using

Row data adjust = Row feature vector^{-1} × Final data

Row feature vector^{-1} is the inverse of row feature vector. Take all eigen vectors in the feature vector, the inverse of feature vector is equal to the transpose of the feature vector. This is true because the elements of matrix are in the unit eigen vectors of our dataset. The equation becomes as follows:

Row data adjust = Row feature vector^{T} × Final data

We have to add the mean of the original data to get back the original data.

Row data adjust = (Row feature vector^{-1} × Final data) + Original mean

For the completeness, re-create the dataset and create the graph. Figure 9.15(b) shows the recreated plot for a single eigen vector (First column of Table 9.2). Comparing with Figure 9.14(a), we can see the variations happened.

Applications

The major applications of PCA in machine learning is to visualise high dimensional data (For example, Analogy space), find hidden relationships (For example, topics in articles), compress information, avoid redundancy, reduce algorithm complexity, outlier detection an noise reduction in a dataset.

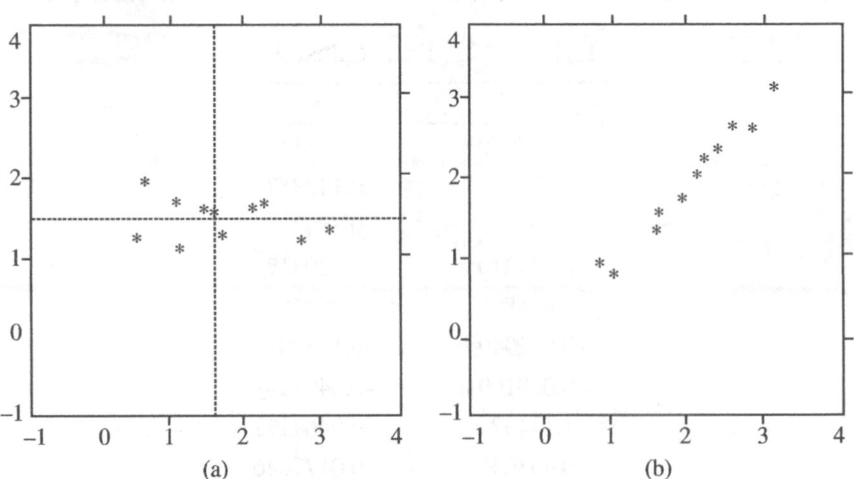

Figure 9.15 (a) Table of data by applying the PCA analysis using both eigen vectors, and a plot of the new data points (b) There construction from the data that was derived using only a single eigen vector.

9.4 Semi-supervised Learning

The human learning is from labelled and unlabelled data. So, we can say that the human learning is semi-supervised. Semi-supervised learning is a class of machine learning technique that uses small quantity of labelled data with large quantity of unlabelled data for the training. Semi-supervised learning falls between supervised and unsupervised learning techniques. Using the mixture of labelled and unlabelled data for training is cost effective and produces good results. There is a practical value for semi-supervised learning because the cost associated with labelling process of the training data is expensive and infeasible. But it works like a human learning. Figure 9.16 shows labelled and unlabelled data classification by an expert. The involvement of expert causes expenses in the labelled data.

Figure 9.16 Labelled and unlabelled data.

In semi-supervised learning, given labelled training data $L=\{x_i, y_i\}_{i=1}^{l}$, unlabelled data $U=\{x_j\}_{j=l+1}^{l+u}$ and learning a function f is used to predict labels for the future test data. Usually number of data in $U \gg L$. Learning a function that applied on the test data, hence semi-supervised learning is inductive. A semi-supervised learning model is given in Figure 9.17.

For a semi-supervised learning, we have labelled data $\left(\{x_i, y_i\}_{i=1}^{n}\right)$ and unlabelled data $\left(\{x_i\}_{i=1}^{n}\right)$. The goal is to achieve a better prediction model. Some of the semi-supervised learning algorithms are self-training, generative methods, mixture models, graph-based methods, co-training and semi-supervised SVM.

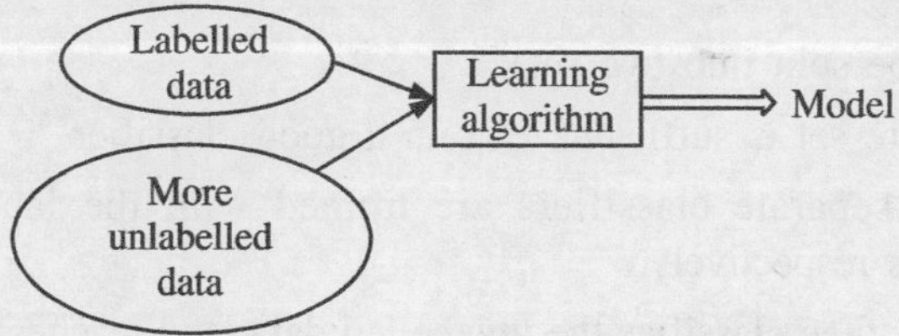

Figure 9.17 Semi-supervised learning.

9.4.1 Self-training

Train classifier on labelled data use classifier to classify unlabelled data. Add predicted unlabelled with high confidence to training set followed by retraining the classifier. The advantages of self-training includes—It is a simplest form of semi-supervised learning method that can be applied to other existing classifiers. It is frequently used in real-time tasks in natural language processing. Mistakes can re-inforce themselves which is a major disadvantage of self-training. That is, the model is unable to correct its own mistakes which is the main problem of self-training. Model's errors are amplified, if the unlabelled data used for training are confident but wrong. This effects the model confidence into a poor predictor by its performance. The procedure of self-training is given below.

Input: labelled data $\{x_i, y_i\}_{i=1}^{n}$, unlabelled data $\{x_i\}_{i=1}^{m}$

1. Initially, let $L = \{x_i, y_i\}_{i=1}^{l}$ and $U = \{x_j\}_{j=l+1}^{l+u}$
2. Repeat the following steps:
   (a) Train f from L using a supervised learning model.
   (b) Apply f to the unlabelled instance in U.
   (c) Remove a subset S from U; add $\{(x, f(x)) | x \in S\}$ to L.

9.4.2 Co-training

Another semi-supervised training method is co-training. There are certain assumptions made for this training. The features can be split into two sets—each subfeature set is enough to train a classifier and subfeature sets are conditionally independent in the class.

Two classifiers are trained by using the subfeature sets of the labelled data and then classifying unlabelled data using each classifier. Using the predicted labels of the unlabelled data, each classifier trains the other classifier. The co-training works well when conditional independence holds and is able to work when one classifier correctly labels a data that the other classifier misclassified. Features are randomly split into two parts, if no natural feature split is present.

As per the assumptions, feature split $x = [x_1, x_2]$ exists, and is sufficient for a classifier. According to Mitchell in 1999, co-training procedure is given as follows:

Assumptions

1. Features can be split into two sets.
2. Each subfeature set is sufficient to train a good classifier
3. Initially two separate classifiers are trained with the labelled data on the two subfeature sets respectively.
4. Each classifier then classifies the unlabelled data, and teaches the other classifier with the few unlabelled examples (and the predicted labels) with feel most confidence.
5. Each classifier is re-qualified with the additional training examples given by the other classifier, and the process repeats.

Co-training procedure is given as follows:

Co-training algorithm

Input: Labelled data $L = \{x_i, y_i\}_{i=1}^{l}$ and unlabelled data $U = \{x_j\}_{j=l+1}^{l+u}$

Each instance has two views, i.e. $x = [x_1, x_2]$ with a learning momentum k.

Let $L_1 = L_2 = \{(x_1, y_1), (x_2, y_2), \ldots, (x_l, y_l)\}$

Repeat the used up unlabelled data:

Train view-1 f_1 from L_1, view-2 f_2 from $L - 2$

Classify unlabelled data using f_1 and f_2 separately

Add f_1's top k-most confidence predictions $(x, f_1(x))$ to L_2

Add f_2's top k-most confidence predictions $(x, f_2(x))$ to L_1

Remove these from the unlabelled data

The views are different and sufficient for training a classifier. Figure 9.18 shows a simple example for different views X_1 and X_2.

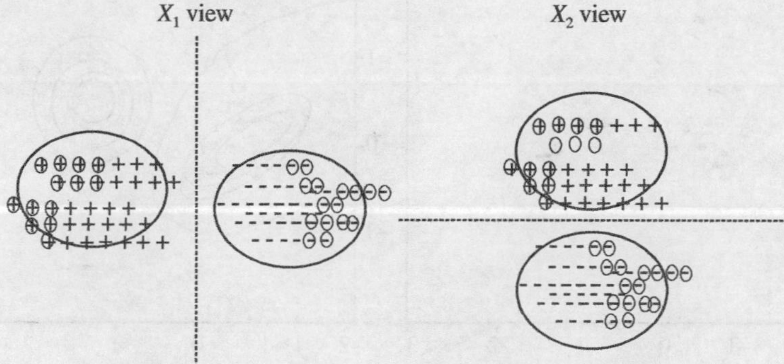

Figure 9.18 X_1 view and X_2 view.

9.4.3 Generative Methods

The Gaussian Mixture Model (GMM) is identified as a mixture model, $p(x, y) = p(y)p(x|y)$, where $p(x|y)$. For a large unlabelled data, we can identify mixture values and then we requires one labelled example per mixture value to identify the mixture distribution. Generative models are a semi-supervised method that identifies mixture components. Expectation Maximisation (EM) algorithm is generally used to solve but may fall into local maxima. The unlabelled data may upset learning if the local minima are far away from the global maxima. Based on the components per class, train generative mixture models. Expectation maximisation algorithm converts labelled examples into fixed length Fisher score vector that is respective of model parameters.

A generative model having a conditional probability of observable X and a given target y, is given as $P(X|Y = y)$. Assume that, $(X|Y)$ is Gaussian, then expectation maximisation algorithm can be used to find parameters of classes. Expectation maximisation algorithm finds the hidden variables in the clusters that are associated with labelled examples, which is not actually hidden. Suppose a variable set $x = (x_1, \ldots, x_M)$ and class variable y is given.

The probability density function $p(x, y)$ is a generative model since we can generate more samples artificially. For example, we have certain labelled data (X_l, Y_l) in Figure 9.19(a). Using supervised learning, only with labelled data, the classification is given in Figure 9.19(b) using a decision boundary. In semi-supervised, mixture model, add unlabelled data as shown in Figure 9.20(a) and the most likely model gives decision boundary as shown in Figure 9.20(b).

Types of generative models are Gaussian mixture model, hidden Markov model, probabilistic context-free grammar, Bayesian network, averaged one-dependence estimators, latent Dirichlet allocation, Boltzmann machine, variation auto-encoder, generative adversarial network and flow-based generative model. The following chapter discusses these models in detail. A semi-supervised learning technique may incorporate any of these techniques along with the labelled and unlabelled dataset.

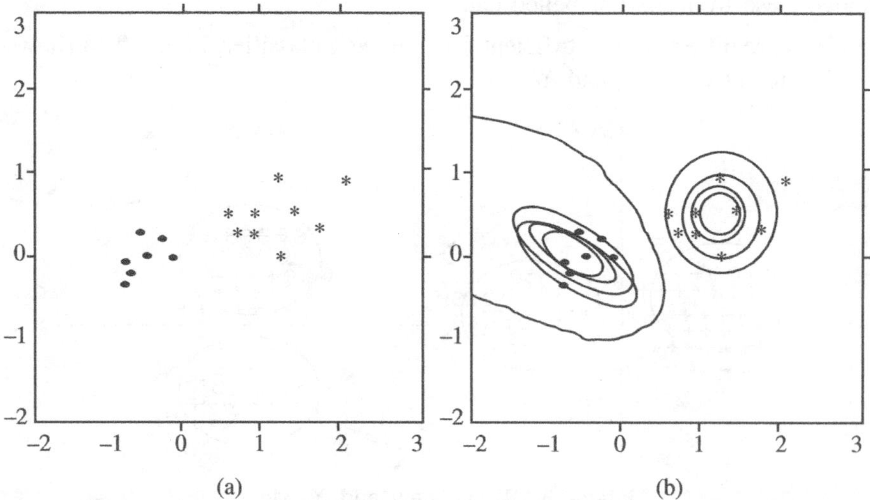

Figure 9.19 (a) Labelled data (b) Model learned from labelled data.

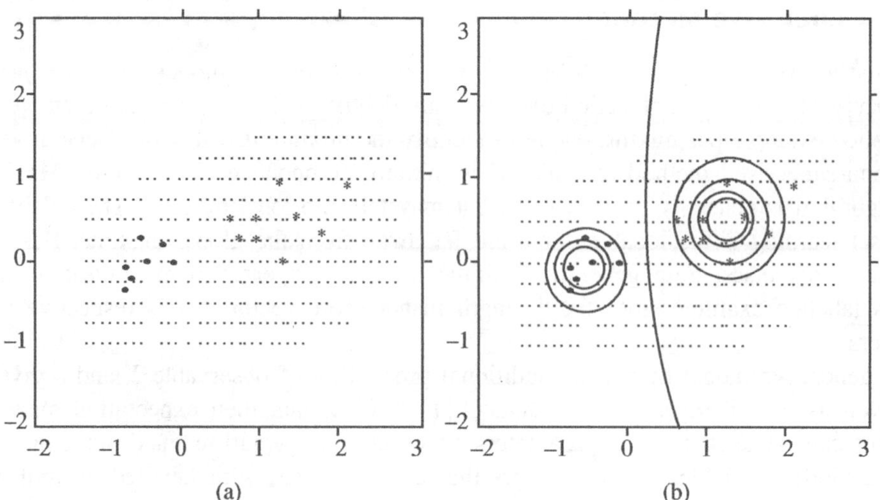

Figure 9.20 (a) Labelled data (star) and unlabelled data (small dots) (b) Model learned from labelled and unlabelled data.

9.4.4 Graph-based Methods

Basic idea is to construct a graph connecting similar data points. The intuition is that similar data points have similar labels and the information propagates from labelled data points.

A basic assumption of this method is that similar unlabelled data have similar labels. The similarity in the graphs modelled a local neighbourhood relations between data points. We have labelled dataset L and unlabelled dataset U with number of data l and u respectively. In the graph regularisation, the number of nodes are $X_l \cup X_u$ (Binary labels $y \in \{0, 1\}^n$). The edges are computed from features. For example, for an unweighed graph (0, 1) weights computed by k-nearest graph. For a fully connected graph, weight decays by distance, $w_{ij} = \exp(\|x_i - x_j\|^2/\sigma^2)$. It is assumed that nodes connected by heavy edges tend to have similar label.

For a large weight w_{ij}, data points i and j are more similar than the similarity between their labels. $f_i = f_j$,

$$\min_f \Sigma_{i \in l}(y_i - f_i)^2 + \lambda \Sigma_{ij \in l, u} w_{ij}(f_i - f_j)^2 \tag{9.1}$$

First component of the above equation gives loss on labelled data (mean square, 0 – 1) and the second component gives graph based smoothness prior on labelled and unlabelled data. The graphical model for handwritten characters is shown in Figure 9.21.

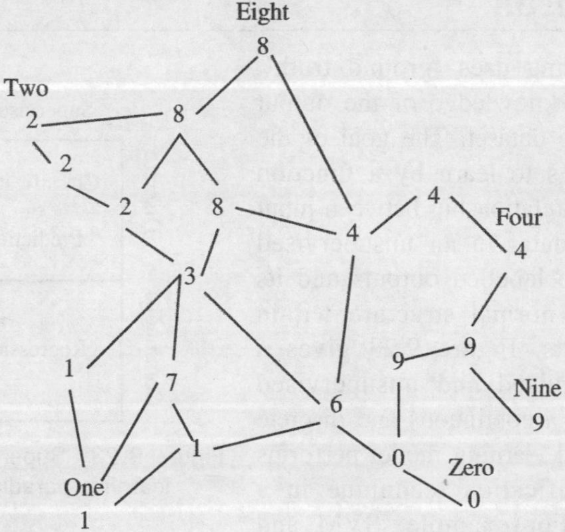

Figure 9.21 Graph-based model for handwritten characters.

9.4.5 Semi-supervised SVM

Transductive SVMs (Semi-Supervised Support Vector Machines or S3VM) optimise over both margin and labels of unlabelled points, instead of finding maximum margin between labelled points. Basic assumption of this SVM is unlabelled data from different classes are separated with large margin. Transduction is the prediction of the labels Y_u of the unlabelled data X_u. It is an inductive approach that uses labelled data and trains a supervised classifier. Transductive approach considers all points that label the unlabelled data points according to clusters that they belong to. The transductive algorithm needs to be repeated again by adding unseen data points. The SVM and S3VM is shown in Figure 9.22. In S3VM, unlabelled data guides the linear boundary away from thick regions, hence the classification is better than SVM.

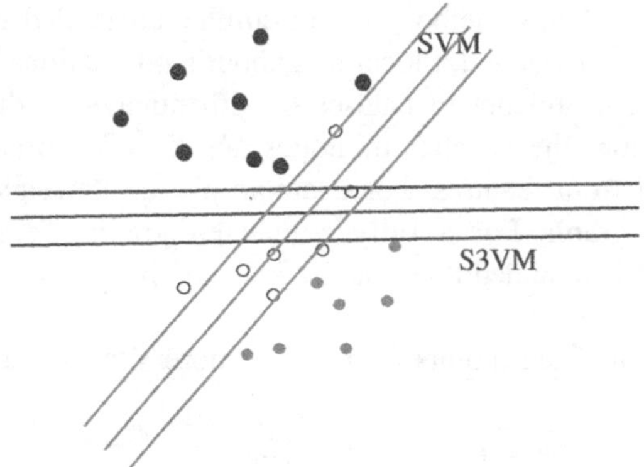

Figure 9.22 SVM and S3VM.

9.5 A Comparison

The supervised learning uses 'ground truth', which means, a prior knowledge of the output values of the available dataset. The goal of the supervised paradigm is to learn by a function that identifies the best relationship between input data and the output data. In an unsupervised paradigm there are no labelled outputs and its goal is inferring the normal structure within the set of data points. Figure 9.23 gives a comparison of supervised and unsupervised learning methods over a continuous and discrete dataset. The supervised learning model performs as a prediction/classification technique in a continuous dataset. For example, SVM and Multi-layer perceptron perform as a classifier on continuous dataset. In a discrete dataset, the supervised learning technique performs as a regression model. The unsupervised learning technique performs as a dimensionality reduction method for a discrete dataset. For a continuous dataset, the unsupervised learning acts as clustering of data points. A comparison of supervised and unsupervised method is given in Figure 9.24. The predictive model and the machine leaning algorithm are different based on the problem selected. If the data is not labelled then unsupervised learning techniques can be adapted to solve a problem. In a supervised technology the expected labels are predicted but in unsupervised learning the clusters are formulated or better representation suggestion is given.

Data \ Learning	Supervised	Unsupervised
Discrete	Classification or Prediction	Clustering
Continuous	Regression	Dimensionality reduction

Figure 9.23 Supervised and Unsupervised learning paradigm—a comparison.

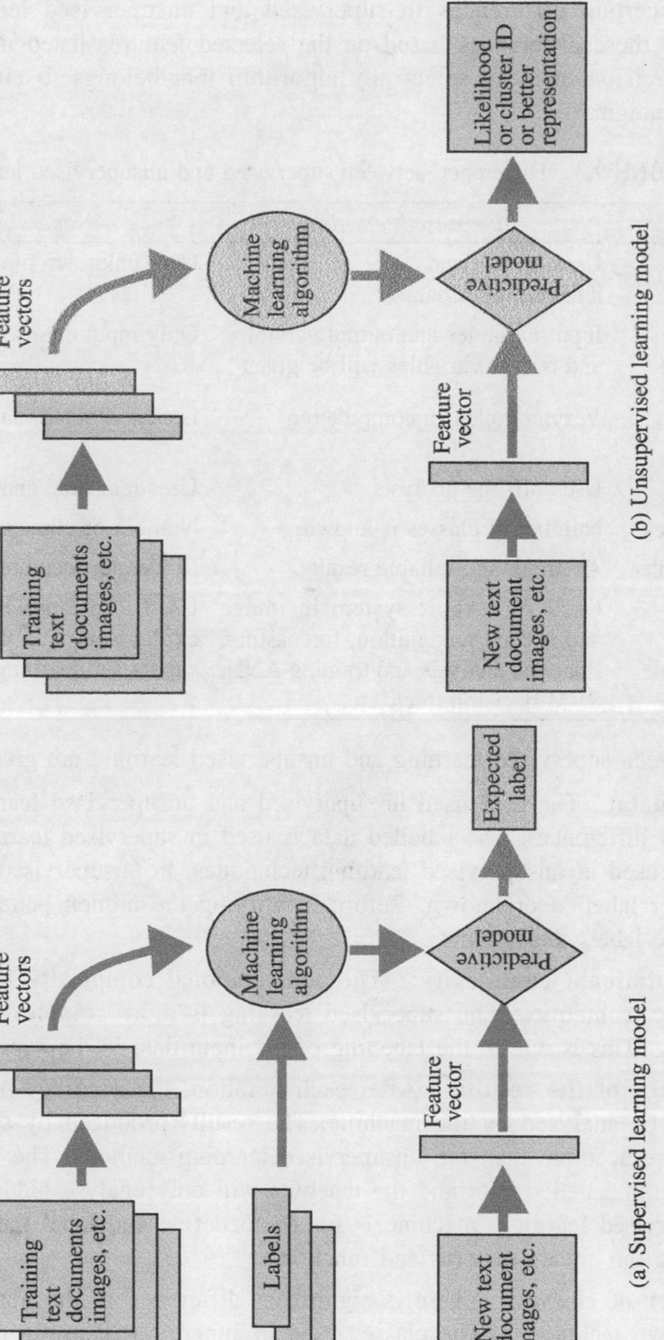

Figure 9.24 Supervised and unsupervised learning.

There are certain differences in supervised and unsupervised learning techniques. Table 9.3 shows these differences based on the selected features listed in the first column. Based on the decision one can select any algorithm that belongs to either supervised or unsupervised paradigm.

TABLE 9.3 Difference between supervised and unsupervised learning

	Supervised learning	*Unsupervised learning*
Input data	Uses known and labelled input data	Uses unknown input data
Method	Input variables and output variables, and output variables will be given	Only input data will be given
Computational complexity	Very complex in computation	Less computational complexity
Real-time	Uses off-line analysis	Uses real-time analysis of data
Number of classes	Number of classes is known	Number of classes is not known
Accuracy of results	Accurate and reliable results	Moderate, accurate and reliable results
Uses	Used for export system in image and speech recognition, forecasting, financial analysis and training ANN, SVM decision trees, etc.	Used to pre-process the text, during exploratory analysis or to pre-train supervised learning algorithms

Differences between supervised learning and unsupervised learning are given as follows:

1. **Input data:** The data used in supervised and unsupervised learning is one of the primary differences. The labelled data is used in supervised learning but unlabelled data is used in unsupervised learning techniques. In unsupervised learning, the data is either labelled or known. Before identifying the hidden patterns and functions, machine labels the raw data.
2. **Computational complexity:** The computational complexity is different in both learning techniques. The supervised learning is more complex than unsupervised learning. This is due to the labelling of the input data in the supervised learning.
3. **Accuracy of the results:** After each iteration, the accuracy of results produced should be analysed by the machine. The results produced by supervised learning are more accurate than the unsupervised learning methods. The reliable results are because of labelled data and the machine will only analyse hidden patterns. But in unsupervised learning, machine is forced to define and label the input data before entering the hidden patterns and functions.
4. **Number of classes:** There is significant difference in the number of classes in these two techniques. The classes used in supervised learning are known and the goal is to identify the unknown clusters. Since, unsupervised learning has no prior knowledge so that the number of classes is also unknown. The users who use the unsupervised learning technique are not aware of any information regarding the raw data and expected results.

5. **Real-time learning:** The unsupervised learning is well placed in real-time models but supervised learning is suitable for off-line model. The designers are involved in the preparation and labelling of the off-line input data and the analysis of hidden patterns and functions can be done online. The entire input data is analysed and labelled in the presence of the designers which help them how to classify the raw data. This is the significant merit of the unsupervised learning.

   In summary we can say,

   Supervised learning set = {(sample [acquisition], class [expert])}

   Unsupervised learning set = unlabelled samples

   Semi-supervised learning set = some labelled samples + many unlabelled samples

9.6 Case Studies

9.6.1 Language Classification

Dravidian languages have been attested since the 6th century BCE. There are nearly 85 Dravidian language families' genetically related languages spoken by about 217 million people. These people are spread over the world, but focussed to southern part of India and parts of eastern and central India as well as in north-eastern Sri Lanka, Pakistan, Nepal, Bangladesh, Afghanistan, Iran, and overseas in other countries such as Malaysia and Singapore. The most extensively spoken Dravidian languages are Kannada, Malayalam, Tamil, and Telugu. Many small Dravidian-speaking scheduled tribes, are available across our country, most of them do not have a proper script. Kaldava, Kodava, Tulu, etc. are some examples of such language. It is not that much hard for a native speaker to identify his language but is hard to recognise the other Dravidian languages, even when these languages originated from the same root.

To identify a language three focusses are important: script, speech and usage. Scripts are symbols that convey a message in written form for visual pattern. Speech signals are sounds which represent hearing pattern. The usage is fused in both representations. Motivation of this study includes the investigation of age of a script, which was used years before. So, usages of scripts are very important.

Stochastic model plays an important role in elucidating many areas of the natural and engineering sciences. They can be used to analyse the variability inherent in biological and medical processes, to deal with uncertainties affecting managerial decisions and the complexities of psychological and social interactions, and provide new perspectives, methodology, models, and intuition to aid in other mathematical and statistical studies.

Use of supervised learning techniques is good to solve this problem. A stochastic model is a good model to solve this problem. HMM is a stochastic model for solving prediction problem by its power as a statistical learning method.

9.6.2 Software Reliability Prediction

Software Reliability is an important factor to measure the quality of software as compared to all other attributes like availability, maintainability, functionality, etc. Software reliability is defined as the probability of failure-free operation of software under the stated conditions for a stated period of time. The reliability of software is measured for a given environment. The 'period of time' in this definition may represent any number of actual data items, such as number of executions; number of lines of code traversed or clock time.

Reliable software is a major requirement of both the developers and the users. Software reliability engineering is an essential component of Total Quality Management (TQM) in the context of software. Software reliability models give the behaviour of a system based on its failures. Software reliability models mainly focus on software reliability estimation and software reliability prediction. Software reliability estimation uses the data from the current development effort and estimate reliability at either present or future time. Software reliability prediction uses historical data and predicts reliability at some future time. Prediction such as time-to-next failures, mean time to failure, total number of faults remaining at time t are the examples of measurements derived from the reliability models and historical failure data.

A proliferation of software reliability model has emerged as people try to understand the characteristics of how and why software fails and try to quantify software reliability. Decades of research has resulted in many software reliability models, but how to quantify software reliability still remains unsolved.

As many models are already there and many more are emerging, none of the models are widely acceptable. There are many drawbacks for the current models. Mainly, none of the models can capture a satisfying amount of complexity of the software. Also, certain unrealistic constraints and assumptions have to be made for the quantifying process. Therefore, there is no model that can be used in all situations. Some models may work well in some software but may be completely off-track for other kinds of problems. The failure data collected for software reliability modelling done at various phases like software architecture, design and testing, etc. will also make much difference in the software reliability prediction. If the reliability and performance can be modelled from architecture itself, then it can surely improve the reliability measure.

Both statistical and soft computing techniques have been applied in developing software reliability models. The statistical method completely relies on some assumed external parameters. A model achieved through the soft computing methodologies is considered as non-parametric. Neural networks, Fuzzy logic, Decision tree, Genetic algorithm, etc. are some of the soft computing paradigms useful for the prediction. Since the soft computing is tolerant of imprecision, uncertainty, partial truth and approximation, it can be adopted as a better forecasting tool. This is the methodology that is proposed to adopt in the current research.

9.6.3 Decision Support System for Health Informatics

Developments in healthcare delivery are constant; in recent years, informatics in health has emerged as a new field to support emerging and ongoing developments. At the same time, further evolution of the field of health informatics is reflected in the introduction of concepts at the macro or health systems delivery level with major national initiatives related to electronic health records, data standards, and public health informatics. Health informatics is a discipline at the intersection of information science, computer science and healthcare. It deals with the resources, devices and methods required to optimise the acquisition, storage, retrieval, and use of information in health and biomedicine. To involve the acquisition and manipulation of healthcare related data, health informatics is an emerging area. Health informatics tools include not only computers but also clinical guidelines, formal medical terminologies, and information and communication systems.

A decision support system can be approached from two major disciplinary perspectives, those of information system science and artificial intelligence which offer different conceptualisations of decision support system. The use of decision support systems in healthcare results in changes in healthcare practices, processes and outcomes. The aim of this development is to improve healthcare delivery. Users in healthcare are asking about useful systems. That is, systems that provide users with information and knowledge that support their work and actions in their working environment. These changes will continue to shape health services in the twenty-first century. By making full and creative use of the technology to tame data and to transform information, Health informatics will foster the development and use of new knowledge in healthcare.

Soft computing is tolerant of imprecision, uncertainty, partial truth and approximation. In effect, the role model for soft computing is the human mind. Soft computing has developed as a discipline and method for the diagnosis in health informatics. The principle of soft computing is to exploit the tolerance for imprecision, uncertainty, partial truth, and approximation to achieve tractability, robustness and low solution cost. The clinical data may consist of missing, incorrect and sometimes incomplete values set, so using soft computing is the better alternative to handle such data. Soft Computing techniques, especially those that are based on exploiting human knowledge and experience, are extremely useful to model complex decision making procedures.

Potential problems in this field include designing machine learning models for predicting organ transplantation survival. Investigating intelligent methods and research techniques with application in medical imaging and medical diagnosis. Investigating and implementing intelligent system for the pattern recognition and classification problems from the medical diagnosis area such as kidney and liver transplantation. Identifying areas where tools are needed to transform uncertain and incomplete data and insights into useful knowledge. To be more prescriptive about effective decision making by using intelligent systems. Assisting and guiding practitioners in improving their core knowledge of effective decision support, etc.

9.7 Exercises

1. At what point in the PCA process can we decide to compress the data? What effect does this have?
2. For an example of PCA and a graphical representation of the principal eigen vectors, research the topic 'eigen faces', which use PCA to do facial recognition. How is it used?
3. What is a feature vector? How will it prepare?
4. What is stratification? What is cross-validation? What is 'leave-one-out' cross-validation? Why is cross-validation useful and what are its drawbacks? Explain, using an example, how it attempts to address the problem of overfitting.

5. How will you select supervised technique while designing a machine learning problem?
6. Under what conditions one selects semi-supervised learning methodology?
7. How will you select *k* in a *k*-fold cross fold validation?
8. Is cross fold validation an enough method in a supervised learning method?
9. Which is more important for you, model accuracy or model performance?
10. How do you handle an imbalanced dataset?
11. How do you ensure a non-overfitting model while designing?
12. How do you evaluate a logic regression model?
13. How do you handle missing or corrupted data in a dataset?
14. How will you fix a source dataset which is used for training and testing?
15. How will you prepare independent dataset?
16. What is the trade-off between bias and variance?
17. What is selection bias? What other types of biases could you encounter during sampling?
18. When should you use classification over regression?
19. What is the difference between test dataset and validation dataset?
20. What is bias-variance decomposition of classification error in ensemble method?
21. Explain the difference between principle component analysis, Kernel principle compact analysis and independent component analysis?
22. What are the areas in robotics and information processing where sequential prediction problem arises?
23. List out real-world supervised learning terminology.
24. A dataset is given that contains 1000 columns and a classification problem. You are requested to reduce the dimension of this data so that the computational time of the model is reduced, because the machine has memory constraints. What would you do?
25. Is rotation necessary in PCA? If yes, why? What will happen if you do not rotate the components?
26. A dataset is given that has missing values which spread along one standard deviation from the median. Check the percentage of data remaining unaffected. What is its percentage? Why?
27. How would you go about choosing an algorithm to solve a business problem?
28. Are you ready to work on time series dataset? You are requested to build a high accuracy model in either supervised or unsupervised terminology? It should work fairly on all kinds of data. How will you select a model? If you select a time series regression model and have a higher accuracy than decision tree model. If it happens, why?
29. How will you select a model which has a low bias and high variance? Which algorithm should you use to tackle it?

30. For a new project, food delivery company by online cooking to save money. The company people are not able to deliver food on time and hence the customers are unhappy. To keep them happy, they end up delivering food for free. Which machine learning algorithm can be used to solve this problem?
31. A dataset contains many variables, and some of them are highly correlated. If you use PCA, would it remove correlated variables first? Why?
32. After analysing your regression model, suffering from multicollinearity, how do you check? Without losing any information, can you still build a better model?
33. Explain your idea to select important variables from your working dataset.
34. Check out the difference between covariance and correlation.
35. A dataset consists of variables having more than 30% missing values. For example, out of 50 variables, 10 variables have mission values higher than 30%. How will you deal with this?
36. In what conditions one chooses any of the semi-supervised learning technologies?
37. When and why does transductive learning have worse performance than inductive learning?
38. What are the major challenges in semi-supervised learning?
39. Is rotation necessary in PCA? If yes, Why? What will happen if you do not rotate the components?
40. You are given a dataset, which contains many variables, some of which are highly correlated and you know about it. Your teacher has asked you to run PCA. Would you remove correlated variables first? Why?
41. You came to know that your model is suffering from low bias and high variance. Which algorithm should you use to tackle it? Why?
42. Differentiate between semi-supervised and transductive learning.
43. Suppose you are running a majority classifier on the following training set. The training set is shown in Figure 9.25. It consists of 10 data points. Each data point has a class label of either 0 or 1. A majority classifier is defined to output the class label that is in the majority in training set, regardless of the input. If there is a tie in the training set, then always output class label 1.

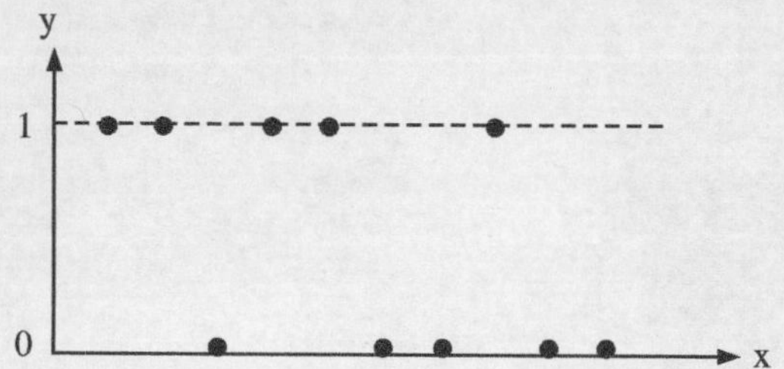

Figure 9.25 Training set for cross validation.

(a) What is the training error? (report the error as a ratio)

(b) What is the leave-one-out cross validation error? (report the error as a ratio)

(c) What is the two-fold cross validation error? Assume the left 5 points belong to one partition while the right 5 points belong to the other partition (report the error as a ratio).

44. For each of the following technical issues, discuss how it might arise in this setting and how you would address it. It is fine if a different machine learning algorithm is used in each of your three answers, but for each answer focus on atmost one specific algorithm.

(a) Medical treatments and laboratory tests change over time, so medical records from 100 years ago are not the same as from 10 years ago.

(b) The amount of information is not the same for each patient. For example, some people have undergone medical test *X* many times, while others never have had this test.

(c) There might be formalised domain knowledge about disease D. You may assume 'formalised' means written in some artificial intelligence knowledge representation and you are free to choose the particular formalism in which it is written.

45. What is overfitting, and how can you avoid it?

46. What is 'training set' and 'test set' in a machine learning model? How much data will you allocate for your training, validation, and test sets?

47. How do you handle missing or corrupted data in a dataset?

48. How can you choose a classifier based on a training set data size?

49. What are the three stages of building a model in machine learning?

CHAPTER

10

Reinforced Learning

Have you ever tried to play 'throw and throw' with your pet dog? If so, you are a champion in implementing reinforced learning. You throw a disk, if the pet gets disk back to you, you give him a biscuit else you wait till he takes it (Figure 10.1). Unknowingly you have taught your pet a reward punishment system.

A typical learning always starts by interaction with the environment. Small kids are always fascinated by candle lights and try to catch it even if the parent warns them. They learn that it will burn their hand only when he touches it. By doing so they gain wealthy knowledge on the cause and effect of such process. Starting from the childhood, throughout our lives, interactions with the environment have gained us a lot of experience and that remains as a major source of knowledge. This is what we call learning from environment. It is the foundational idea that is employed in Reinforcement Learning (RL). Developments of reinforcement learning algorithms are associated with dynamic programming methodology, a general approach to optimal control. Reinforcement evolved from many sub-branches like machine learning, optimal control, reward system, operation research, etc. Figure 10.2 shows evolution faces of reinforcement learning. It evolved from the participants of Engineering, Mathematics, Computer science, Economics, Physiology, and Neuroscience.

Figure 10.1 Dog training.

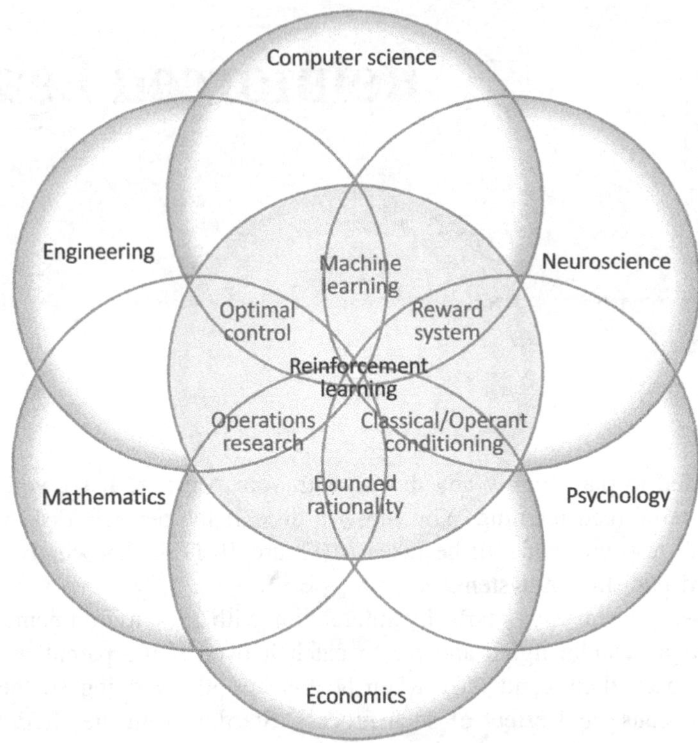

Figure 10.2 Many faces of reinforcement learning.

To design a model in supervised learning, labelled data is required and in unsupervised learning, unlabelled data is required. That is, presence of an expert is involved in supervised learning and the expert is not required in unsupervised learning. These are two typical learning terminologies.

How do we practice the bicycle riding? Is it supervised or unsupervised learning? In which we examine how a person (agent) can learn from success and failure, reward and punishment. The reinforce learning helps a person to ride the bicycle without falling over. Riding the bicycle, and performing a series of actions result in a bicycle being tilted 45 degrees to the right. Turning of the handle bar left or right suddenly crashes the system. The reinforcement learning, system has learned that the state of bicycle being titled 45 degrees to the right is bad. To repeat trial using 40 degrees to the right and by performing enough of these trial-and-error interactions with the environment. The reinforcement system will ultimately learn how to prevent the bicycle from ever falling over.

A major difference between supervised learning and reinforced learning is shown in Figure 10.3. The training information contains evaluations ('rewards' / 'penalties') in the reinforcement learning but desired (target) outputs are dependent on the learning process. The outputs from the reinforced learning are not simply output values but 'actions'.

So, the reinforcement learning have some key features as follows:

- Learning from interaction
- Goal-oriented learning

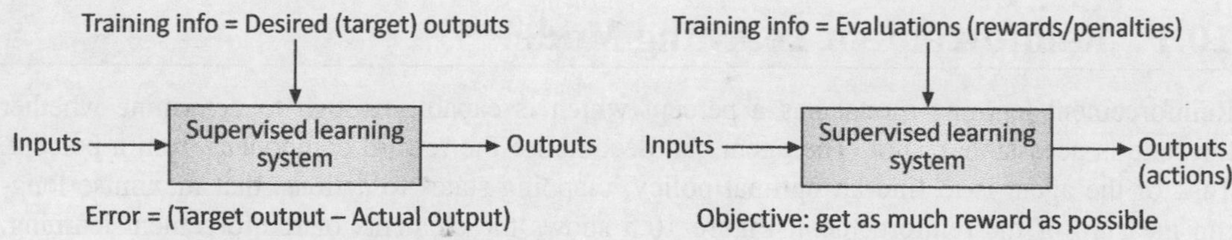

Figure 10.3 Difference between supervised and reinforced learning.

- Learning about, from, and while interacting with an external environment
- Learning what to do, how to map situations to actions, so as to maximise a numerical reward signal.

A question arises, how can an agent learn behaviours when it does not have a teacher to tell it how to perform.

The agent has a task to perform and it takes some actions in the world. At some later point, it gets feedback telling it how well it did on performing the task. The agent performs the same task over and over again. The agent gets positive reinforcement for tasks done well and the agent gets negative reinforcement for tasks done poorly. In reinforcement learning, goal is to get the agent to act in the world so as to maximise its rewards. The agent has to figure out what it did that made it get the reward/punishment. It is called credit assignment problem. Reinforcement learning learns how to behave successfully to achieve a goal while interacting with an external environment. Reinforcement learning approaches can be used to train computers to do many tasks, like backgammon, chess playing, job shop scheduling and controlling robot limbs.

Reinforcement learning is a type of machine learning where learning occurs by interacting with the environment. Figure 10.4 shows schematic of reinforcement learning model. Reinforcement learning stands unique among other machine learning approaches, it can be mentioned as a goal-directed learning process. Reinforcement learning will have an agent which learns from cause and effect. These agents select the actions from their past experiences and through choices, which can be treated as a trial and error method. The agent will receive a reinforcement signal (reward) which shows the success of an action. The agent selects an action which gives maximum rewards. In short, we can conclude that the basic goal of reinforced learning algorithm is selection of actions that maximise reward (the return) of the agent.

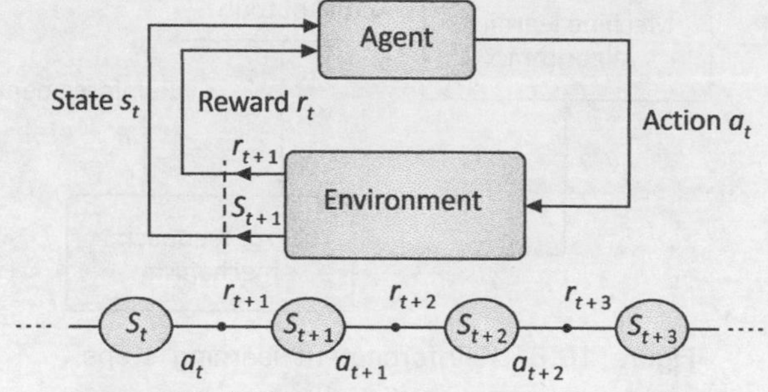

Figure 10.4 Reinforcement learning model.

10.1 Reinforcement Learning Model

Reinforcement learning model has a percept which is capable enough to determine whether the state is accessible or not. The agent can decompose the reward component from a percept. Task of the agent is to find an optimal policy, mapping states to actions, that maximise long-run measure of the reinforcement. Figure 10.5 shows the elements of reinforcement learning. That is, the following are elements in reinforcement learning along with their jobs:

- Policy: what to do
- Reward: what is good
- Value: what is good because it predicts reward
- Model: what follows what

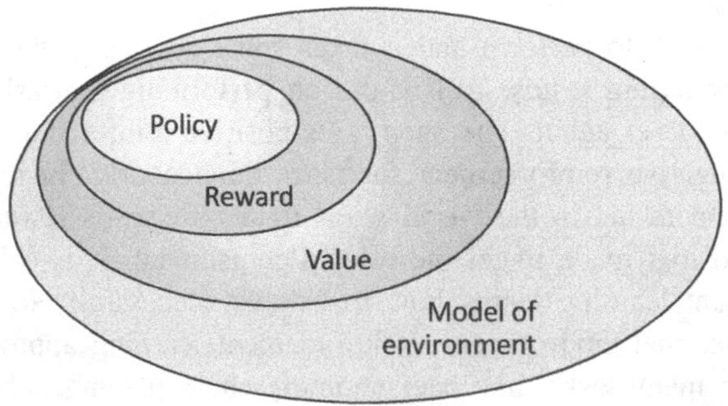

Figure 10.5 The agent-environment interface.

Agent and environment interact with a discrete time step $t = 0, 1, 2, ..$ The agent observes state at t, $s_t \in S$ which produces action at step t, $a_t \in A(s_t)$, get resulting reward, $r_{t+1} \in R$ and resulting next state, s_{t+1}.

Reinforced learning is same as supervised learning, but with a restriction that the actual output may be true only for some inputs. For other input, the output might be correct to some extent, for example, '60% correct'. Thus, with the reinforced learning, a critic information, but not the exact information is available. Figure 10.6 depicts how reinforced learning works.

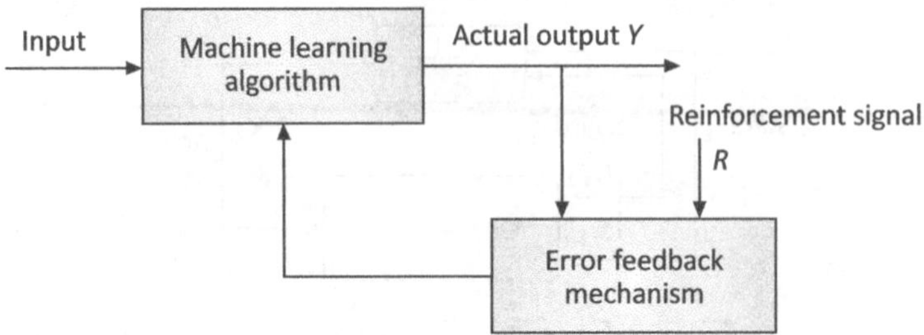

Figure 10.6 Reinforcement learning steps.

Learning model basically has five components as follows:

- Set of environment states S
- Action sets A
- Rules of transition between the states
- Rules determining the reward
- Rules that show the agents observation

As we have seen earlier, the reinforcement learning agent interacts with the environment in discrete steps. For every discrete time t, the agent receives an observation O_t, for which it gets the reward a_t. It then chooses an action a_t from the set of actions available which is subsequently sent to the environment. The environment moves to a new state s_{t+1} and the reward a_{t+1} associated with the transition (s_t, α_t, s_{t+1}) is determined. The main aim of a learning agent is to get the maximum reward which is possibly available. The agent can randomise its custom selection and can in fact opt for any action as a function of the history.

In order to get the notion of regret, the agent's performance is compared to one which works optimally. Difference in their performance is termed as the notion of regret. For optimal behaviour the agent must reason all the long-term consequences of its action. It is just saying like, for becoming a successful bike stunt hero, you need to learn the basics of stunt riding. Even though in such a situation you are negatively rewarded first by considering you as a stunt hero, but in future you will gain positive rewards. Thus, we may say that reinforcement learning is suited to problems which include a long-term reward trade-off. Various applications where it is commonly used are in the area of robot control, telecommunication, backgammon and checkers.

Reinforcement learning is mainly aimed at solving the so called Markov Decision Problem (MDP). We may also term this as 'the problems that are being solved by reinforcement learning will be a Markov decision problem or a variant of Markov decision problem'. Next section discusses the Markov decision problem.

Dynamic programming

Richard Bellman developed dynamic programming in 1950's as a general method for optimising multi-stage decision process. Here the word programming stands for planning. It was a technique mainly used for solving problems with overlapping subproblems. These subproblems arise from a recurrence relating a solution to given problem with solutions to its smaller subproblems of the same type. Dynamic programming suggests such a solving method for each of the smaller subproblems only once and recording the results in a table from which we can obtain solution to the original problem.

10.1.1 Limitation of Reinforcement Learning

There are various challenges in the recent reinforcement learning research. First, when the problems become complex it is difficult to store the values of each state in memory. To solve this we can introduce techniques for value estimation like neural network and decision trees. But these estimates with imperfect values will have a straight consequence in the results.

Hence, the usage is very much reduced. Moreover, the problems are usually seen to be modular; similar behaviours are found to reoccur often, thus in order to prevent learning everything all over, modularity can be proposed. Hierarchical approaches are commonly used. However, doing this automatically is found to be difficult. Due to limited knowledge of the states and situation, we will not be able to fully determine about the current state. This will also affect the performance of the algorithm.

10.1.2 Applications of Reinforcement Learning

One reason why reinforcement learning is popular is that, it serves as a theoretical tool for studying the principles of agents learning to act. Reinforcement learning is used vastly by various researchers for constructing autonomous systems that train themselves to improve performance by interacting with the environment. We can summarise the applications of reinforcement learning as follows:

- Robotics: navigation, robo-soccer, walking, juggling
- Control: factory processes, admission control in telecommunication, resource control in multimedia networks
- Games: backgammon, chess, othello, checkers
- Operations research: warehousing, transportation, scheduling
- Others: adaptive treatment design, biological modelling
- Combinatorial optimisation: VLSI placement and routing, elevator dispatching

10.2 Markov Decision Problem

We have already mentioned the various components that the reinforcement learning does possess. Similarly, Markov Decision Problem (MDP) also has a set of framework elements as follows:

- State of the system
- Actions
- Transition probabilities
- Transition rewards
- A policy
- A performance metric

State

A 'states' describes the present condition of a system with the help of attribute sets or parameters. It can either be a time dependent or independent system. If the system is a time dependent it is called as a dynamic system. That means the state changes as the time moves.

Let's look an example from real life, consider a queue in front of a supermarket. The state of such a system can be defined by the number of people in the queue. Thus, it is sure that the state changes with time, when a customer is being served and moved out or when a new person enters into the queue, state changes. Transition of states is a regular matter while considering a Markov decision problem. Consider there is only one serving window and only one queue. Here, let x define the state by the no of customers in a queue, whenever a new customer arrives the probability will be x. But consider the case when a customer leaves, then the probability changes from x to $x + 1$ with the remaining probability. When does this happen? It occurs when a customer departs from the system after the service completion.

Actions

It may be defined as the changes that need to be done when the system attains a particular condition. For the above mentioned example, we may say that whenever the number of customers exceed a particular number, remaining people need to be diverted to a new counter, that is opened. So here we may have two actions, namely 'Opening' a new counter and 'Not opening' a new counter.

Transition probability

Consider the situation where one state changes to another under the influence of an action. Let a be the action at a state i. By this action a it changes to another state j. We can denote this as $p\ (i, a, j)$ which denotes the probability of going from state i to j under the action a. This is termed as transition probability. Three states and two actions in a Markov decision problem will result in nine transition probability per action.

Immediate rewards

The system always gets an immediate reward during transition from one state to another. It is denoted by $r\ (i, a, j)$. Policy: The system changes its state frequently, the action that needs to be selected in every state is defined by the term policy. There are cases where no action needs to be done for some states. But the states which make some actions to be chosen are called as decision-making states.

Performance metric

Every policy will be associated with a performance metric which judges the performance of the policy. Our aim is to have a policy which gives the best performance matrix. One such performance matrix is average reward of a policy. We will discuss about it in the latter parts.

Time of transition

It is the time taken by the Markov decision problem for transition from one state to another. Normally it is assumed to be one. It does not have any meaning like 1 hour, 1 minute or 1 second. It is just a fixed quantity by analyst.

Consider name of the policy that is following is $\prod$. Then $\prod(i)$ will denote the action done by this policy for state i. We say a transition has occurred whenever a jump in the

Markov chain takes place. It is even possible to have transition from a state to itself. Let X_s be the state of the system before the sth transition. Then, the following quantity, in which $X_1 = i$, is called the average reward of the policy $\prod$ starting at state i.

$$p_i = \lim_{k \to \infty} \frac{E[\Sigma_{s=1}^{k} r(x_s, \pi(x_s) x_{s+1}) \mid x_1 = i]}{k} \tag{10.1}$$

Average reward shows the sum of total immediate rewards earned, divided by the number of transitions, which has been calculated over a very long time period (that is k assumes a large value). The major objective of average-reward Markov decision problem is to find out that policy which maximises average metric (performance reward) of the policy.

Curses

We have generally two concepts like, 'curse of modelling' and 'curse of dimensionality'. In cases where we have large number of governing variables, it is quite difficult to derive the exact values. It is called curse of modelling. For large scale systems, we are unable to store all the states or values, this is termed by curse of dimensionality.

We can use dynamic programming for solving Markov decision problem, but it requires all the transition probability and rewards for solving it. Thus, it may eventually suffer from any of these sores and here the reinforcement learning plays its role. It is able to generate optimal solutions to large and complex Markov decision problems. In other words, reinforcement learning is able to make inroads into problems which suffer from one or more of these two curses and cannot be solved by dynamic programming.

10.3 Q-learning

As we have mentioned earlier, reinforcement learning-methods is a goal directed scheme. It is used for learning to reach a goal in step by step or to learn avoiding a disturbance. There are several ways to implement the learning process. One among this is Q-learning.

Q-learning was introduced by Watkins during 1989 as a machine learning method. Q-learning is a form of reinforcement learning where the agent learns to assign values (Q-values) to state-action pairs rather than finding a mapping from states to state values. Instead of having an associated value function, Q-learning makes use of the Q-function. A Q-value is associated with each action for each state. Definition of a Q-value is the sum of possibly discounted reinforcements that are received when performing the associated action and then subsequently following the given policy. In background of Q-learning, the value of a state is defined to be the greatest Q-value in the given state.

Information about the Q-values of each state-action pair can be used by the agent to select the action for each state. The problem with this is, the agent will not have any sort of information about the Q-values initially. So, the primary goal of the agent is to settle on an optimal Q-value function, which assigns appropriate values for all state/action pairs. Yet another issue is that the Q-value depends on current and future reinforcements. The agent will only have information regarding the immediate reinforcement. For this, agents make use of two principles, which are considered as the essence of reinforcement learning.

- If an action in a particular situation produces an adverse effect, then learn not to do the same action in a similar situation. Similarly, an action which produces a favourable result, then learn to do and repeat the same action in a similar forth-coming environments.
- If all of the action is producing adverse effect then learn to avoid that state. That is, do not take actions in other states that would lead you to be in that bad state. If any action in a given state causes something good to happen, then learn to like that state.

The reinforcement learning is made to happen by this second principle. Even if there is no immediate reinforcement associated with particular actions, reinforcement learning permits the agent to learn positive and negative values for particular actions for each state.

10.3.1 Q-learning Algorithm

The problem model consists of an agent, states S and a set of actions per state A. By performing an action $a \in A$, the agent moves from state to state. Rewards are provided to the agent in each stage. The whole purpose is to maximise the total reward by learning which action needs to be selected for each state.

The algorithm works on the function which evaluates the Quality of a state/action combination given as follows:

$$Q : S \times A \rightarrow R \tag{10.2}$$

Q will be a constant value fixed by the designer (before learning). When state changes, rewards are provided to the agents and new values are calculated for each state (s from S) and action (a from A). The algorithm works by simple value iteration. New values are added to the old one to make some corrections depending on current information. This is illustrated in Figure 10.7.

R_{t+1} is the reward observed after performing a_t in s_t, $\alpha_t(s, a)(0 < \alpha \leq 1)$ the learning rate (may be the same for all pairs). γ_t is the discount factor, such that $0 \leq \gamma < 1$.

The above formula is equivalent to:

$$Q(s_t, a_t) \leftarrow Q(s_t, a_t)(1 - \alpha_t(s_t, a_t)) + \alpha_t(s_t, a_t)[R_{t+1} + \gamma^{\max}_{\alpha_{t+1}} Q(s_{t+1}, a_{t+1})] \tag{10.3}$$

The algorithm terminates when state s_{t+1} is a final state. The procedure is given in Algorithm 17.

$$Q(s_t, a_t) \leftarrow \underbrace{Q(s_t, a_t)}_{\text{Old value}} + \underbrace{\alpha_t(s_t, a_t)}_{\text{Learning rate}} \times \left[\underbrace{R_t + 1 +}_{\text{Reward}} \underbrace{\gamma}_{\text{Discount factor}} \underbrace{\max_{a_t+1} Q(s_t + 1, a_t + 1)}_{\text{Max future value}} - \underbrace{Q(s_t, a_t)}_{\text{Old value}} \right]$$

Figure 10.7 Q-learning equation.

Algorithm 17 Q learning algorithm

for each action pair (s, a), initialise the table entry $Q(s, a)$ to zero
Observe the current state s
do forever
    Select an action a and execute it
    Receive immediate reward r
    Observe the new state s'
Update the table entry for $Q(s, a)$ as follows
    $Q(s, a) = r = \gamma^{\max_{a'}} Q(s', a')$
    $s = s'$

10.3.2 Q-learning Example

Consider a modified maze problem. Here we need to travel from box s_1 to s_6 in the best way available. The initial condition and the various travel paths are shown in Figure 10.8. The reward of 100 is given if it gets to state s_6, and 0 otherwise. Relative value of delay vs immediate reward is 0.5. Table 10.1(a) shows its rewards.

Starting from s_1, the various available actions are a_{12}, a_{14} opting a_{12}.
From the general formula, $Q(s, a) = r = \gamma^{\max_{a'}} Q(s', a')$, Update $Q(s_1, a_{12})$.

$Q(s_1, a_{12}) = r + 0.5 \times \max(Q(s_2, a_{21}), \max(Q(s_2, a_{25}), \max(Q(s_2, a_{23}) = 0$

So, we are now in s_2 and trying for the next move. There are three available actions such as a_{21}, a_{23}, and a_{25}.
We opt a_{23}. So, let us now update $Q(s_2, a_{23})$ according to our formula,

$$Q(s, a_{23}) = r + 0.5 \times \max(Q(s_3, a_{32}), \max(Q(s_3, a_{36}) = 0 \qquad (10.4)$$

Now, position is s_3, and two available actions are there, namely a_{32} and a_{36}. We opt for a_{36}, and this is the final state and we can update $Q(s_3, a_{33}) = r = 100$. Then our initial values are in Table 10.1(b),

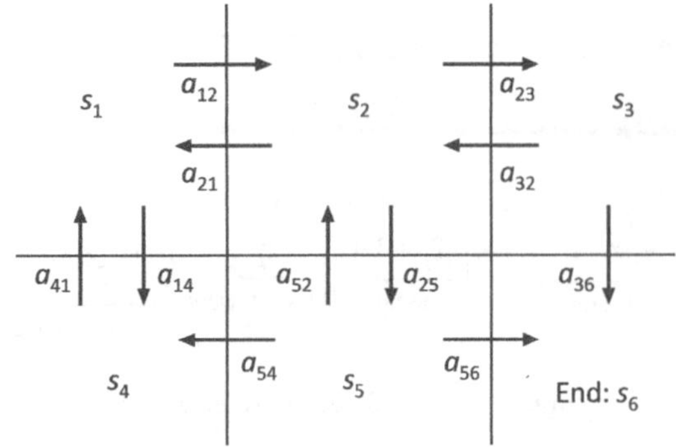

Figure 10.8 Q-learning example.

TABLE 10.1 (a) States-reward (b) Initial table, (c) Final reward table

States	*Reward*	*States*	*Reward*	*States*	*Reward*
s_1, a_{12}	0	s_1, a_{12}	0	s_1, a_{12}	25
s_1, a_{14}	0	s_1, a_{14}	0	s_1, a_{14}	25
s_2, a_{21}	0	s_2, a_{21}	0	s_2, a_{21}	12.5
s_2, a_{23}	0	s_2, a_{23}	0	s_2, a_{23}	50
s_2, a_{25}	0	s_2, a_{25}	0	s_2, a_{25}	25
s_3, a_{32}	0	s_3, a_{32}	0	s_3, a_{32}	25
s_3, a_{36}	0	s_3, a_{36}	100	s_3, a_{36}	100
s_4, a_{41}	0	s_4, a_{41}	0	s_4, a_{41}	12.5
s_4, a_{45}	0	s_4, a_{45}	0	s_4, a_{45}	50
s_5, a_{54}	0	s_5, a_{54}	0	s_5, a_{54}	25
s_5, a_{52}	0	s_5, a_{52}	0	s_5, a_{52}	25
s_5, a_{56}	0	s_5, a_{56}	0	s_5, a_{56}	100

New game starts from state 2, with three available actions and we select a_{23}. The process continues as mentioned above but the values change while updating the table, now

$$\begin{aligned} Q(s_2, a_{23}) &= r + 0.5 \times \max(Q(s_3, a_{32}), \max(Q(s_3, a_{36}) \\ &= 0 + 0.5 \times 100 \\ &= 50 \end{aligned}$$

Thus, after much iteration, we get the rewards in Table 10.1(c).

10.4 Temporal Difference Learning

Temporal Difference (TD) learning is a supervised learning process which predicts a state or quantity of a signal from the past values. The name 'temporal difference' derives from its use of changes or differences, in classifications over successive time steps to drive the learning process. We can update the prediction at any given time slot to bring it closer to the prediction of same quantity for the next time slot. TD algorithms are widely employed in the area of reinforcement learning, for predicting the total amount of reward expected over the near future.

Let us begin explaining this section by explaining the term 'value function'. Value function is a state action pair function which estimates the effectiveness of an action in a particular state. The general notation that we are going to follow is mentioned as follows:

$V\pi(s)$ — denotes the value at a state s under the policy π.

$Q\pi(s, a)$ — denotes the actions value a at a particular state s under the policy π. The reason for estimating the value functions is to accurately choose an action that provides the best total reward possible, after being in that given state.

For estimating the function values we make use of the temporal difference learning scheme. If we are calculating the value function without estimation, the agent needs to wait until the last reward is obtained. After finishing the task it needs to backtrack the path that was taken to reach the end state, and will have to update the function values accordingly. This is a tedious task. This may be represented by the equation: $V(s_t) \leftarrow V(s_t) + \alpha[R_t - V(s_t)]$, where s_t is the state visited at time t, R_t is the reward after time t and α is a constant.

The temporal learning on the other hand estimates the final reward and state action values are updated for every step in the way. It can be expressed as: $V(st) \leftarrow V(st) + \alpha[r_{t+1} + \gamma V(s_{t+1}) - V(s_t)]$, where r_{t+1} is the reward observed at time $t + 1$.

TD learning is also called as bootstrapping method as the values are updated partially on each step depending on the estimate and not on the final reward.

10.4.1 On-policy and Off-policy Learning

On-policy temporal difference technique discovers the value of the policy which is used for making decisions. The value functions are obtained and the values are updated by executing actions which are determined by some policy. All these policies are usually soft and non-deterministic. It opens the path for more exploration to the current policies. The action which gives high rewards are chosen as the policy, thus making the policies not very strict. Three commonly used policies are, ε-soft, ε-greedy and softmax.

Greedy: Most of the time, action with highest estimated reward is chosen, called the greediest action. The action with a small probability is selected at random. The action is selected uniformly, independent of the action-value estimates. This method ensures that if enough trials are done, each action will be tried an infinite number of times, thus ensuring that optimal actions are discovered.

- ε-soft: It is very similar to -greedy. The best action is selected with probability $1 - \varepsilon$ and the rest of the time a random action is chosen uniformly.
- Softmax: One drawback of ε-greedy and ε-soft is that they select random actions uniformly. The worst possible action is just as likely to be selected as the second best. Softmax remedies this by assigning a rank or weight to each of the actions, according to their action-value estimate. A random action is selected with regards to the weight associated with each action, meaning that the worst actions are unlikely to be chosen. This is a good approach to take where the worst actions are very unfavourable.

Off-policy method learns different policies for behaviour and estimation. Here also, as we mentioned in on policy method, the behaviour policy is soft, so there is sufficient exploration going on. Off-policy algorithms can also update the estimated value functions using hypothetical actions, those which have not actually been tried. This is in contrast to on-policy methods which update value functions based strictly on experience. What this means is that off-policy algorithms can separate exploration from control, and on-policy algorithms cannot. In other words, an agent trained using an off-policy method may end up learning tactics that it did not necessarily exhibit during the learning phase.

10.4.2 Advantages of TD Prediction Methods

- TD methods do not require a model of the environment, only experience.
- TD methods can be fully incremental. You can learn before knowing the final outcome.
- Less memory.
- Less peak computation. You can learn without the final outcome.
- From incomplete sequences.

10.5 Learning Automata

An automation in simple words can be defined as a process of making things processes to happen automatically. It is a machine or control mechanism which is designed for automatically following a predetermined sequence of operations or instructions. Learning automata started by Tsetlin in the early 1960s in the Soviet Union and was considered his outstanding work. In a survey paper published in 1974, the term 'learning automation' was used by Narendra and Thathachar. Learning automata is an approach to learning, especially reinforcement learning. It selects one of its actions related to its past experiences and rewards or punishments from the environment.

Learning automata is, defined as:

The concept of learning automation grew out as a fusion of works from psychologists, statisticians and operational researchers. The psychologists worked for modelling the observed behaviour where the statisticians took effort to model the choice of experiments by considering the past observations, the operation researchers attempted to implement the optimal strategies in the context of the two-armed bandit problem. The system theorists group made rational decisions in random environments.

Learning paradigm in the learning automation may be stated as a finite number of actions performed in a random environment. For a specific action performed to the environment, that provides a random response which may be either favourable or unfavourable. The primary objective of automation is to decide on the fact, whether any choice of the action at a particular state needs to be guided by past experience and responses. The decisions can be made with a knowledge concerning the environment. It may have time varying characteristics. Even the decision maker becomes a part of hierarchical decision structure but is unaware of his role in the hierarchy. For an uncertain situation, the output of environment depends on the way the other agents work which will remain unknown to the decision maker. Figure 10.9 shows general architecture of the learning automata. Figure 10.10 and Figure 10.11 show the automation and environment respectively.

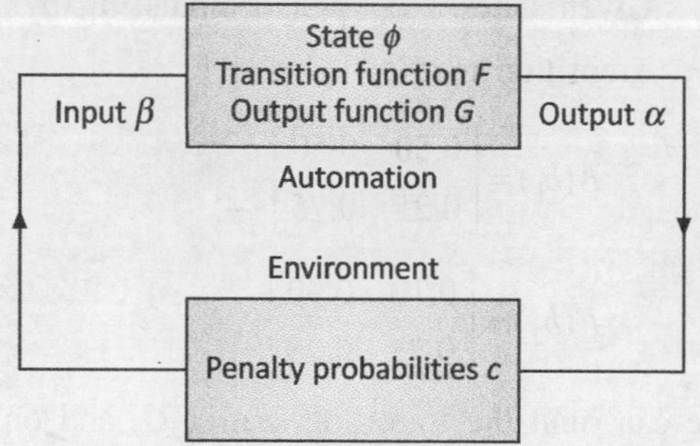

Figure 10.9 The automation and environment.

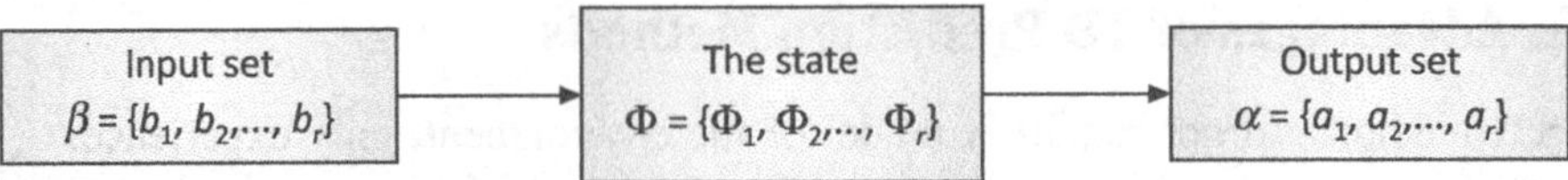

Figure 10.10 The learning automation.

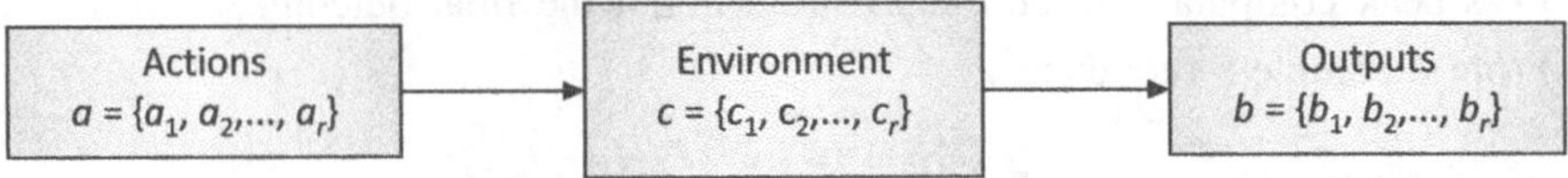

Figure 10.11 The environment.

The automation is a quintuple (Φ, α, β, F, H) where

- Φ is the set of internal states. At any instant n, the state $\Phi(n)$ is an element of the finite set $\Phi(n) = \{\Phi_1, \Phi_2, ..., \Phi_s\}$
- α is a set of actions also defined as the outputs of the automation. The output at the instant n is denoted by $\alpha(n)$, which is an element of the finite set $\alpha = \{a_1, a_2, ..., a_s\}$.
- β is a set of responses also called as the inputs from the environment. The input from the environment $\beta(n)$ is an element of the set β which can be either a finite set or an infinite set such as an interval on the real line: $\beta = \{b_1, b_2, ..., b_s\}$.
- F: $\phi \times \beta \rightarrow \Phi$ is a transition function that maps the current state and input into the next state. F can be deterministic or stochastic.
- H: $\phi \times \beta \rightarrow \alpha$ is an output function that maps the current state and input into the current output. the automation is referred to as state-output automation, if the current output depends on only the current state.

Here environment is a triple: $\{a, c, b\}$ where

- Set of inputs $a = \{a_1, a_2, ..., a_r\}$
- Set of outputs $b = \{b_1, b_2, ..., b_r\}$
- Set of penalty probabilities $c = \{c_1, c_2, ..., c_r\}$

Consider an example,

Given states: ϕ_1, ϕ_2 and inputs: b_1, b_2

Transition matrices:

$$F(b_1) = \begin{bmatrix} 0.50 & 0.50 \\ 0.25 & 0.75 \end{bmatrix}$$

$$F(b_2) = \begin{bmatrix} 0.10 & 0.90 \\ 0.75 & 0.25 \end{bmatrix}$$

We can omit the transition matrix G, and only consider the transition matrix F because the automation is a state-output automation. As the transitions never depends on the environmental responses, there is only one state transition matrix F where the elements are f_{ij}. The transition graph for $\beta = b_1$ and $\beta = b_2$ are shown in Figure 10.12.

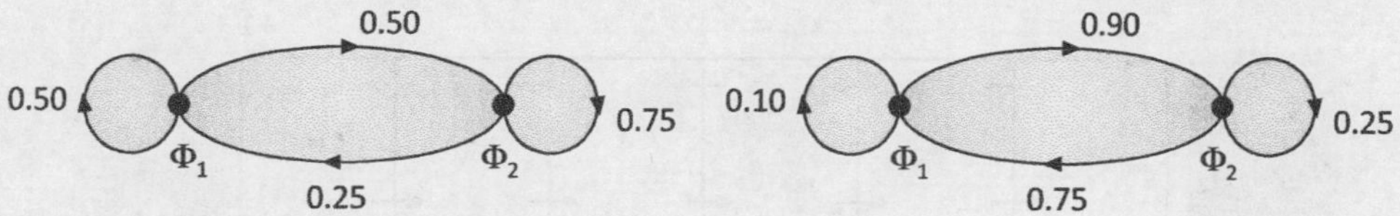

Figure 10.12 Transition graph.

There are two major types of learning automata, namely deterministic and stochastic. For deterministic automation

$$f_{ij}^{b} = 1, \quad \text{if } \Phi \rightarrow \Phi \text{ for an input } b; \; = 0 \text{ otherwise} \tag{10.5}$$

For stochastic automation

$$g_{ij} = 1 \quad \text{if } G(\Phi_I) = a_j; \; = 0 \text{ otherwise} \tag{10.6}$$

Applications of learning automata are as follows:

- Network routing
- Priority assignment in queueing system
- Task scheduling
- Multiple-access networks
- Image compression
- Pattern classification

10.6 Case Studies

10.6.1 Super Mario Game

Super Mario again!! Nobody will ever forget the champ Mario and the little princess. Yet again the princess is being kidnapped by the dragon, Bowser. Mario is now travelling through the Mushroom Kingdom in order to rescue the Princess Toadstool from the antagonist dragon Bowser. This time Bowser has made a special 1D trap for Mario. The trap is so created that for every wrong step (decision) that Mario makes, he gets punished and for every positive move he gets rewarded (reinforced). Mario is now in a one dimensional world of cell where he can either go left or right in single steps. In the left end cell, dragon is hidden to kill Mario and in the right end cell, he has locted up the princess. Here hero needs to make correct steps and get rewarded and must reach the right end cell to save the princess.

In the beginning our agent Mario receives no information at all from the environment but can sense which cell he is in. He also has no idea what cell he gets to by going left or right and what sort of reinforcement he will receive. To start with, our agent moves a single step towards the right, as he is going in the same direction where the princess is locked; he gets a reinforcement of +1. This indirectly says that Mario is moving in the correct direction. If the first step that our agent took was towards the left, then a punishment would have been waiting for our hero (that is, a negative reinforcement). Mario checks for the reinforcement each time and makes positive steps in each move and finally gets to the right end cell and unlocks the princess and saves her (Figure 10.13).

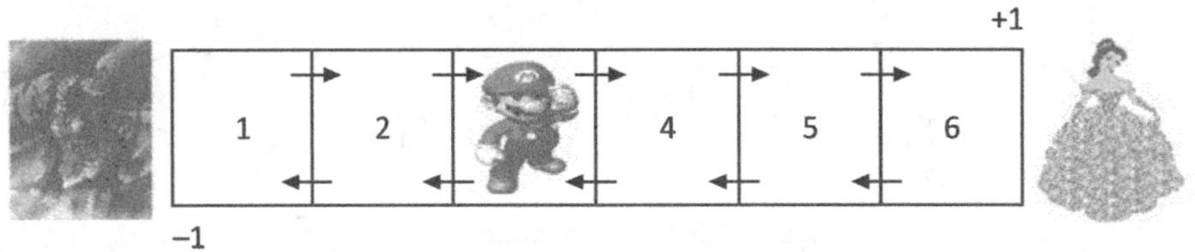

Figure 10.13 Reinforcement learning example.

On the other side, let us say this can be treated as a situation where reinforced learning is used. The goal of reinforcement learning is to maximise the rewards by figuring out those actions which turn good for reinforcement. That is, the agent is learning a policy, a mapping from states to actions. In this case, our agent is learning the route to his destination from the reinforcement that he obtained from the environment. We can divide the agent's policy into two components, i.e. how good the agent thinks an action is for a given state and how the agent uses what he knows to choose an action for a given state. That is, for every positive reward the agents holds strong and repeat the actions which he did in the previous step. Thus, he learns a new rule and a new method. This kind of learning is normally termed as reinforced learning technique.

10.6.2 Cross Junction Problem

Have you not been at a traffic signal waiting for a signal change? At times, we will be forced to wait for 40 or 50 seconds even if there is absolutely no vehicles in the other direction. This can be treated as multi-agent reinforcement problem. Consider Figure 10.14, here five agents are deployed in multi-path junction. Each agent gets eight different combinations to cross the junction. So, which one do you prefer if you are in the car?

From a computational point of view, this can be seen as an optimisation or a decision problem. The decision problem can be stated as 'will the car reach the other side of the junction through minimum signal blocks'? The optimisation problem can be stated as 'What will be the minimum time function taken for crossing the junction'? Both ways, the system needs to undergo a reinforced learning strategy.

When driven by an expert driver, he may be able to crack the problem very simply, but how can we make the system learn? In case of an automatic car whenever the car gets an option to take left, right or go straight then it will have an internal calculation by considering some parameters. Whenever the car takes a left, a matrix LM will be updated with a corresponding value and it will be mapped to a hash function for every correct assumption. Similarly the hashing function gets a negative mapping in case of wrong direction. Thus, the three matrix corresponding to left, right and straight—LM, RM and SM will be respectively updated.

Once the system learns through the hashing function, it can be seen that the car goes through the correct path to reach the destination through shortest path.

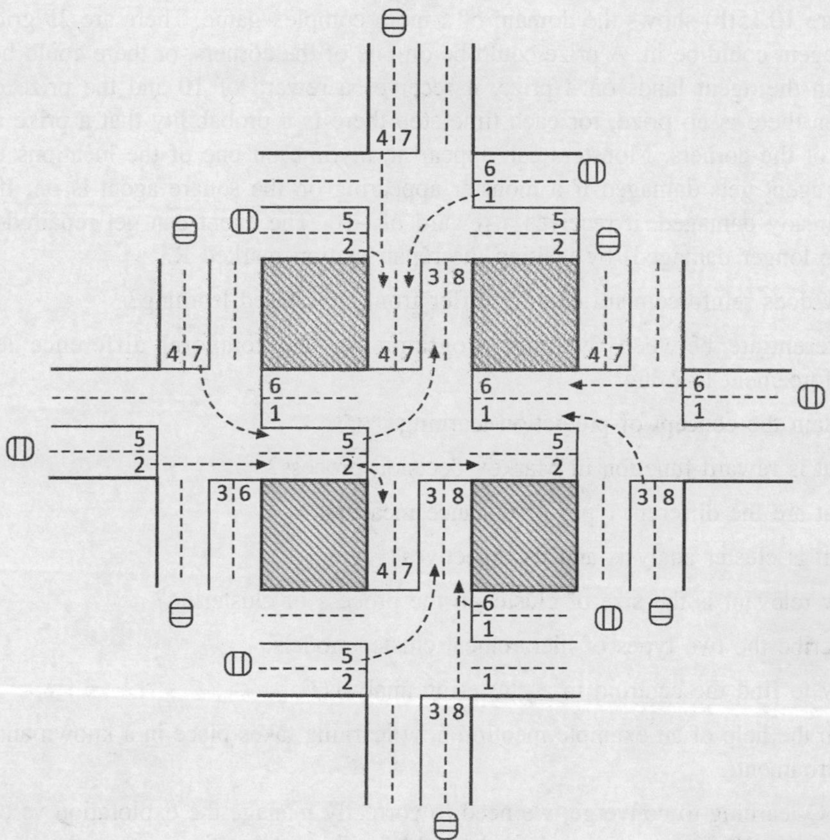

Figure 10.14 Traffic signal problem.

10.7 Exercises

1. Consider the tiny reinforcement learning problem shown in Figure 10.15(a). There are six states the agent could be in, labelled as s_0, ..., s_5. The agent has four actions: UpC, Up, Left, Right. That is, the agent knows all before it starts. It does not know how the states are configured, what are the actions to do, or how rewards are earned.

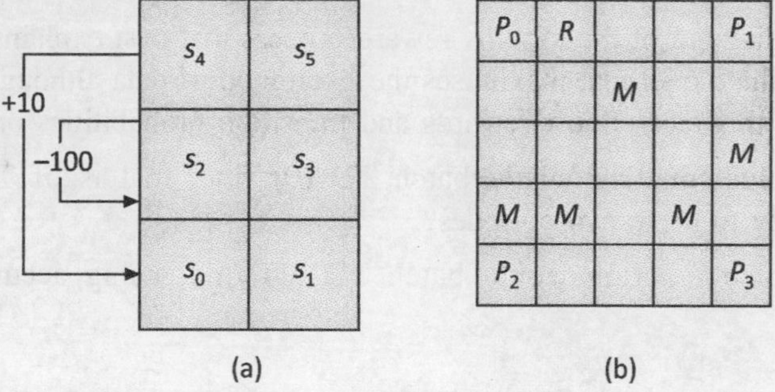

Figure 10.15 (a) The environment of a grid game (b) Domain of a more complex game.

2. Figure 10.15(b) shows the domain of a more complex game. There are 25 grid locations the agent could be in. A prize could be on one of the corners, or there could be no prize. When the agent lands on a prize, it receives a reward of 10 and the prize disappears. When there is no prize, for each time step there is a probability that a prize appears on one of the corners. Monsters can appear at anytime on one of the locations marked M. The agent gets damaged if a monster appearing on the square agent is on. If the agent is already damaged, it receives a reward of –10. The agent can get repaired (that is, it is no longer damaged) by visiting the repair station marked R.
3. How does reinforcement learning differ from supervised learning?
4. Differentiate between dynamic programming and temporal difference learning in reinforcement learning.
5. Explain the concept of prediction learning.
6. What is reward function in Markov decision process?
7. What are the different types of distance measures?
8. What is cluster analysis and its objectives?
9. How relevant is the size of cluster in the process of clustering?
10. Describe the two types of hierarchical cluster models.
11. How to find the centroid in a clustering analysis?
12. With the help of an example mention how learning takes place in a known and unknown environment.
13. For Q-learning to converge, we need to correctly manage the exploration vs exploitation trade-off. What property needs to be hold for this exploration strategy?
14. Write an example of using Q-learning with bid data?
15. Consider an un-discounted Markov reward process with two states A and B. The transition matrix and reward function are unknown, but you have observed two sample episodes: $A + 3 \rightarrow A + 2 \rightarrow B - 4 \rightarrow A + 4 \rightarrow B - 3 \rightarrow$ terminate $B - 2 \rightarrow A + 3 \rightarrow B - 3 \rightarrow$ terminate.

    In the above episodes, sample state transitions and sample rewards are shown at each step, example, A + 3 *rightarrow* A indicates a transition from state A to state A, with a reward of '+3'

    (a) Draw a diagram of the Markov reward process that best explains these two episodes (that is, the model that maximises the likelihood of data although it is not necessary to prove this fact). Show rewards and transition probabilities on your diagram.

    (b) What value function would batch TD (0) find, that is, if TD (0) was applied repeatedly to these two episodes?

    (c) What value function would batch TD (1) find, using accumulating eligibility traces?

16. Consider an MDP with states {4, 3, 2, 1, 0}, where 4 is the starting state. In states $k \leq 1$, you can walk(W) and $T(k, W, k - 1) = 1$. In states $k \geq 2$, you can also jump(J) and $T(k, J, k\ 2) = T(k, J, k) = 1/2$. State 0 is a terminal state. The reward $R(s, a, s') = (s - s')\ 2$ for all (s, a, s'). Use a discount of $\gamma = 1/2$.
    (a) Compute $V^*(2)$ and Compute $Q^*(4, W)$
    (b) Consider the same MDP, but with infinite states {4, 3, 2, 1, 0, –1, . . } and no terminal states. Like before $(k, J, k - 2) = T(k, J, k) = 1/2$ and $T(k, W, k - 1) = 1$ and $R(s, a, s') = (s - s')2$. Compute $V^*(2)$.
    (c) In the infinite Markov decision problem above, an agent acts randomly at each time step. With probability, it walks. With probability $1 - p$, it jumps. What is its expected utility in terms of p, starting from state 4?
17. Suppose a Q-learning agent, with fixed α, and discount, in state 34, did action 7, received reward 3, and ended up in state 65. What value(s) get updated? Give an expression for the new value.
18. Place an agent in any one of the rooms (0, 1, 2, 3, 4) and the goal is to reach outside the building (room 5). Can this be achieved through artificial intelligence? If yes, explain how it can be done.
19. This grid world MDP operates like to the one we saw in class. The states are grid squares, identified by their row and column number (row first). The agent always starts in state (1, 1), marked with the letter S. There are two terminal goal states, (2, 3) with reward +5 and (1, 3) with reward –5. Rewards are 0 in non-terminal states. (The reward for a state is received as the agent moves into the state.) The transition function is such that the intended agent movement (North, South, West, or East) happens with probability 0.8. With probability 0.1 each, the agent ends up in one of the state's perpendicular to the intended direction. If a collision with a wall happens, the agent stays in the same state.

CHAPTER 11

Nature Inspired Learning

Nature has mystified man since immemorial time. Nature gifts solutions to all problems but needs selected learning. Several natural phenomena have been adapted in computational science to create optimisation algorithms. Nature inspired design models are commonly adapted by human for engineering designs. This is intelligent approach in design science, models and their mathematics. Figure 11.1 shows some nature inspired design models applicable to engineering and technology design.

Evolutionary process of natural selection has created many emergent behavioural patterns in organisms that have optimising overtones. These have been separately identified and used in nature inspired algorithms. Various algorithms like evolutionary algorithms and swarm algorithms have been developed on the basis of naturally occurring mechanism. The major nature inspired algorithms are spread over the behaviour of species which live in a nest or an environment. Intelligent behaviour of these species can be turned-out by an algorithm that can help to solve complex problems. The field of nature inspired computing has been endowed with many optimising algorithms. These algorithms are distinct in the source of inspiration as well as in the methodology by which it has been mapped into the computational domain. Most of the algorithms designed till now principally deal with an agent or a multitude of agents that exhibit collective intelligent behaviour.

Figure 11.1 Engineering design adapted from nature models.

Nature exhibits intelligent ways to decipher complex optimisation problems. Nature inspired computing is a relatively strong paradigm that has profound applications in the optimisation arena. Swarm intelligence, Evolutionary computation, Cellular automata, and Neural computation are among the sought-after subdomains of nature inspired computing in the engineering framework.

In recent years there has been found a surge in the research of bio-inspired computing. Many novel meta-heuristic approaches inspired from a natural or man-made metaphor include imperial colonisation, league championships, gravity, electromagnetism, river formation, bacterial foraging and another subclass inspired by animal behaviour like ants, bees, bats, wolves, cats, termites, frogs, eagles, dolphins and much more. But many of the above-specified methods except a few are without any strong mathematical background.

It is also very difficult to identify if a new meta-heuristic is different from a familiar one even though the metaphor that inspired them was quite different. Popularly studied and analysed swarm based methods are Ants, Birds, and Bees. Figure 11.2 shows the swarms which it builds the colonies.

Research has not yet analysed the natural swarm behaviour optimising multiple objectives simultaneously or optimisation at different levels to be applied for multi-objective or multi-level optimisation problems with the maximum spread, diversity and minimum computational speed. A popular way of handling Multi-objective optimisation is the use of evolutionary technique. But it is a black box optimisation with parameter fine tuning done using trial and error method and does not account for self-adaptation. Meta-heuristic approaches work well when compared with traditional methods when there exists no clear understanding of the problem at hand. There does not exist a single method that optimises a problem effectively and efficiently as these objectives are contradictory. They also suffer from the curse of dimensionality problem as it takes quadratic, cubic or polynomial time with an increase in number of parameters. The limitations of the popular evolutionary approaches like Genetic Algorithm (GA) lies with fine tuning all the parameters like mutation rate, elitism percentage, crossover parameters and fitness normalisation/selection parameters. Genetic algorithm uses trial and error to obtain comprehensible solutions.

Figure 11.2 Ant, ants flocking and their colony, bee, bee flocking and their colony.

The nature inspired models can be classified as: bio-inspired models, swarm based models, and evolutionary based models (Figure 11.3). While designing the nature inspired algorithms the following is required to note. First, identification of analogies between the nature and mathematical problem. The next step is understanding the computer modelling of realistic behaviour. Finally engineering the model in a simplified manner in tune with the application. Figure 11.4 gives a brief idea of how to make a nature inspired model.

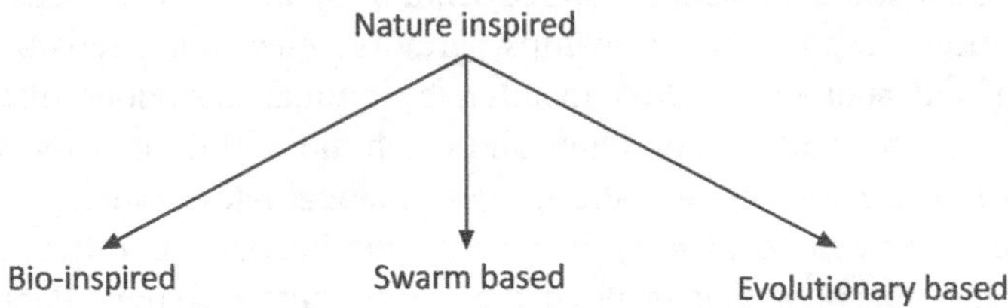

Figure 11.3 Types of nature inspired algorithms.

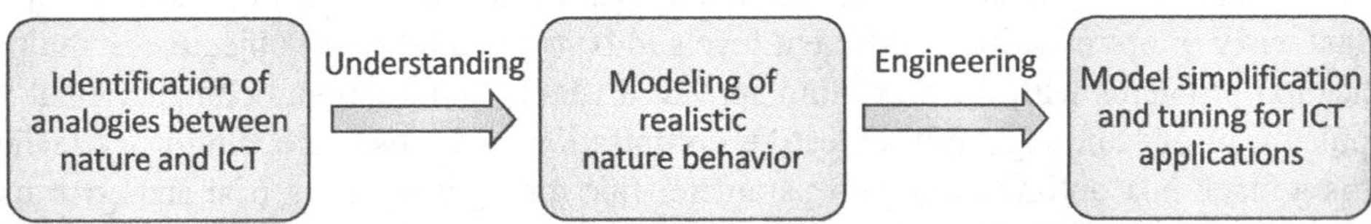

Figure 11.4 Transformation of nature inspired algorithms.

11.1 Nature Inspired Computing

Nature Inspired Computing (NIC) is a relatively recent research domain where observations found in nature are modelled mathematically to solve complex computational problems. The research has contributed to the emergence of new areas like evolutionary computing, artificial neural networks, swarm intelligence, etc. Recent advances in the field of computational analytics and big data have emphasised the concept of acceptable or satisfactory solutions to large scale computational problems. This has also contributed to the ground breaking research in the field of nature inspired computing. Nature inspired computing is based on the concept of emergence, self-organisation and decentralised behaviour of different species found in nature to accomplish different action for their livelihood.

Objective of nature inspired computing is to model algorithms to address complex problems based on the behaviours observed in nature. The autonomous agents in a nature inspired computing systems are coined as effectors and detectors. Detector agents of similar type receive the information and stimulus from the environment and other agents. Effectors contribute to the change in environment by changing the internal states and the surrounding.

Nature inspired systems are based on the principle of emergence. The prominent feature of systems which shows the property of emergence arising from the interactions between the individual agents is uncertainty. Some of the natural systems where the property of emergence can be observed is birds, ants, human societies, biological immune systems, etc. Three inherent properties of nature inspired computing which made it desirable for large scale computing problems are the parallelism, simplicity and the localised interaction. Any nature inspired computing model follows a general method where a group of autonomous agents following simple rules, coordinate in a decentralised manner to achieve an intelligent act.

Large scale computational problems have the features like varying properties of target functions, presence of highly nonlinear constraints, the presence of a large number of variables—mixed, discrete, discontinuous variables, complex search domain which makes each problem unique in nature. No differentiability, multi-modality, functions with significant null space effect, functions with a large number of local optima and computational complexity are some of the significant properties of target objective functions. The recent growth in this research area is due to the vast nature of the application domains and different types of problem requirements.

Recent years have found a surge in the research in nature inspired computing. Many novel meta-heuristic approaches inspired from a natural or man-made metaphor include imperial colonisation, league championships, gravity, electromagnetism, river formation, bacterial foraging and another subclass inspired by animal behaviour like ants, bees, bats, wolves, cats, termites, frogs, eagles, dolphins and many more falling under nature inspired computing. But many of these methods except a few are without any strong mathematical background. It is also very difficult to identify if a new meta-heuristic is different from a familiar one, even though the metaphor that inspired them was quite different.

Ability to fine tune parameters according to the specific problems and ability to handle problems of different nature makes the field of nature inspired computing promising. Evolutionary computing and swarm intelligence are popular nature inspired methods successful in engineering applications.

There are models in literature based on the perceptions of human mind which has the property of tolerance to impression calculations and uncertainty, robustness and cost effective computational process. The approaches like fuzzy logic, roughest theory, perception based computing, wisdom computing are some of the popular ones.

There are a class of methods inspired from the evolutionary theory. It is based on the adaptation of the agent with time. The methods which come under this category are Genetic algorithm, Neuro computing, DNA computing, Membrane computing and Cellular automata.

There is another class inspired from the natural swarms which interact locally following simple rules to achieve a collective task in a decentralised manner. There are swarm intelligence methods inspired from ants, bees, fish schooling, termites, firefly algorithm, bacterial foraging, etc.

Swarm intelligence based method develops a solution to complex computational problem obtained by modelling the natural computation approach of the swarms, where a group of agents explores the search area in parallel, to arrive a set of acceptable solution. There is another class of solutions under the category of trajectory based nature inspired methods like intelligent water drops method, simulated annealing, stochastic diffusion search, gravitational search where the trajectory of a particular agent is followed to reach the optimal solution.

There is another field called as geo-sciences which relate to the study of the planet earth, atmosphere, oceans, hydrosphere. The natural process like plate tectonics, volcano eruptions, tides, oceanic waves, earthquakes are considered for the study. Several nature inspired computational intelligence techniques inspired from the geological study of planet earth are adapted to real-time scenarios such as in robotics, path planning, destination prediction, image classification and object identification.

11.2 Bio-inspired Models

Life is derived from biology. The real-life models are derived from biological species. Human being is such a biological model with diversified character sequence. But there are many species in nature that show peculiar character behaviour. Swarm based algorithms are work based on the group behavior of these species which is contributed individually. In addition to the common behaviour of these species, individual performance gives solution to the world. For example, ants working as a group while canine (dogs) performing individually. So, both are intelligent agents but behaviours are different. Their movements can be derived as a solution to the complex optimisation problem.

A high level advantage of these solutions are lesser mathematical derivation compared to other mathematical or statistical models. But the solutions are limited to solve certain application in the real-world problems. In nature inspired technique, usually mathematical derivation is made after completion of the solution identification. As we know that one of the first applications of hidden Markov model was speech recognition, starting in the mid-1970s. Even the Markov model was suggested in 1900s and many modifications and algorithms are added but the first application was originated after 70 years. This is because of the solution maker's understanding of a model or even high-level statistics and derivation involved.

Consider the growth of a plant. Its branches are towards sunlight. Suppose, we have placed a set of similar plants in different environmental conditions. The plant grows towards a high intensive light source or we can say that the growth is dependent on the light. This is a nature phenomenon that the branches are identified in optimised path itself. This phenomenon can make use of an algorithm that will support complex problem solving. Suppose we have to consider insect groups, fireflies. These fireflies are work individually and deliver a result while in a group. Here light of the firefly acts as a biological behaviour that can be developed as a solution matrix to a complex problem. The intensity of its light is an attractive nature phenomenon used by fireflies that can be developed as an algorithm to the complex problems. Consider another example, of nature sniffing behaviour of canine. The track path of a canine from a starting point to the ending point while sniffing is marked as the shortest path. Obstacles in the travel path of the canine are covered during its traversal. In the movement of canine, the starting point can be the source and ending point can be the destination. This can be extended to solve shortest path finding problem and can be used to solve many complex tasks.

The above solutions formed are not accidental, but derived from their behaviour of species agents in the nature. It is followed by giving proof like mathematical derivations. That is, first, identify the problem that needs a solution which is far away and which the species in nature can solve easily through their biological behaviour. Based on the problem statement and behaviour of the species, a cross matching can be checked. If it is a perfect match or if there is a modified match that can give a solution, then we can fix the nature behaviour and the species. An initial solution is made from this behaviour followed by proposal of its mathematical derivation/formulation and equations. Revising the initial procedure accordingly may lead to the solution. This process continues in cycles giving a perfect solution to the problem.

11.2.1 Bio-inspired Computing

Biological inspired computing techniques originate from the natural behavior of biological systems. The field avails contributions from biology, computer science and mathematics. It falls as a subset of nature inspired computing. Biological concepts inspired from our surroundings like natural immune system, brain, life, natural agents like ants, birds, bees, molecular activity, evolution, etc. are modelled mathematically to solve complex real-time tasks. The feature of practical combinatorial optimisation problems like uncertainty and the nonlinear parameter space, limits the use of deterministic methods for deriving the perfect solution. Adaptable and self-organizing features of the biological systems make it suitable for handling real-time problems with complex parameter space.

Generally, bio-inspired computing takes a bottom up approach in a decentralised manner. While tuning, it meets the problem specification. These methods have been successful in different fields of parallel computing, networking, data analyties, power systems and optimisation. As the complexity of the problem increases, the number of possible solutions also increases. The potential solution can be categorised as worst, passable and best.

Biologically inspired methods follow a heuristic approach in generating random solutions to the problems. They are found superior to other deterministic methods when the uncertainty and randomness of the problem space increases. Meta-heuristics form an alternative approach to tackle complex search and optimisation problems when mathematical methods fail to produce optimal solutions. Heuristic and meta-heuristic approaches propose clever ways for scanning the possible solutions to arrive at the attractive choice for the specific problem in affordable computational time

Generic approach for problem modelling involves the adaptation of the specific bio-inspired model to match with specific problem requirement, with parameter and constraint specification done in a customised environment. Two popular approaches under the category of evolutionary computing and swarm intelligence based methods are the forerunners under biologically inspired computing.

Swarm intelligence studies the emergence of different agents to achieve a collective intelligent act. It uses a population of similar agents exploring the search area to generate a set of potential solutions. Features of natural swarms are simplicity, scalability, fault tolerance and decentralised computing. These properties make it robust while searching the solution space for feasible answers from the huge search space.

Swarm refers to an unorganised group of similar social animals which moves to perform the designated task. Swarm Intelligence is a field which explores ideas arising from behaviours of social animals to achieve and intelligently act. Intelligence arises from the simple interactions among and between the individuals which are unorganised.

Evolutionary computing is derived from the biological evolution process of different species. They are successful for search and optimisation problems. Evolutionary approach always maintains a population of potential solutions, which are made to propagate to further generations, where the principle of survival of the fittest is preserved. The operations of mutation and crossover are applied on a pair of potential solution to develop the next generation of solutions which continues until a satisfactory result is obtained.

Social behaviour of similar species help them interact to find food or escape predator attack. Fish schools escape an attack, where individual members remain alert about the intruders trying to attack. Bee swarms help each other in detecting better foraging sources. Ants follow perfect path to the food source. Birds flock together for feeding efficiency and to successfully defend predator attacks. Lion optimisation algorithm mimics the behaviour of lions and their characteristics of cooperation. Behaviours of these kind of natural swarms are emulated to develop intelligent computational models to solve the routing, searching and optimisation problems.

Meta-heuristic optimisation techniques based on simulating the analogical reasoning behaviours of biological beings, perform an informed or blind search to solve complex optimisation problems. The choice of a particular algorithm for a specific problem is a significant process in developing the optimised solution. These optimisation algorithms are becoming very popular because they are based on simple ideas, ease of implementation, no gradient information of data is needed, finding optimal neighbourhood for getting it adapted to different applications.

11.3 Evolutionary Models

Population based algorithms constitute a collection of complete solution that is a collective of partial solutions and solution attributes. There are multiple approaches while selecting individuals that constitute a new population. This is by selecting newly generated individuals and 'elite' individuals. At the time of initialization, a diverse population is selected through a randomization and take an explicit criterion. The populations are evolved in 'generations'. A new offspring is created by combining features of current parents. Usually an offspring is developed from two parents. Figure 11.5 gives the generation by evolutionary algorithm. The offspring is evolved by variation operation such as 'mutation' or 'recombination'. The size of population is fixed at most of the time. Multiple copies of a given parent may involve in the selection process. A best one carried out in the next generation. During the iteration the values become better. For example, Figure 11.6 shows a 0th and *n*th population.

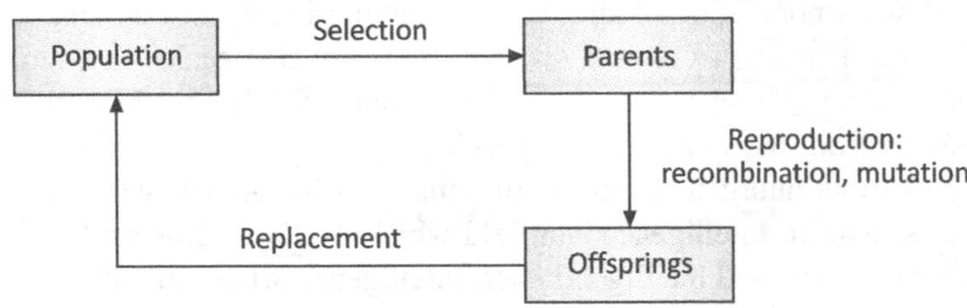

Figure 11.5 A generation in evolutionary algorithm.

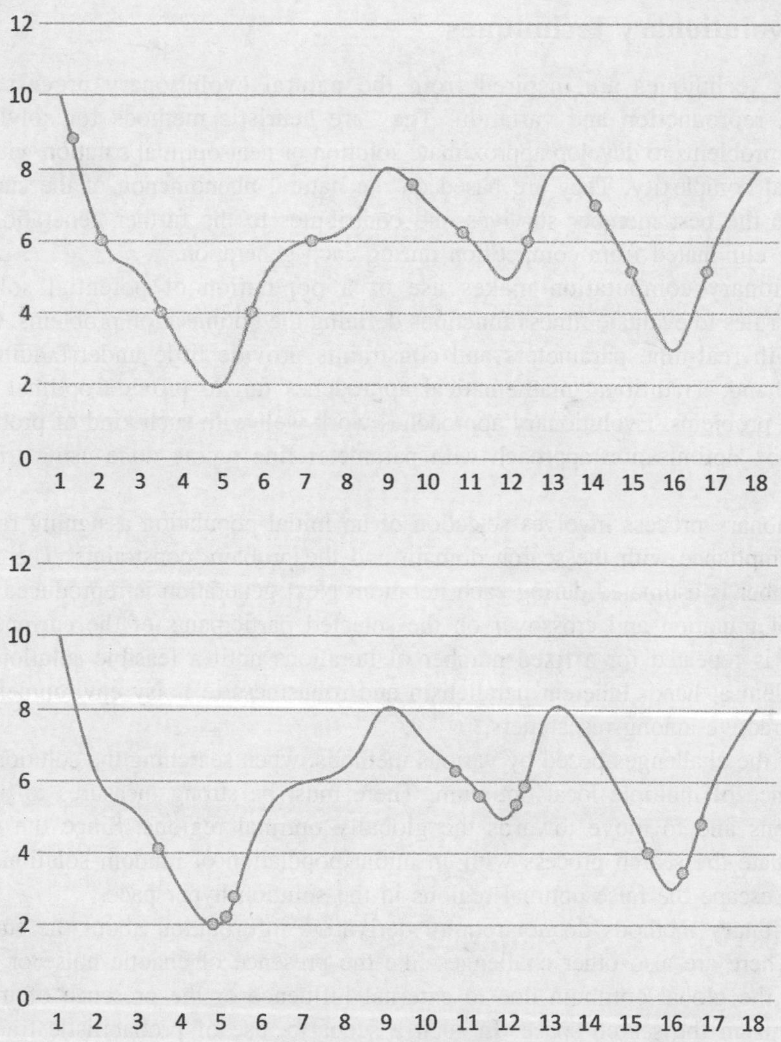

Figure 11.6 0th and *n*th population example.

Basic evolutionary computation steps

```
P ← Generate Initial Population
Evaluate (P)
while termination conditions not met do
    P' ← Recompute(P)
    P" ← Mutate (P')
    Evaluate (P)
    P ← Select (P" ∪ P)
endwhile
```

11.3.1 Evolutionary Techniques

Evolutionary techniques are inspired from the natural evolutionary process comprising of selection, reproduction and variation. They are heuristic methods for solving complex optimisation problems to develop approximate solution or near optimal solution with reasonable computational complexity. They are based on the natural phenomenon of the survival of the fittest, where the best member survives and contributes to the further generations and unfit members are eliminated from competition during each generation.

Evolutionary computation makes use of a population of potential solutions with probabilistic rules to evaluate fitness functions defining the optimisation problems. Optimisation problems with real-time parameters and constraints provide little understanding about the problem at hand. Traditional mathematical approaches fail to provide optimal solutions to such kind of problems. Evolutionary approaches work well with such kind of problems, which is a black box optimisation approach with parameter fine tuning done using trial and error method.

Evolutionary process involves selection of an initial population assigning random initial values in compliance with the search domain and the problem constraints. The fitness value of each member is estimated during each iteration. Next generation is reproduced by applying operations of mutation and crossover on the selected participants of the current population. The process is repeated for a fixed number of iterations until a feasible solution is obtained for the problem at hand. Inherent parallelism and robustness to noisy environments make the approach attractive among researchers.

One of the challenges posed by various methods, when searching the solution hyperspace is the presence of multiple local optimum. There must be strong measures to overcome the local solutions and to move towards the globally optimal regions. Since the evolutionary methods initiate the search process with an initial population of random solutions, the search is guided to escape the false optimal regions in the solution hyperspace.

Evolutionary methods do not require derivative information about the function to be optimised. There are also other challenges like the presence of chaotic noise or the dynamic variation of the global optimum due to external influence or the presence of multiple local optimal points in the search space. In such a scenario, use of probabilistic transition rules, guide the search process, rather than it being a random search operation. Those individuals with a higher fitness value are considered in the competition for taking the generations forward. Winners of each generation are paired by applying crossover or mutation operation to produce a more attractive individual for the next iteration.

The various computational paradigms that come under the evolutionary techniques are genetic algorithm, genetic programming, evolutionary strategies and evolutionary programming. Genetic algorithm imitates the natural evolution found successful in classification and optimisation problems. An initial population is randomly selected in the solution hyperspace, where the individuals propagate in the successive iterations to move towards the optimal goal. The individual members are denoted using encoded strings, where the dimensionality and type of each member vary with the specific problem.

Evolutionary programming applies a top down approach which simulates the adaptive behaviour of the species during evolution, while genetic algorithm simulates the genetic behaviour of various species. Mutation is applied to develop offspring. The best members of parents and offsprings are ranked and the top members enter the next generation.

Evolutionary strategies follow a slightly different approach, where the problem solutions are considered rather than population itself. It uses a set of strategy parameters that stochastically control the amount of mutation applied on individual members. The mutation rate evolves with the population being optimised.

11.4 Swarm Models

'Swarm intelligence is artificial intelligence based on the collective behaviour of decentralised, self-organised systems'.

Swarm intelligence emerged from the collective behaviour of natural agents in realising a complex task. A swarm is a collection of agents that depicts an emergent behaviour to elucidate a complex problem without any centralised coordination. The intelligent behaviour of different types of swarms, such as ants, birds, bees, bacteria, and fish, has already been used to solve many combinatorial optimisation problems. Ant Colony Optimisation (ACO), Particle Swarm Optimisation (PSO), Artificial Bee Colony (ABC) optimisation are well-received techniques in the field of swarm intelligence. Swarm intelligence is an emerging field of bio-inspired artificial intelligence based on the behavioural models of natural swarms where intelligence emerges by the collective actions of decentralised and self-organising agent to solve an intricate task. The collective intelligence is accomplished by natural agents by following simple rules. The agents interact locally in a decentralised manner, leading to the emergence of an intelligent action collectively that is unknown to the individual members. Primary capabilities of swarm intelligence focus on load balancing, clustering, optimisation, and routing. Complex tasks from truck routing to military robots are later realised using swarm intelligence based algorithms.

The swarm intelligence has two fundamental concepts—self organising and division of labour. The self-organisation of swarms includes positive feedback, negative feedback, fluctuations and multiple interactions of agents. The swarms work by division of labour by simultaneous task performance by cooperating specialised individuals that enables the swarms to respond to changed conditions in the search space.

11.4.1 Self-organisation

The importance of swarm intelligence where the features like self-organisation and decentralised computation make it suitable for real-time applications. The natural swarm behaviours solving multiple objectives are analysed to propose computational frameworks to tackle different kinds of real-time problems.

The basis of self-organisation is as follows:

1. Positive feedback (amplification) promotes better solutions by allocating more agents.
2. Negative feedback for counter-balance and stabilisation wherein avoids the situation all individuals converge to the same behaviour or to the same state.

3. Amplification of fluctuations (randomness, errors, random walks).
4. Multiple social interactions.
5. Continuous conflict between positive feedback and negative feedback.

Basic principle of swarm intelligence is as follows:

1. **Proximity principle:** The swarm is capable of doing elementary space and time computations related to its neighbourhood. Computation is considered as a behavioural response to each agent to some environmental stimulus raised during local interactions between swarm members. Some of the fundamental behavioural computations by natural swarms include foraging, nest building and predator defence.
2. **Quality principle:** A swarm should respond to quality factors of food and safety of a location.
3. **Principle of diverse response:** Resources should not be concentrated in a narrow region. The distribution should be designed so that each agent will be maximally protected facing environmental fluctuations.
4. **Principle of stability:** The population should not change its mode of behaviour every time the environment changes. Such changes consume energy which may not produce a worthwhile return on the investment.
5. **Principle of adaptability:** The swarm is sensitive to the changes in the environment that result in different swarm behaviour.

11.4.2 Properties of Swarm Intelligent Systems

Scalable: Intelligence arises due to the individual interaction of each agent with its neighbours. Each individual member's behaviour is influenced only by its neighbour. Thus, by increasing the size of the swarm one can increase the productivity without really needing to redesign the entire system.

Parallel: Individual members of the swarm can perform different functions at various locations at the same time. This behaviour can accomplish different independent parts of a complex task simultaneously. Parallel processing can have a significant effect on the flexibility as well as the computational complexity.

Fault tolerance: Self-organised and decentralised nature of swarms leads to fault tolerant systems as the agents are interchangeable and none is in total control of the entire system. Individual failures can be easily replaced by a fully functional agent without degradation in performance.

Adaptability: Swarm can dynamically accommodate any kind of changes in the environment very promptly.

Decentralised: There is no single point of failure in a swarm subsystem as nobody is in full charge of the system. Agents choose their actions and then carry them out independently.

11.5 Swarm and Evolutionary Algorithms

Evolutionary computing and swarm intelligence are popular nature inspired methods successful in engineering applications. Evolutionary computing emulates natural evolution process while swarm intelligence evolves after natural swarm behaviour in accomplishing a complicated task successfully. Successful and well-studied algorithms in swarm intelligence are inspired after natural swarms like Ants, Birds, Bees and Bacterial colony.

Evolutionary approaches are stochastic methods based on the Darwinian theory of evolution by means of natural selection. Popular approaches are genetic algorithms, evolution strategies and evolutionary programming which operate on a set of solutions which are altered by two operators—mutation and crossover (selection and variation). Evolutionary computing methodology is popularly applied to search and optimisation domains. This approach consists of a population consisting of individual members where each individual represents a potential solution. Each individual of the current population is assigned a fitness value as a measure of its strength. A random sampling is done to generate a mating pool based on fitness measure of each member. Variation and recombination operator is applied to develop the next generation to carry the population forward till the desired termination criteria are met.

Swarm intelligence is a paradigm which considers intelligence as a behaviour emerging through local interaction and cooperation of large numbers of swarms belonging to the specific category. Swarm intelligence uses adaptive strategy and its application mainly concentrating on optimisation, clustering, load balancing and routing problems. Well studied algorithms emerge from the natural behaviour of ants, birds, bees, bats and bacteria. Basic swarm intelligence methods used for optimising single objective problems are adapted to solve multiple objectives. The basic behaviour found in the natural swarm colony is adapted to suit the formulation of multi objective optimisation. But those methods lack optimal convergence with increase in a number of objectives. Also, the size of non-dominated solution set increases to make the task of the decision maker tedious.

11.6 Important Nature Inspired Algorithms

In computer science, logic is very important but in mathematics procedure and proof are important. A meta-heuristic or a high-level procedure is generated while designing an optimisation algorithm. The designers of nature inspired algorithms considered these for a problem. Table 11.1 shows different nature inspired algorithms that built procedure based or meta-heuristic based algorithms used in different applications.

These algorithms are spread over optimisation solution building by searching over a large set of feasible solutions. The meta-heuristics can often find good solutions with less computational effort than optimisation algorithms, iterative methods, or simple heuristics.

In early 80s, the algorithms where derived from the biological fields. Neural networks and genetic algorithms are such classes. In the later years, swarm based algorithms were evolved from nature. For example, ACO algorithm from natural behaviours of ant, foraging and PSO is from movements of birds. These algorithms are heuristics based algorithms. In late 80s, the meta-heuristic algorithms were developed to solve complex problems. So, the

ACO, PSO, etc. were re-modified, many times, as a meta-heuristic algorithms. Figure 11.7 shows the timelines of these algorithms. A huge change in the development of algorithms was after 2005. Most of these algorithms are meta-heuristic driven. Present developments in the nature inspired algorithms are hyper-heuristics based. That is, a combination of nature inspired techniques can be developed to solve complex problems.

TABLE 11.1 Important nature inspired algorithms

	Inspired model	*Proposed by and year*
1.	Simulated annealing	Kirkpatrick et al., 1983
2.	Ant colony optimisation	Dorigo, 1992
3.	Particle swarm optimisation	Kennedy and Eberhart, 1995
4.	Harmony search	Geem, Kim and Loganathan, 2001
5.	Artificial bee colony algorithm	Karaboga, 2005
6.	Bees algorithm	Pham, 2005
7.	Glow-worm swarm optimisation	Krishnanand and Ghose, 2005
8.	Shuffled frog leaping algorithm	Eusuff, Lansey and Pasha 2006
9.	Imperialist competitive algorithm	Atashpaz-Gargari and Lucas, 2007
10.	River formation dynamics	Rabanal, Rodrguez and Rubio, 2007
11.	Intelligent water drops algorithm	Shah-Hosseini, 2007
12.	Firefly algorithm	Xin-She Yang, 2008
13.	Gravitational search algorithm	Rashedi and Nezamabadi, 2009
14.	Cuckoo search	Yang and Deb, 2009
15.	Bat algorithm	Yang, 2010
16.	Spiral optimisation algorithm	Tamura and Yasuda, 2011
17.	Flower pollination algorithm	Yang 2012
18.	Cuttlefish optimisation algorithm	Eesa, Mohsin & Orman, 2013
19.	Artificial swarm intelligence	Rosenberg, 2014
20.	Smell detection agent algorithm	Chandra, 2014
21.	African Buffalo optimisation	Beneoluchi and Nizam Mohmad, 2015
22.	Duẹlist algorithm	Biyanto, 2016
23.	Killer whale algorithm	Biyanto, 2016
24.	Rain water algorithm	Biyanto, 2017
25.	Mass and energy balances algorithm	Biyanto, 2017
26.	Hydrological cycle algorithm	Wedyan et al., 2017
27.	Phototropic algorithm	Chandra and Anand, 2018

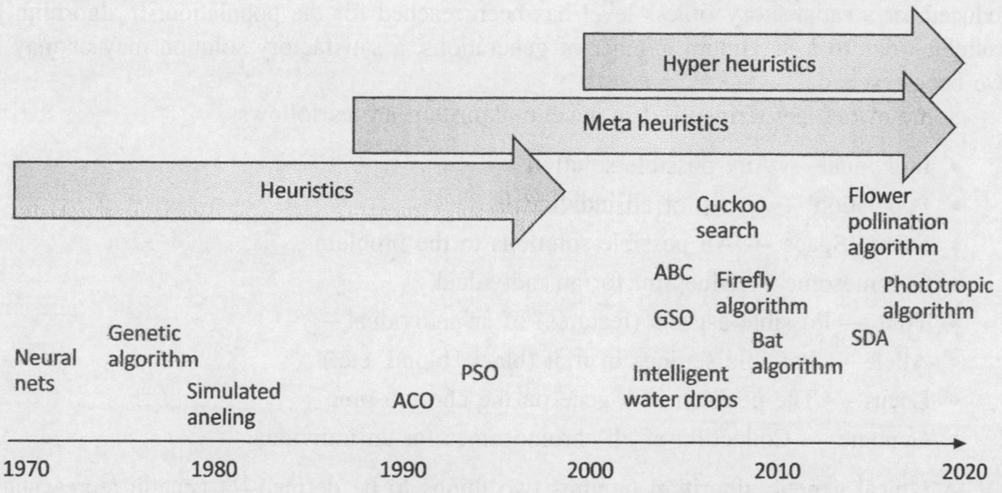

Figure 11.7 Evolution of nature inspired algorithms.

11.6.1 Genetic Algorithms

Genetic algorithm is a directed search algorithm based on the mechanics of biological evolution developed by John Holland from University of Michigan in 1970's. The algorithm is inspired by genetic process of life to understand the adaptive processes of natural systems. This algorithm provides efficient, effective techniques for optimisation and machine learning applications that are widely used today in business, scientific and engineering circles.

It is a class of iterative optimisation methods that use the principles of evolutionary biology. When referring to genetic algorithms, some of the terminologies are usually associated with optimisation changes in the biological terms. Solutions in a multi-objective problem are referred to as individuals, and the set of individuals is known as a population. The fitness of an individual refers to the values of the objective functions.

A genetic algorithm is a search technique used in computing to find true or approximate solutions to optimisation and search problems. This technique is categorised as global search heuristics using a particular class of evolutionary techniques inspired by evolutionary biology such as inheritance, mutation, selection, and crossover (also called recombination). Genetic algorithms are implemented as a computer simulation in which a population of abstract representations, called chromosomes or the genotype or the genome of candidate solutions (called individuals, creatures, or phenotypes) to an optimisation problem, evolves toward better solutions. Traditionally, solutions are represented in binary as strings of 0s and 1s, but other encodings are also possible. The evolution usually starts from a population of randomly generated individuals and happens in generations. In each generation, the fitness of every individual in the population is evaluated, multiple individuals are selected from the current population (based on their fitness), and modified (recombined and possibly mutated) to form a new population. The new population is then used in the next iteration of the algorithm. Commonly, the algorithm terminates when either a maximum number of generations has been produced, or a satisfactory fitness level has been reached for the population. If algorithm has terminated due to a maximum number of generations, a satisfactory solution may or may not have been reached.

Some of the key terms used in genetic algorithm are as follows:

- Individual — Any possible solution
- Population — Group of all individuals
- Search Space — All possible solutions to the problem
- Chromosome — Blueprint for an individual
- Trait — Possible aspects (features) of an individual
- Allele — Possible settings of trait (black, blond, etc.)
- Locus — The position of a gene on the chromosome
- Genome — Collection of all chromosomes for an individual

A typical genetic algorithm requires two things to be defined—a genetic representation of the solution domain, and a fitness function to evaluate the solution domain. A standard representation of the solution is as an array of bits. Arrays of other types and structures can be used in essentially the same way. The main property that makes these genetic representations convenient for their easily aligned parts due to their fixed size, which facilitates simple crossover operation. Variable length representations may also be used, but crossover implementation is more complex in this case. Tree-like representations are explored in genetic programming.

The genetic algorithm is a probabilistic search algorithm that iteratively transforms a set or a population of mathematical objects (typically fixed-length binary character strings), each with an associated fitness value, into a new population of offspring objects using the Darwinian principle of natural selection and using operations that are patterned after naturally occurring genetic operations, such as crossover (sexual recombination) and mutation.

Basic steps in genetic algorithm

```
Generate initial population;
Compute fitness of each individual;
Repeat /* New generation /*
  for population size/2 do
      Select two parents from old generation;
      /* biased to the fitter ones */
      Recombine parents for two offspring;
      Compute fitness of offspring;
      Insert offspring in new generation
  endfor
Until population has converged
```

11.6.2 Ant Colony Algorithms

Ants are eusocial insects living as society. Ant society has division of labour in communication with individuals. The communication is achieved by 'stigmergy' of the ants. Stigmergy is a mechanism of indirect coordination, through the environment, between agents or actions. Stigmergy can be operational by coordination through indirect interaction which is more appealing than direct communication which reduces or eliminates communication between the agents. Stigmergy is used to get the problem solved bit by bit. The individuals communicate with each other, affecting what each of them does on the task. The individuals leave markers or messages.

Usually ants show unsophisticated behaviour but collectively they can perform complex tasks. The ants have highly developed sophisticated sign-based stigmergy, communicate using pheromones. The ants lay trails of pheromone that can be followed by other ants. Out of more than one pheromone trails, the ant has a choice of two pheromone trails to follow which should be stronger. Figure 11.8 shows the cooperative behaviour of ants based on pheromone trails. Pheromones are chemical substance produced and released into the environment by the ants (or other species like a mammal or an insect) affecting the behaviour or physiology of others of its species. This trail attracts other ants and serves as a guiding path. The pheromones will evaporate in timely manner.

A social insects like ants use their pheromone to communicate travel path among them. The initial trails of pheromone as they return to the nest with food are shown in Figure 11.8(a). There are two ants chosing different paths from their nest (source) to the food (destination). If we keep check, there are two separate paths created by these ants through depositing pheromones as shown in Figure 11.8(b). When one ant reaches the food source, using the same pheromone trails, it returns to the nest. The second ant also travelled back to the nest after visiting the food source (Figure 11.8(c)). The first ant guides the other ants to the food source through the path it travelled (Figure 11.8(d)). The second ant reaches the nest after a long travel compared to the first ant as shown in Figure 11.8(e). The pheromone evaporation rate is high in the second path because of its length. Finally, all ants choose the first path based on the high pheromone deposit. Figure 11.8 (f) shows the final shortest path chosen by the ants.

In such cooperative behaviour, the ants select a shortest path which will be an optimum path from the source to the destination. This phenomenon trail can be evolved as an algorithm used to solve optimisation problems.

Study of realistic ant and its colony is the major motivation of the Ant colony algorithms. The inspiration is based on ant foraging from one point to another using pheromone trails. The main idea is that the self-organising principles which allow the highly coordinated behaviour of ants is derived to artificial agents (ants) to solve complex problems. The behaviour of real ant colony system like foraging, division of labour, brood sorting, and cooperative transport is imitated to form an artificial ant colony system.

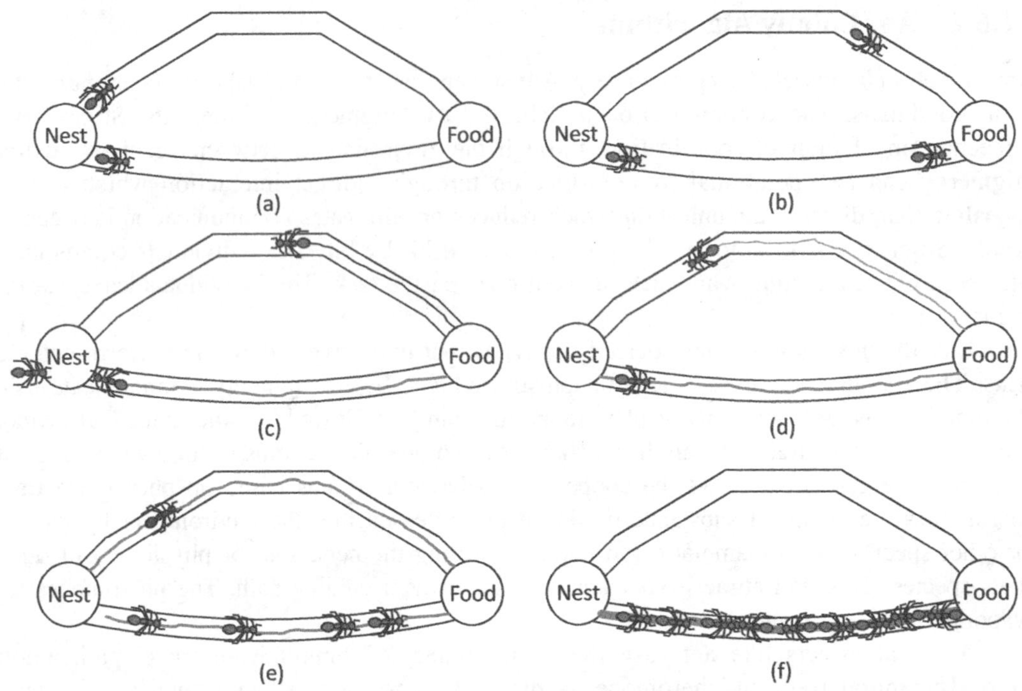

Figure 11.8 Cooperative search by pheromone trails.

As we know that ant's pheromone deposits while its move, are used as a communication material. This pheromone is considered as an artificial pheromone material while designing an ant colony system. The ants are agents in this system which is communicating and giving solutions. There are decision rules used by ants during the system construction process. Pheromones are biological material secreted by an ant that primarily maintain the line in a pheromone trail. While moving the ant deposits certain amount of pheromone. There should be a direction of move, i.e. either left, right or straight. This can be represented mathematically as a probable value. Figure 11.9 shows such a move by an ant by $P(C) < P(B) < P(A)$. That is, the ant may move either left, straight or right.

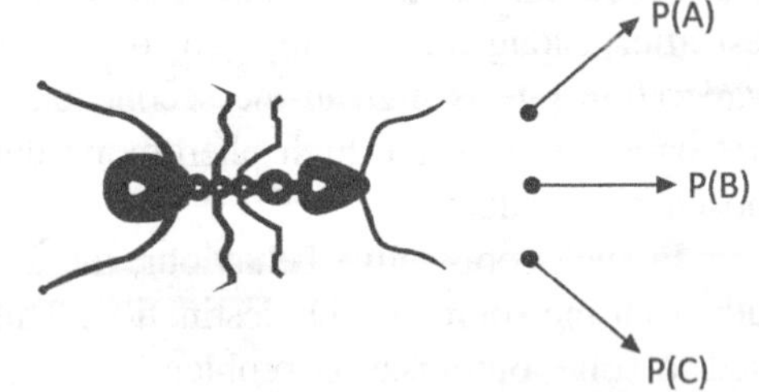

Figure 11.9 Pheromone trails movements.

Ants are making a straight line pheromone trail that connects a food source to their nest. An elementary behaviour of real ants can be used to explain how they can find the shortest path. The broken paths were reconnected if an unexpected obstacle has interrupted the initial path. This is by pheromone trails of the ants. When an obstacle is met by an ant which disables to continue to follow the pheromone trail, the ant has to choose between turning right or left. For such situations, we can expect half the ants to choose to turn right and the other half to turn left. By chance, a group of ants that chooses the shorter path around the obstacle will more rapidly reconstitute the interrupted pheromone trail. Based on the greater pheromone deposit the ants will receive a greater amount of pheromone per time unit. So, a larger number of ants will use that path, i.e. the shorter path. Based on the time stamp, the pheromone trails start to evaporate that leads to reduce its attractive strength. The path with higher pheromone intensity will be chosen by most of the ants which becomes the shortest path. In more time, ants travelling down a path and coming back in a same path, will deposit more pheromone with lesser evaporation rate. Thus, a higher density pheromone path is more frequently compared with other paths. Step by step movements of ants are given in Figure 11.10(a) through Figure 11.10(h).

The first of its kind proposed by Marco Dorigo in 1991 is called as Ant System (AS) which is based on swarming behaviour of natural ants. Ant colony optimisation algorithms are based on meta-heuristic derived from ants feedback systems: pheromones. In this algorithm, ants move in the search space by marking the current best solutions (positions) at a particular instant of time. They always coordinate their movement by considering the previous search positions trying to optimise their findings.

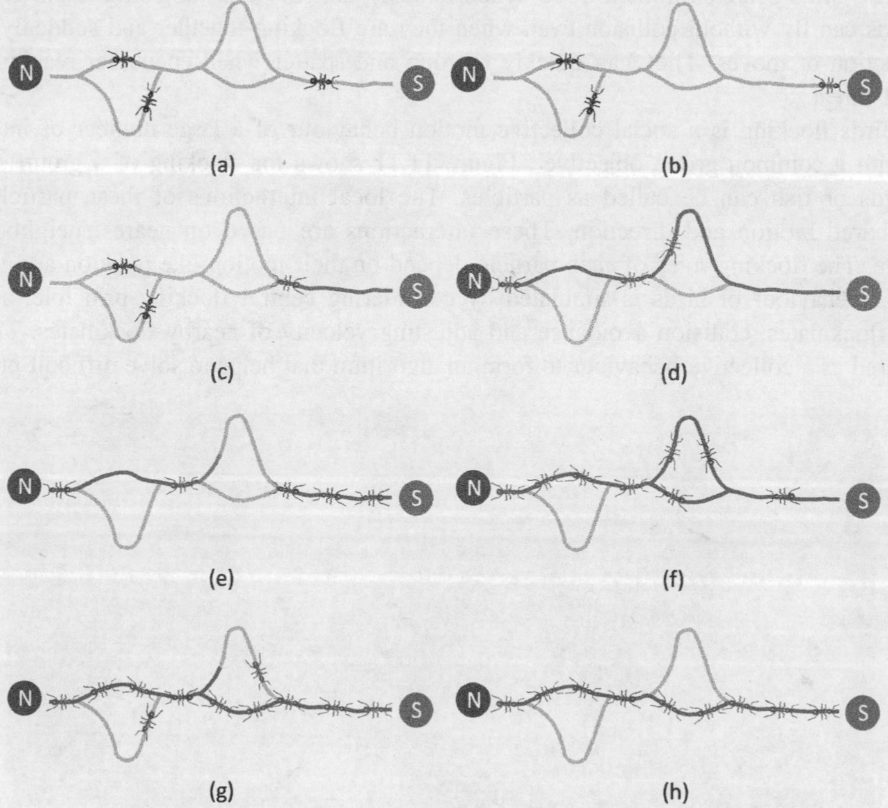

Figure 11.10 Shortest path around an obstacle by ants by pheromone trails. (a) Search randomly around the nest (b) Find food source (c) Laying down pheromones trail (d) Keep following the trails (e) Reinforcing the trail with more pheromones (f) Uncover more possible paths (g) Reducing the pheromone strength by evaporation (h) Identified shortest path.

The principle of distributed computation, positive feedback, and constructive greedy heuristic used by ants are computationally exploited to find feasible solutions for different applications. The computational model is successful on classical and asymmetric single objective problems like travelling salesman problem, job-shop scheduling, and quadratic assignment problems. Many algorithms inspired by the ant colony behaviour emerged in the literature. Elitist ant system, ANT-Q, Max – Min ant System, Rank based ant system, ANTs are popular techniques for global optimisation problems. Ant colony based algorithms are very popular in specific areas of optimisation, scheduling, classification and load balancing.

11.6.3 Particle Swarm Algorithms

The important sense for flocking behaviour is eye vision. The birds and fish are eyes on both sides of their heads. This phenomenon allows them to see the objects on each side at the same time. Large size of bird's eyes and their arrangement of heads gives a wide range field view. For example, without turning the head the woodcocks and pigeons can see 360-degree field view. The birds search for their food synchronously and move a long migration. It is seen that birds can fly without collision even when they are flocking together and suddenly change the direction of moves. They can quickly regroup and scatter when encounter reaching some external threat.

'Birds flocking is a social collective motion behaviour of a large number of interacting birds with a common group objective'. Figure 11.11 shows the flocking of a group of birds. The birds or fish can be called as particles. The local interactions of these particles have some shared motion and direction. These interactions are based on nearest neighbourhood principle. The flocking rules of each particle depend on their motion like position and velocity. Flocking behaviour of birds is simulated by considering central flocking principle, adjusting nearby flockmates, collision avoidance and adjusting velocity of nearby flockmates. These are considered as a collective behaviour to form an algorithm that helps to solve difficult problems.

Figure 11.11 Birds flocking.

Every optimisation technique focuses on finding the best solution from a set of feasible solutions. Particle Swarm Optimisation(PSO) is a swarm based optimisation technique first introduced by Kennedy and Ebenhart in 1995. This technique was inspired by the animal social behaviour such as flock of birds, school of fish etc. and mimics the way they find their food sources. PSO is initialized with a group of random particles (solutions) and then searches for optimum by updating generations.

Behaviour of particle swarms are collection of flying particles (swarm) giving changing solutions by its search area. Movement of particles towards a promising area to get the global optimum value. The particles working in a group shows three simple behaviours (Figure 11.12). The three behaviour of particles are as follows:

- Separation: avoid crowding local flock mates
- Alignment: move towards the average heading of local flock mates
- Cohesion: move toward the average position of local flock mates

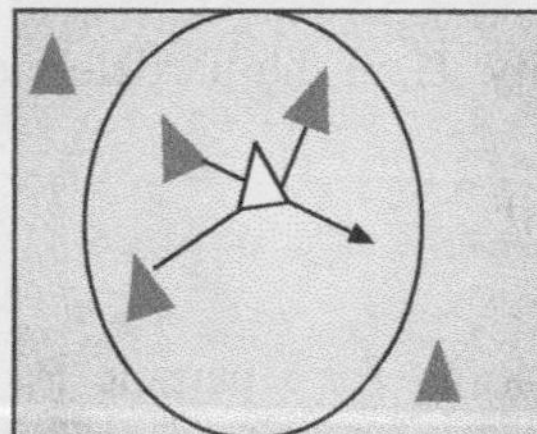 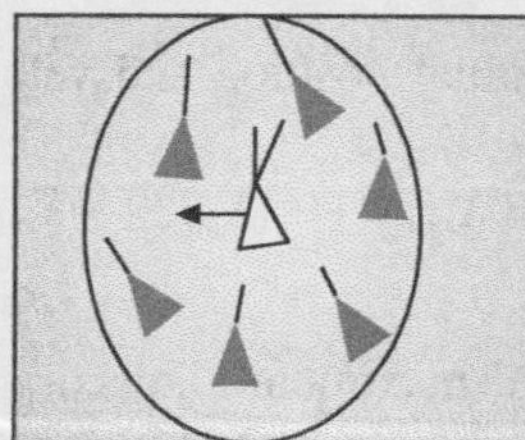 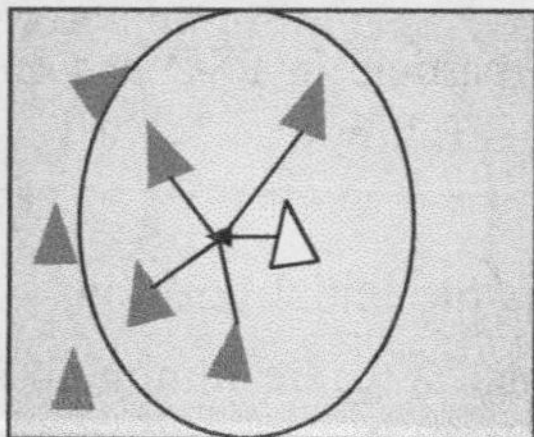

Figure 11.12 Behaviours of particles.

Particle movements

Each particle keeps track its best solution, personal best (*pbest*) and the best value of any particle, global best (*gbest*). Each particle adjusts its travelling speed dynamically corresponding to the flying experiences of itself and its colleagues. Each particle in a group is searching an optimum value by adjusting its velocity. Each particle remembers the position it was in where it had its best result so far (its personal best). Each particle modifies its position according to its current position, its current velocity, the distance between its current position and *pbest* and the distance between its current position and *gbest*. Figure 11.13 gives the movement of each particle in a group by adjusting their *pbest* and *gbest* using their simple behaviours.

Particles are working as a group hence they are co-operating. They exchange information about what they have discovered in the places they have visited. That is a particle has neighbourhood associated with it. They know the fitnesses of those in its neighbourhood, and uses the position of the one with best fitness. This position is simply used to adjust particle's velocity (*pbest*).

Each particle P in the swarm is represented by the following characteristics:

x_i : The current position of the particle P.

v_i : The current velocity of the particle P.

y_i : The personal best position of the particle.

$y^{\cap}$: The neighbourhood best position of the particle.

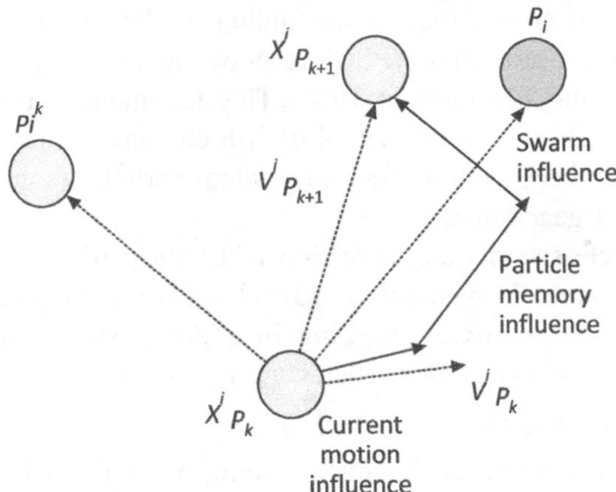

Figure 11.13 Movement of each particle in a co-operative environment.

Each particle in PSO has an initial position and velocity. Here the particle's position is initialised as follows.

$$x_0 = X_{min,j} + r_j(X_{max,j} - X_{min,j}) \tag{11.1}$$

where, x_0 is the initial position.

X_{min} and X_{max} are minimum and maximum allowed positions for a particle P.

j refers to the dimension r refers to a random value between 0 and 1.

A particle's position represents a possible solution to the optimisation problem. Goodness of the position is determined by a user-defined fitness function denoted by f. The fitness function varies depending on the optimisation problem. During its flight a particle updates its position and velocity based on its own experience and experience of its neighbours. In every iteration, each particle is updated using two best values. The first one is the best fitness it has achieved so far. This value is called personal best position of the particles denoted by y_i. The personal best value of a particle at time step t is updated as follows:

$$y_i(t + 1) = y_i(t); \quad \text{if } f(y_i(t)) \geq f(x_i(t + 1)) \tag{11.2}$$

$$y_i(t + 1) = x_i(t + 1); \quad \text{if } f(y_i(t)) < f(x_i(t + 1)) \tag{11.3}$$

Another best value is the best value obtained so far by any particle in the population.

This best value is called global best represented by $y^{\cap}$.

$$y^{\cap} = \max\{f(y_0(t)), f(y_1(t)) \;.......\; .f(y_s(t))\} \tag{11.4}$$

After finding the two best values, the particle updates its velocity and position by Eqs. 11.5 and 11.6. The velocity of particle v_i is updated as follows:

$$v_{i,j}(t+1) = wx_{i,j}(t) + c_1 r_{1,j}(t)[y_{i,j}(t) - x_{i,j}(t)] + c_2 r_{2,j}(t)[y_i^{\cap} - x_{i,j}(t)] \tag{11.5}$$

where w is the inertia factor.

c_1 and c_2 are the acceleration coefficients, $0 \leq c_1, c_2 \leq 2$.

r_1 and r_2 are two separately generated uniformly distributed random numbers in the range of [0, 1]

The position of particle x_i is updated as follows:

$$x_i(t + 1) = x_i(t) + v_i(t + 1) \tag{11.6}$$

11.6.4 Bee Colony Algorithms

Honeybee or bee is a flying insect constructing large colonies. Bees are closely related to swarms like ants. Bees are adapted for feeding on nectar and pollen. The cooperating behaviour of the bees leads to the creation of large nests. Each bee in the nest has separate jobs or based on the jobs they are grouped together.

Honeybees are social creatures that live in colonies with a queen bee, worker bees and male drones. They forage for food in groups and are also responsible for plant pollination. Honey bees regulate their foraging activity based on colonial demands and environmental conditions. Honey bee foraging provides numerous contributions to the environment with honey accumulation and plant pollination being a few. Foraging activity by the numerous bee agents is a perfect example of an optimised behaviour in a decentralised manner.

The Artificial Bee Colony (ABC) optimisation algorithm is based on the natural honey foraging behaviour of bees where the location of a food source with respect to a bee colony represents a solution and the algorithm iterates to an optimal position to solve many types of unimodal and multi-modal functions. Behaviours based on the mating flight of queen bees, evolution of queen bee are also explored for the purpose of optimisation which evolves in a static domain with unperturbed environmental conditions.

Bees in nature

The bee colony is based on cooperation like any of the swarm algorithms. A major difference between other swarm based algorithm and ABC algorithm is the different types of agents working together for solution(s). The division of labour is major agenda in the colony(hive). The communication of bees is exploited according to quality and distance from the hive. Exchange of information among bees is the most important occurrence in the formation of collective knowledge of the bees. Communication among bees related to the quality of food sources occurs in the dancing area. This communication is called waggle dance which delivers the quality of food and amount of energy usage for the movement from hive to the food source. Selection of food source is based on proximity to the nest, richness of the food and ease of extraction.

In a bee colony, there are employed bees and unemployed bees (scouts and onlookers). The employed bees are associated with a particular food source who carry and share information to other bees. Role of unemployed bees is looking for a food source to exploit. A hive of honey bees can extend itself over long distances in multiple directions. Figure 11.14(a) shows a hive and range of bees they can travel (10 kilometres). Here A, B and C are food sources with different quality and quantity. Scout bees search for food randomly from one flower patch to another as shown in Figure 11.14(b). The bees who evaluated the different patches (A, B or C) return to the hive and perform a communication through waggle dance. Quality threshold taken by the bees is measured as a combination of some elements, such as sugar content. They deposit their nectar or pollen onto the 'dance floor' to perform a 'waggle dance'. The dance floor is nothing but the entrance of hive for the bees. Communication of the bees (waggle dance) contains the following information:

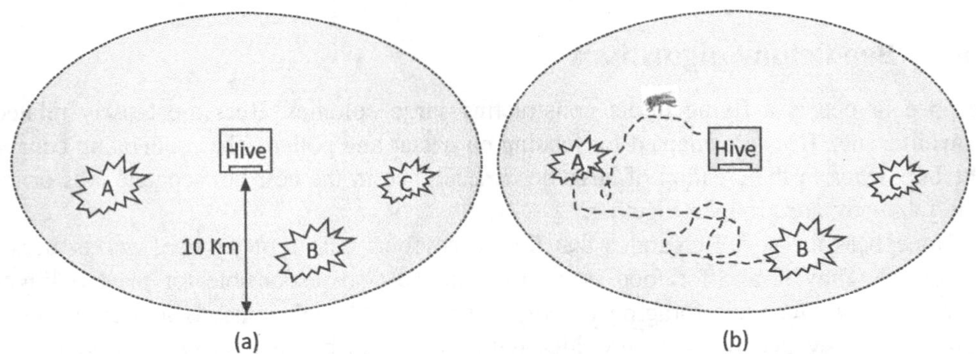

Figure 11.14 (a) A colony of bees and three food resources A, B and C (b) Scout bees search food in flower patches.

- Direction of flower patches (angle between the sun and the patch)
- Distance from the hive (duration of the dance)
- Quality rating or fitness (frequency of the dance)

This information helps the hive to send its bees precisely to the exact food source. Follower bees go after the dancer bee to the patch to gather food efficiently and quickly. Same patch will be advertised in the waggle dance again till it is good enough as a food source and more bees will be recruited to that source. This depends on the food level. So, more bees visit flower patches for plentiful amounts of nectar or pollen. Based on the fitness of bees, patches can be visited by more bees or may be abandoned. The process is given in Figure 11.15. Here S represents Scout bees, R represents Onlooker bees, UF represents Uncommitted Follower, EF_1 represents a shared information and EF_2 represents the continuation of work of the bee.

ABC basics

ABC algorithm was proposed by Dervis Karaboga in 2005. ABC algorithm was developed based on inspecting the behaviours of real bees on finding nectar and sharing the information of food sources to the bees in the hive. This algorithm is used to solve multidimensional and multi-modal optimisation problems.

Bees forage, honey, pollen, water, and resins for their survival. The computational model proposed here consists of multiple sources of various resources foraged by bees, multiple bee colonies, and bee swarm. Multi-modal foraging is carried out by the members of each colony without any centralised coordination. The objective of optimising profitability at their respective hives. The optimisation function is defined as a maximisation problem to maximise the profitability at a hive and is defined based on the amount of energy consumed by each member while foraging. Robustness of the proposed strategy is evaluated by mapping it over multi-commodity transportation problem with multiple optimisation parameters. As compared with other nature inspired techniques, such as genetic algorithm and ant colony optimisation which is used in a static scenario, the self-organising nature of honey bees provides a solution to the dynamic transportation problem model with multi-dimensional parameters.

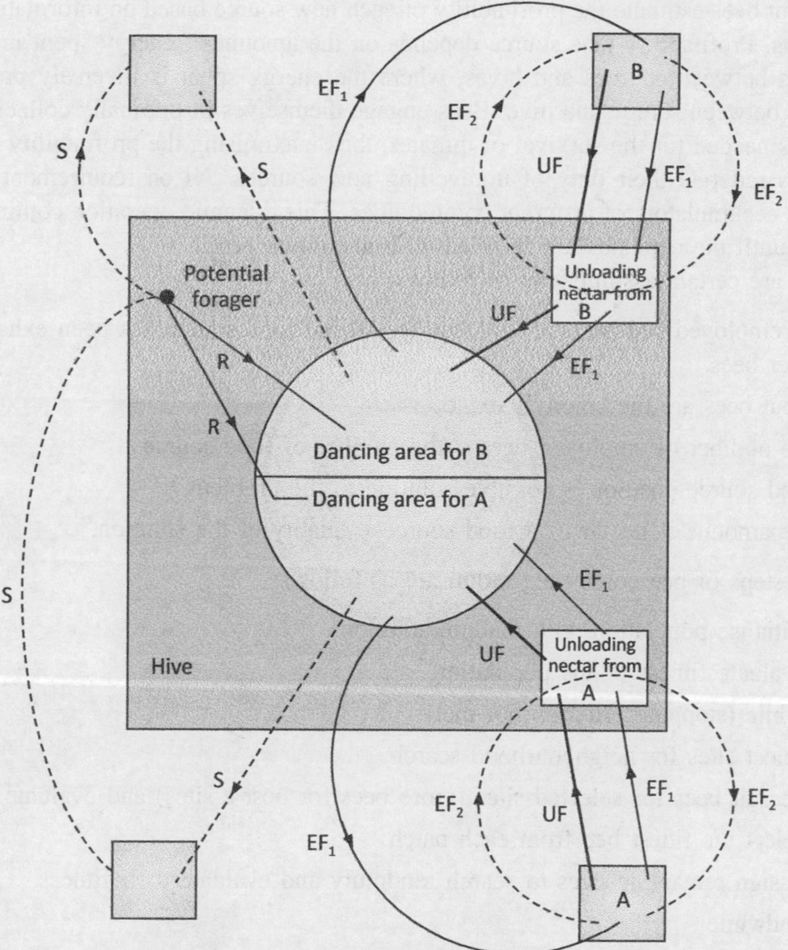

Figure 11.15 Movements of bees in a hive.

The worker or forager bees of a colony are responsible for the collection of different commodities required for their survival: nectar, pollen, water, and propolis (resins). Nectar is the main source of energy; it is converted to honey and stored for future use. Pollen provides the inmates with necessary protein, vitamins, and traces of minerals; excess pollens are stored in the hive for future use. Water is collected as required and not stored; bees use it to dilute honey before consumption and to keep the hive cool when the temperature soars. Propolis is a resinous material obtained from the bark of trees; it is used to seal the holes of the hive to insulate it in winter. The forager bees in a colony are classified as scout bees and reticent bees.

Scout bees, which from 5 to 10 percent of the total population, roam to discover new sources of various commodities. These scouts report the discovery of competent sources to the reticent bees at the hive by using a special dance named waggle dance. The quality of food source, distance of the source from hive, and the direction of source with respect to the colony are among the information conveyed through waggle dance.

Reticent bees estimate the profitability of each new source based on information conveyed by the scouts. Profitability of a source depends on the amount of energy spent in transporting commodities between sources and hives, where the energy spent is inversely proportional to the distance between source and hive. Bees engage themselves in optimally collecting different commodities needed for the survival of inmates, thus maximising the profitability at each hive, while scouts reiterate their duty of unravelling new sources. As on requirement bees switch their role as accumulators of different commodities. This dynamic operation continues in every bee colony until the demands are fulfilled with maximum profit.

There are certain assumptions as follows:

- An employed bee becomes a scout bee whose food source has been exhausted by the other bees.
- Scout bees are the colony's explorers.
- The number of employed bees = the number of food sources.
- Food source position = possible solution to the problem.
- The amount of nectar of a food source = quality of the solution.

Basic steps of bee colony algorithm are as follows:

1. Initialise population with random solutions.
2. Evaluate fitness of the population.
3. While (stopping criterion not met).
4. Select sites for neighbourhood search.
5. Recruit bees for selected sites (more bees for best e-sites) and evaluate fitness.
6. Select the fittest bee from each patch.
7. Assign remaining bees to search randomly and evaluate their fitness.
8. Endwhile

The ABC algorithm is given in Figure 11.16.

```
Send the scouts onto the initial food sources
REPEAT
    Send the employed bees onto the food sources and determine their nectar amounts
    Calculate the probability value of the sources with which they are preferred by
    the onlooker bees
    Stop the exploitation process of the sources abandoned by the bees
    Send the scouts into the search area for discovering new food sources, randomly
    Momorize the best food source found so far
UNTIL (requirements are met)
```

Figure 11.16 Basic steps of ABC algorithm.

11.7 Case Studies

11.7.1 Population Based Meta-heuristics

Nature is of course a great and immense source of inspiration for solving hard and complex problems in computer science since it exhibits extremely diverse, dynamic, robust, complex and fascinating phenomenon. It always finds the optimal solution to solve its problem maintaining perfect balance among its components. This is the thrust behind bio-inspired computing. Nature inspired algorithms are meta-heuristics that mimic the nature for solving optimisation problems opening a new era in computation. For the past decades, numerous research efforts have been concentrated in this particular area. Still being young and the results being very amazing, broadens the scope and viability of bio-inspired algorithms exploring new areas of application and more opportunities in computing.

Though reasons behind their success are often elusive: their performance often depends crucially, and unpredictably, on design choices and parameters. Furthermore, given a class of bio-inspired algorithms it is unclear on which kind of problems it performs well and on which it performs poorly. This lack of understanding represents a major obstacle for the widespread uptake and usage of bio-inspired computing techniques and for the development of more effective variants.

Meta-heuristic algorithms which make few or no assumptions about a problem and can search very large spaces of candidate solutions have been extensively developed to solve optimisation problems these days. Among these algorithms, population-based meta-heuristic algorithms are proper for global searches due to global exploration and local exploitation ability.

Main population based meta-heuristic algorithms are Adaptive dimensional search, Firefly algorithms, Harmony search, Gaussian adaptation, etc. Rigorous estimation of the expected time required by the algorithms to find a satisfactory solution for various optimisation problems is an important area of research. Such analyses use mathematical techniques drawn and extended from the fields of randomised algorithms, probability theory and computational complexity. The results allow for insights into the working principles of bio-inspired meta-heuristics, enable the assessment of parameter choices and design aspects while contributing to the design of more powerful algorithms.

We have the mathematical methodology developed to reveal where the real power of bio-inspired algorithm is in comparison with the traditional problem-specific algorithms. Main aim is to explain the performance of realistic bio-inspired algorithms for well-known combinatorial optimisation problems through runtime analyses, highlighting the relationships between solution quality and the exploration capabilities of the population to theoretically explain how bio-inspired algorithms can take advantage of the parallelisation available inherently in new technologies to achieve the population diversity required to produce solutions of higher quality in shorter time. Such analyses use mathematical techniques drawn and are extended from the fields of randomised algorithms, probability theory and computational complexity. Second objective is to design meta-heuristics tailored by modifying their source code. Compared to human designed meta-heuristics, these will be better, faster and stronger.

Pilot studies indicate that they will perform better than human designed meta-heuristics. We can design a faster than the manual design process. It is able to make stronger statistical statements about their properties. In short, a machine learning approach is to be employed to design meta-heuristics, explore how features can be used to make predictions, and how effectively we can design meta-heuristics given a problem's features. A new variant of multi-objective algorithm can be developed and is used in packet switched network routing problems to find the best path followed by the packets.

11.7.2 Nature Inspired Optimisation in Database Warehouse

The need for the optimised storage and retrieval of information from large database/data warehouse is growing exponentially with time. The database may be heterogeneous or legacy systems. Also, it may be relational, hierarchical or network databases. A common interface, which is well-suited to accommodate all the databases and consumes optimised storage for the database is difficult to conceive.

A real problem cited here is to establish the need for optimised storage. During the automation of a university management system, one of the modules was student details management. It was disclosed that the database for the students of a single academic year consumed large memory (in Giga bytes) for examination and related activities. The data collected/stored includes personal details, academic details, achievements, i.e. both personal and academic, details of extracurricular activities, students' resume and much more. The database is meant to support queries from various sources like government, industry, and other academic and non-academic institutions. Because of this requirement more details are to be collected/stored which is responsible for the huge size of the database.

The main goal will be to achieve greatest efficiency. The databases may be optimised in terms of file storage, data access speed, query performance, and efficiency of indices. It is supposing, if we concentrate on hardware, networking connections, server performance, coding, indexing, and methods like efficient clustering of data in and among the tables, it may improve the performance of the databases. Nature inspired optimisation techniques are very effective to improve the data store.

Apart from optimising the storage requirements of the databases, we may enhance the system to achieve the following functionalities:

The data of students in each year need to be maintained from a historical perspective. This will add to the storage needs to about TBs. How to optimise the storage is the primary target? A generic system meant for same task may have to take into account the heterogeneity and legacy factors. The interface should bridge the gap between the different implementations of the database. Storage should be optimised irrespective of the databases used (Example, Oracle, MS SQL, etc.). The concept of data warehouse can be applied for the storage of data related to passed out students, which is historical data. This should be organised so as to facilitate online analytical processing. Data structures like data cube can be deployed for efficient organisation of the data. The data warehouse/database of the students from all the colleges in the state may be maintained. The concept of data mining can be applied in an efficient way on this database to discover knowledge/patterns of various kinds like the clusters of students who are not so good in programming skills so that remedial action can be suggested during the course itself. We can find the characteristics in common to these kinds of students. This may be done from the perspective of the industry so that when these students are about to complete the course, they can excel in skills.

These functionalities can be extended to various other options which may be needed by the government, industry and other authorities. Government can have details of professional students at various granularities, which can be used for various purposes. The concept of intelligent databases may also be introduced so as to have an efficient management of data.

11.8 Exercises

1. List five nature inspired optimisation algorithms. Do a search and find at least one real-world problem that has been solved for each of the five algorithms.
2. What are the components of natural selection?
3. What are meta-heuristics? List out meta-heuristics used in our life.
4. How will you arrive a solution using nature inspired algorithms?
5. Can we solve same problems using different nature inspired algorithms? If so, how will you design such algorithms?
6. List out ten practical problems that cannot be solved normal techniques such as NP-hard?
7. How could you cope with division by zero in a program being evolved by a genetic programming approach?
8. What are the five stages when planning to solve a problem using genetic programming? Give a short description of each. Suggest another problem that could be solved by genetic programming.
9. The orthogonal rectangle packing problem requires that rectangles are packed into a bin with the aim of minimising the height of packing. Suggest two representations for this problem when using a genetic algorithm. Present the advantage(s) and disadvantage(s) of each type of representation. Name the items and describe the main features of genetic algorithms.
10. What is the appropriate method to find fitness function for genetic algorithm?
11. When do we use genetic algorithm? How will you know that the solution has reached the global minima?
12. Can genetic algorithm be used in real-time application? Justify.
13. How can you normalise an objective function?
14. What are the ways to choose genetic algorithm parameters (type of selection, mutation, crossover) that make quick convergence?

15. What two requirements should a problem satisfy in order to be suitable for solving it by a genetic algorithm?
16. For the PSO algorithm, what will be the consequence if
    (a) The maximum velocity V_{max} is too large or small?
    (b) The acceleration coefficients c_1 and c_2 are equal or not?
    (c) The acceleration coefficients c_1 and c_2 are very large or small?
17. ACO algorithms are considered very efficient when dealing with dynamic problems. Imagine you have a TSP in which nodes sometimes disappear and reappear again. Can you explain why ACO is a good choice to solve this problem? Think about the pheromone trail.
18. Is running a single ant $m \times k$ times the same as running ants fork iterations? Think what is the role of local pheromone decrease.
19. How can problems of premature convergence and stagnation in the PSO algorithm be prevented?
20. How can the discrete-valued problems be solved by the PSO method?
21. Compare different artificial bee colony and particle swarms algorithms used as a multi-objective optimisation applications.
22. Implement an ABC algorithm that can avoid obstacles in a search space.
23. How are ABC algorithms helpful in function approximation?
24. Implement resource allocation problem using ABC algorithm.
25. Consider a routing problem in a large IP network. What are the main difficulties/ differences when applying an ACO algorithm for the TSP for finding shortest routes in the network?
26. Describe the motivation for particle swarm optimization, its form and the expected advantages and potential drawbacks.

CHAPTER 12

Game Playing

People like games. Physical playing of games are very interesting. It is also a logical playing of games in virtual environments. These are available over video games or over computers. The computer games offer logical steps for a person as an intellectual component when involving. It is easy to represent the state of a game through two or more coins (pieces) or agents. The agents are usually limited to a small number of well-defined moves. Games are treated as search problems, i.e. an oldest endeavours in artificial intelligence. Claude Shannon and Alan Turing had written the first chess programs in 1950s. A machine requires intelligence for doing something like playing games with human. There are a set of simple rules for normal game playing that need to be translated to program. The computer program can correct every move of the game because it already has memory resident. The computer game is viewed as a complex decision problem due to the presence of human opponent.

This can be viewed as a search problem, sometimes a complex tree structure. A player searches a node in the game tree and finally reaches one leaf node for his favourable outcome.

A game can be formally defined as a type of search problem with the following components:

1. Initial state: This includes the board position and next move indication.
2. A set of operators: This defines the legal moves of the player.
3. A terminal test: It determines when the game is completed. The states where the game has finished are called terminal states.
4. A utility function: It is the outcome of a game in terms of a numeric value.

Game theory models strategic behaviour by agents who recognise that their actions affect the actions of other agents. A game consists of a set of players, a set of strategies for each player and the payoffs to each player for every possible list of strategy choices by the players. There are two players or multi-player games (example, chess vs monopoly vs poker), perfect information or imperfect (example, checkers vs card games) and deterministic or non-deterministic (example, go vs backgammon).

Different types of games in artificial intelligence are two-player zero-sum, multi-player, perfect information games, imperfect information games and games of chance. Some types of games that have been developed using artificial intelligence technique are given in Figure 12.1. Two-player games generate game trees which is a strategy that defines complete plan of action for a given player. An optimal strategy is obtained for the games of perfect information enumerating paths of a game tree. In practice, the game tree and its evaluation are for small games.

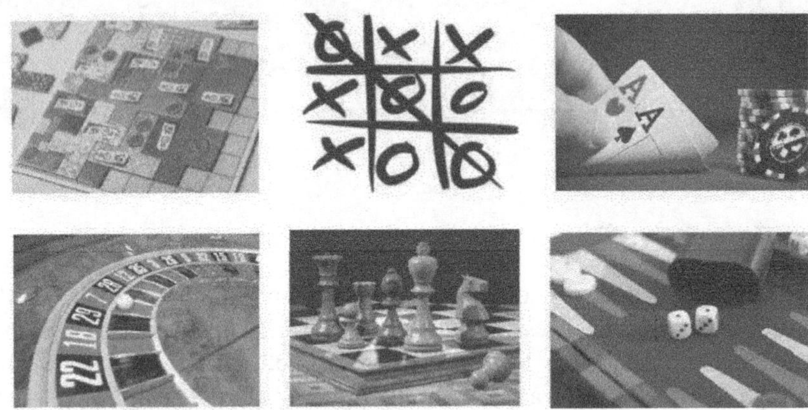

Figure 12.1 Some of the games that have been developed as a software.

To make an evaluation tree for a game is very difficult. Otherwise we can see that such a tree is very huge in size. A move is one of the nodes in this tree selected while playing. Suppose two players are playing the game of Tic-Tac-Toe, then if we wish to draw a graph with the start node having 9 successors and intern having 8 successors each and so on then the entire graph will consist of 9 nodes. In the game of chess if we wish to enumerate all nodes, the graph has 10^{120} nodes. Even if we assume that a node can be generated in $\frac{1}{3}^{rd}$ of a nanosecond, it will take years to generate the entire graph. Similar situation exists for checkers game. For such problems, the graphs along with search methods like BFS, DFS, Best-first-search or A* become inefficient. In most of these cases, it is sufficient to decide the best possible move in a two-player game. To do this, we follow a procedure known as Minimax procedure with two players: MAX (Max) player and MIN (Min) player.

12.1 Minimax Procedure

Minimax or MinMax is a decision rule used in artificial intelligence based games. Most of the games are two-player games that can be visualised in two player status (Figure 12.2). If any one of the ends can be played by computers, then it becomes computer game. In chess, for example, each chessman is black or white. A man who can opt either one of the chessmen for his move. In such cases, there are two players: MAX (Max) player and MIN (Min) player (Figure 12.3). An optimised winning of both MIN and MAX is the objective of this procedure. MAX is supposed to reach the terminal state having highest utility and MIN to the terminal state with lowest utility. The game ends when either MAX or MIN has reached a terminal state. There is a terminal state point obtained that may be awarded or sometimes deducted from the present function value.

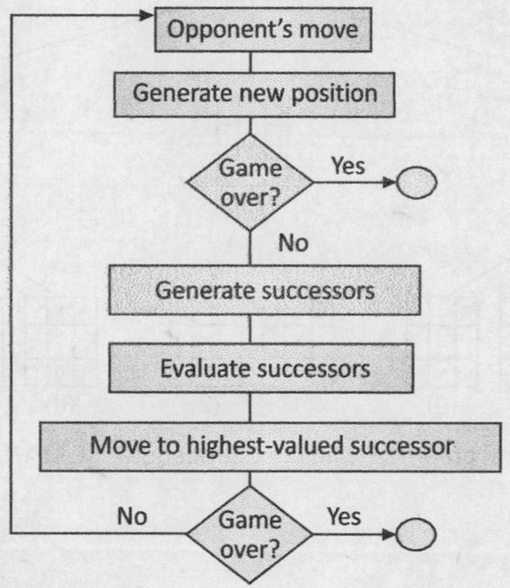

Figure 12.2 Two player game steps.

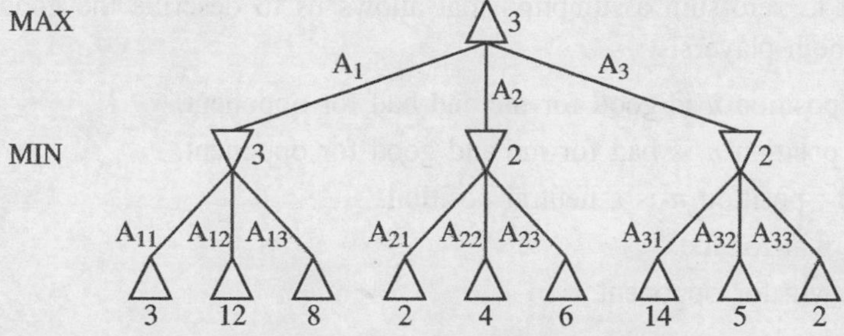

Figure 12.3 Two player game: revisited.

We shall illustrate Minimax procedure with the help of Tic-Tac-Toe game, by finding out the best first move for a MAX player. We also note that a few states are symmetric. For example marking an 'X' at one of the corners has four configurations. But by locating the goal, we note that all are symmetric. Figure 12.4 shows symmetric configurations which are not considered. Figure 12.5 shows the graph up to level two, starting from the empty configuration.

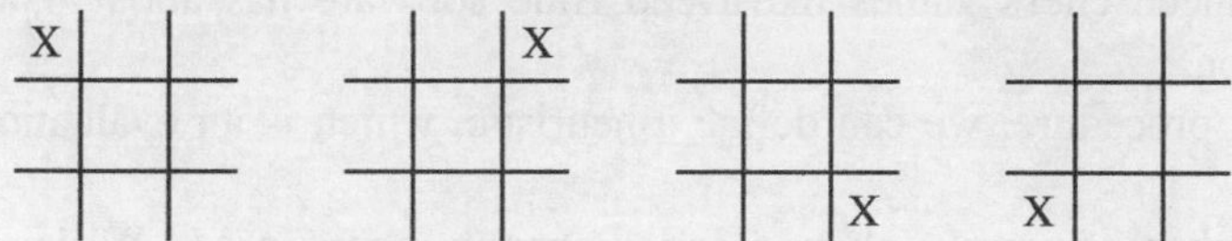

Figure 12.4 Symmetric configuration.

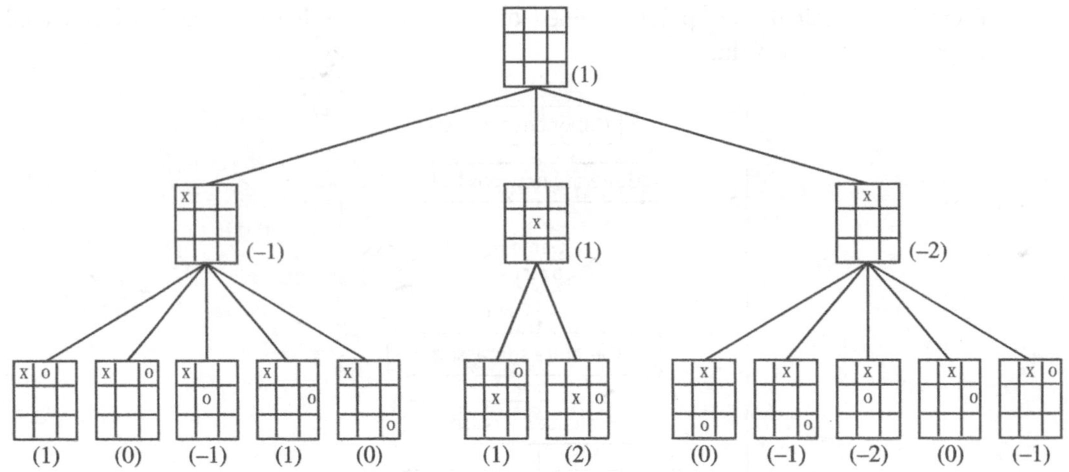

Figure 12.5 Two step graph of Minimax procedure. X-MAX player, O-MIN player.

Evaluation function

The evaluation function or static evaluator for a two-player game assesses the 'goodness' of a game position in contrast with heuristic search where function was a non-negative estimate of the cost from start node to a goal and passing through the given node. A single evaluation function is used to zero-sum assumption that allows us to describe the goodness of a board with respect to both players.

$f(n) > 0$: position n is good for me and bad for opponent.

$f(n) < 0$: position n is bad for me and good for opponent.

$f(n)$ near 0 : position n is a neutral position.

$f(n) \gg 0$: win for me.

$f(n) \ll 0$: win for opponent.

Using the domain expertise, evaluation function is formatted as a heuristic function. For example, Alan Turing's function for chess as follows:

$$f(n) = w(n)/b(n)$$

where $w(n)$ = sum of the point value of white pieces and $b(n)$ is the sum for black pieces. Most evaluation functions are specified as a weighted sum of position features as follows:

$$f(n) = w_1 \times feat_1(n) + w_2 \times feat_2(n) + \ldots + w_n \times feat_k(n)$$

For example, in chess game, piece count, piece placement, squares controlled, etc. are considered. Advanced chess games like Deep Blue software has about 6,000 features in its evaluation function.

In Minimax procedure, we can define a heuristic, which is an evaluation function, given as follows:

$e(p)$ = (number of rows, columns and diagonals open to MAX player) – (number of rows, columns and diagonals open to MIN player)

Among the evaluation function values at level 2, we will find the minimum value and backup the same value to its parent node among the MAX nodes at level 1. We will find the maximum value among them and ascertain that state or configuration is the best move for the MAX player.

To conclude Minimax procedure, we will generate successors up to a certain level and backup the leaf values to their ancestors depending on whether it is a MAX play or a MIN play. In the case of MAX play, we will backup the least value from immediate successor. Among these backup values, MAX player chooses the state, which has the largest backed up value.

Minimax procedure determines optimum strategy for MAX as follows:

1. Entire search tree is generate.
2. A utility function is applied to the terminal states.
3. To determine utility of each node for the upper layer of the tree, utility value of current layer is used.
4. Continue when root node is reached
5. Minimax decision—An assumption is made that MIN will attempt to minimise this utility, so that it maximises the utility for MAX.

The Minimax procedure is given in Algorithm 18. This procedure is good enough only for games with a low branching factor. The complexity of such procedure is high because the game has more number of plies.

In general, the complexity is: $O(m^n)$, where: m = average branching factor in the tree and n = number of plies.

Taking chess as example, on an average, it has 35 branches and 100 moves at the least, resulting in game space of 35^{100}. This is not a realistic game space to search. Also, time being an important factor in this aspect; such game space becomes very undesirable.

Many games have very large search trees. As there is no knowledge of the terminal states, the program has to guess regarding which path to take. Cut-off must be implemented due to time restrictions. So, a function can be used that will, from a given position, return an estimate of the expected utility of the game. This helps the present situation in a way to estimate the next moves value. The quality of evaluation functions determines the performance of a game playing program. To judge the quality of evaluation function, it must agree with the utility functions on the terminal states. It must not take too long (that is, trade-off between accuracy and time cost). The utility function should reflect actual chances of winning. To design the evaluation function, the nature of the game should be identified. Encode the quality of a position in a number representable within the framework of programming language. Then design a heuristic for value to the given position of any object in the game. One of the material advantages of such design is to identify values of the pieces that are judged independent of other pieces on the board. That is, a value is returned based on the material value of the computer minus the material value of the player. For example, a weighted linear function:

$$w_1f_1 + w_2f_2 + \ldots + w_nf_n$$

where w's are weight of the pieces and f's are features of the particular positions. That is, in chess, material value for each piece on the board is worth some value like *Pawns* = 1, *Knights* = 3, etc.

Algorithm 18 MINIMAX algorithm

```
function MINIMAX-Decision (state) returns an action
  v← MAX-Value (state)
  return the action in SUCCESSORS state with value v.

function MAX-Value(state) returns a utility value
  if Terminal-Test(state) then return UTILITY(state)
  v← −∞
  for a, s in SUCCESSOR (state) do
    v← MAX (v, MIN-Value (s))
  return v.

function MIN-Value (state) returns a utility value
  if Terminal-Test(state) then return UTILITY(state)
  v← ∞
  for a, s in SUCCESSOR (state) do
    v← MIN (v, MAX-Value (s))
  return v.
```

Even we are using some evaluation function for a game, which has some disadvantages. Evaluation function can return an incorrect value. That is, if the search is cut-off and the next move results involve a capture, then the returned value may be incorrect. Another problem is Horizon problem. It is a move, that is pushed deeper into the search trees which may result in an oversight by the evaluation function. As the number of possible state is very large, and computers are capable of searching small portion of it, that is, a few plays down the game tree. Thus, in many games, this horizon effect is seen. The computer can search certain number of play. There is a possibility that it will make a move which is detrimental. As the depth of error is not searched the detrimental effect is not visible.

So, how will you overcome such issues? The answer is improving the cut-off of the game tree. This can be visualised by evaluation functions applied to quiescent position. This is a position that is unlikely to exhibit wild swings in value in the near future. Non-quiescent position should be expanding until we reach the decided node. Such an overlooking search is required to identify a node. This extra search is called a *Quiescence search*, which provides more information about any one node in the search tree. Loss of information about the other nodes may be the outcome.

12.2 *Alpha-Beta* Pruning

Straight Minimax procedure requires a two-pass analysis of the search space. First one is desired to play depth and apply a heuristic. Second one propagates values back up in the tree. Minimax procedure pursues all branches in the space, including many that could be ignored or pruned by a more intelligent algorithm. Researchers in game playing developed a class of search technique called *alpha-beta* pruning procedure (Algorithm 19). The *alpha-beta* procedure examines $O(b^{h/2})$ nodes in its average search.

Algorithm 19 $\alpha - \beta$ cut-off algorithm

```
function MAX-VALUE(state, game, α, β) returns the MINIMAX value of state
  Inputs: state, current state in game, game description
  α the best score for MAX along the path to state,
  β the best score for MIN along the path to state
if CUTOFF-TEST(state) then returns EVAL(state)
for each 's' in SUCCESSORS(state) do
  α ← MAX(α, MIN-VALUE(s, game, α, β))
  α ≥ β then returns β
end
return α

function MIN-VALUE(state, game, α, β) returns the Minimax value of state
if CUTOFF-TEST(state) then returns EVAL(state)
  for each s in SUCCESSORS(state) do
  β ←MIN(β, MAX-VALUE(s, game, α, β))
  if α ≥ β then returns α
end
return β
```

The idea for an *alpha-beta* search is simple. The process of eliminating a branch of the search tree starts from consideration without examining it. Rather than searching the entire space to play depth, *alpha-beta* search proceeds in a depth-first fashion. As a part of the search two values *alpha* and *beta*, are created. This is a particular technique to find the optimal solution according to a limited depth search using evaluation functions which returns the same choice as Minimax cut-off decisions, but examines fewer nodes. This got its name from the two variables that are passed along during the search which restrict the set of possible solutions.

The *alpha* value, associated with MAX nodes, and the *beta* value associated with the MIN node. That is, *alpha* value of the best choice so far along the path for MAX and *beta* value of the best choice (lowest value) so far along the path for MIN. For example, a MAX node has *alpha* value 6, then MAX need not consider any backed-up value less than or equal to 6 that is associated with any MIN node. *Alpha* is the worst that MAX can 'score' given a MIN node will also do its 'best'. Similarly, if MIN has a *beta* value of 6, it does not need to consider any MAX node below and has a value of 6 or more. An *alpha-beta* cut-off procedure is given in Figure 12.6.

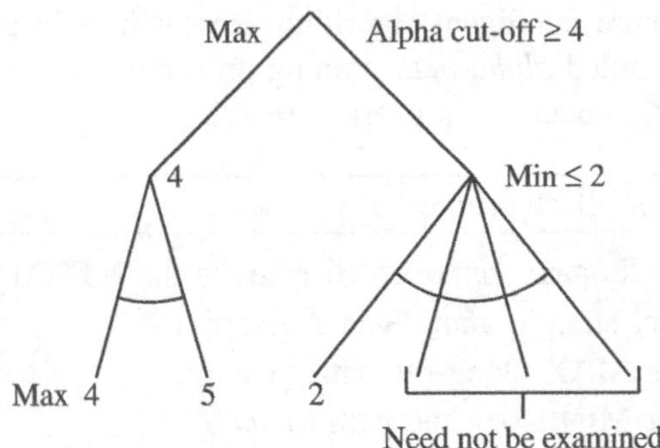

Figure 12.6 $\alpha-\beta$ cut-off example.

Alpha-beta search can be implemented as follows:

1. Set root node *alpha* to negative infinity and *beta* to positive infinity.
2. Search depth first, propagating *alpha* and beta values down to all nodes visited until reaching desired depth.
3. To get the utility of this node, apply evaluation function.
4. If parent of this node is a MAX node and the calculated utility is greater than parent's current *alpha* value, then replace this *alpha* value with this utility.
5. If parent of this node is a MIN node and the calculated utility is less than parent's current *beta* value, replace this *beta* value with this utility.
6. Based on these updated values, it compares the *alpha* and *beta* values of this parent node to determine whether to look at any more children or to backtrack up the tree.
7. In this way, continue the depth-first search, until all potentially better paths have been evaluated.

To begin *alpha-beta* search, we descend to full play in a depth- first fashion and apply the heuristic evaluation to a state and all its siblings. Assume we are applying heuristic in MIN node. The minimum of these MIN values is then backed up to the parent (a MAX node in Minimax tree). This value is then offered to the grandparent of these MINs as a potential *beta* cut-off.

Next, the algorithm descends to other grand children and terminates exploration of their parent if any of their values is equal to or larger than the *beta* value. How will we stop the search process. Two rules are used to terminate the search, based on *alpha* and *beta* value.

1. The search process is terminated when any MIN node has a *beta* value less than or equal to the *alpha* value of any of its MAX ancestors.
2. The search process terminated when any MAX node has an *alpha* value greater than or equal to the *beta* value of any of its MIN node ancestors.

Alpha-beta pruning, thus expresses a relation between nodes at play n and nodes at play $n + 2$, under which the entire subtree rooted at level $n + 1$, can be eliminated from consideration.

Check α–β cutoff in Figure 12.7. Here

A has $\beta = 3$ (A will be not larger than 3)

B is β pruned, since $5 > 3$

C has $\alpha = 3$ (C will be no smaller than 3)

D is α pruned since $0 < 3$

E is α pruned since $2 < 3$ and C is 3

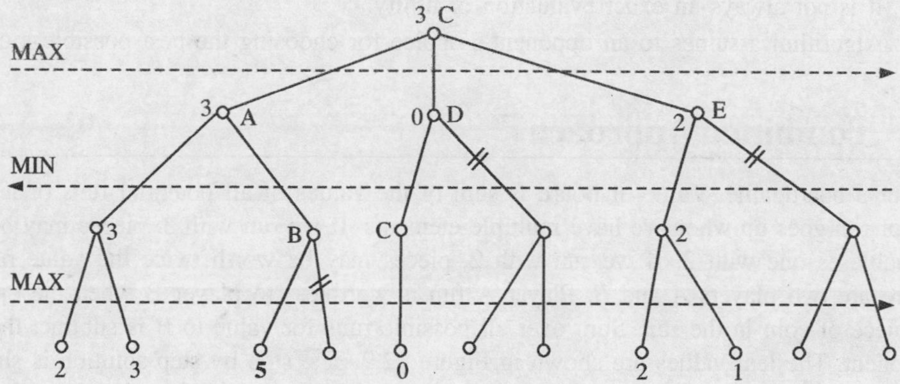

Figure 12.7 *alpha-beta* procedure example.

Example in Figure 12.8 indicates the α–β cut-off. Here

β value = 7

α cut-off is *CI*

β cut-off are *FO*, *IU*, *KY*

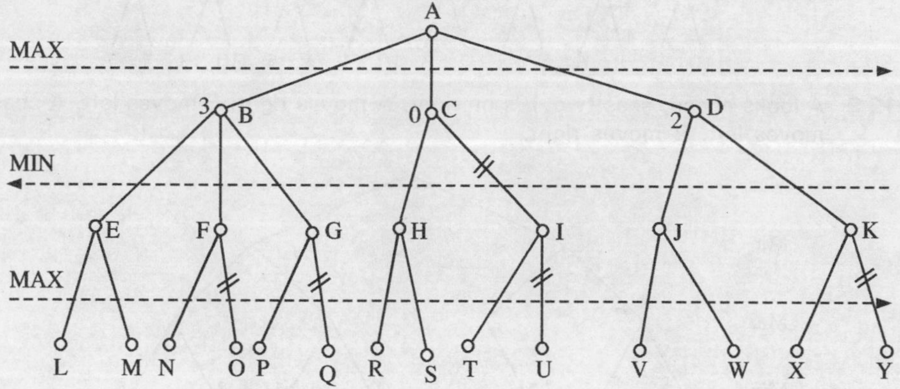

Figure 12.8 *alpha-beta* procedure example.

The effectiveness of *alpha-beta* procedure depends on the order in which the search progresses. If m is the branching factor and n is the depth of the search, best case for *alpha-beta* is $O(m^{n/2})$, compared to the best case of Minimax which is $O(m^n)$.

Major drawbacks of *alpha-beta* procedure includes the following:

- Even if there is only one legal move, *alpha-beta* procedure algorithm will still generate an entire search tree.
- It is designed to identify a 'best-move', not to differentiate between other moves.
- Algorithm overlooks moves that pay something early for a better situation later.
- It is not always an exact evaluation of utility.
- Algorithm assumes to an opponent's choice for choosing the best possible move.

12.3 Combined Approach

Consider a boardgame. Value of board is sum of the values of all potential runs of length 4. Value of run goes up when we have multiple elements. If we run with 3, pieces may be twice as valuable as one with 2. If we run with 2, pieces may be worth twice the value run with 1. There are two players *A* and *B*. Player A run is worthless to player B when the opponent has a piece of coin in the run. Sum over all possible runs for value to B is subtract the value to opponent. The leaf values are shown in Figure 12.9. The step by step solution is shown in Figure 12.10 through Figure 12.12.

Figure 12.9 A looks ahead, sees two 10s on right, A moves right, B moves left, A chases 10, moves left, B moves right.

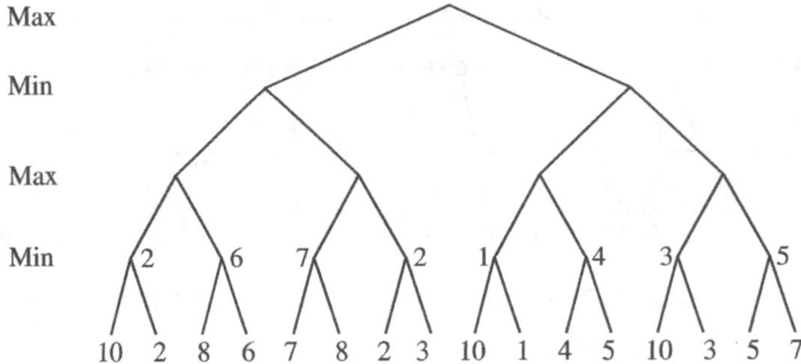

Figure 12.10 A looks at what B's choice would be. Labels node with that value. B wants low values — MIN.

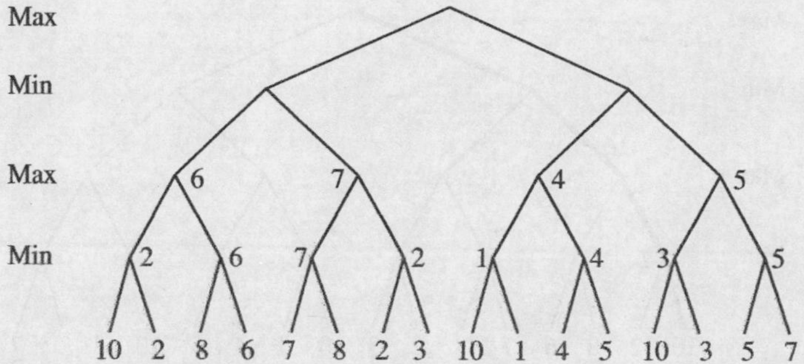

Figure 12.11 A considers which he would prefer. A picks MAX of B's MINs.

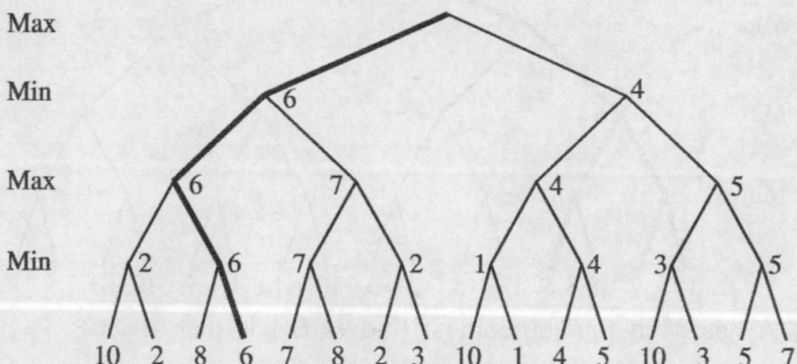

Figure 12.12 Considers B's choices. See A's best choice is 6. Assume both players are rational, and no new information arrives.

How can we improve the search by α–β procedure? Figure 12.13 through Figure 12.19 shows the step by step illustration of how to apply α–β procedure in the current example. Bold lines in the graph is tree walk of each player.

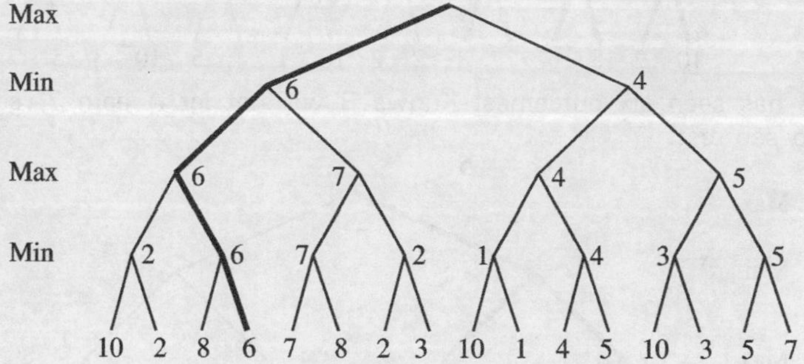

Figure 12.13 A wants MAX (10). B wants MIN (1). A walks tree.

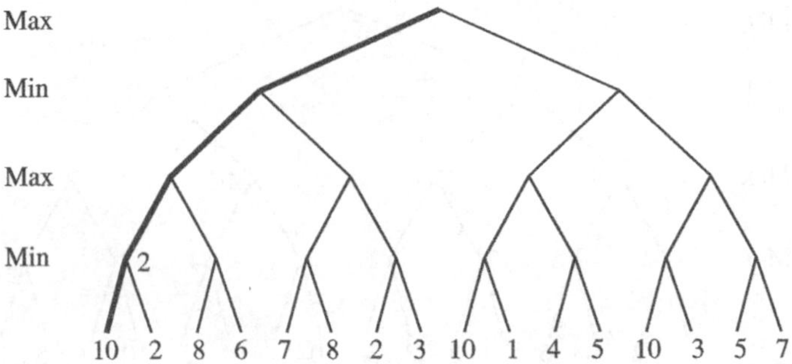

Figure 12.14 A has seen two outcomes. Knows B would pick 2. Labels node with 2.

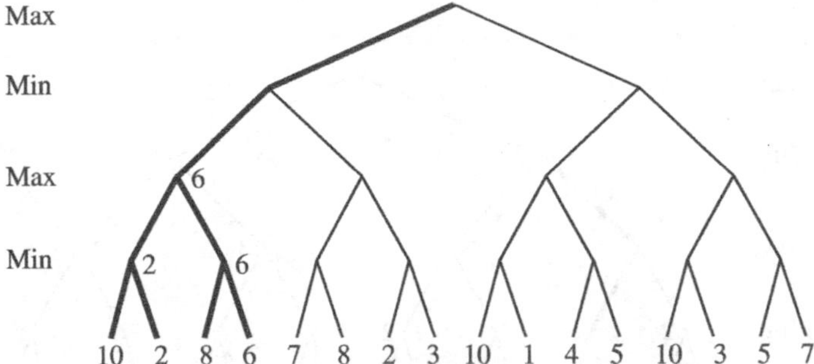

Figure 12.15 A has seen four outcomes. Knows B will pick 6 vs 8. A prefers 6 to 2.

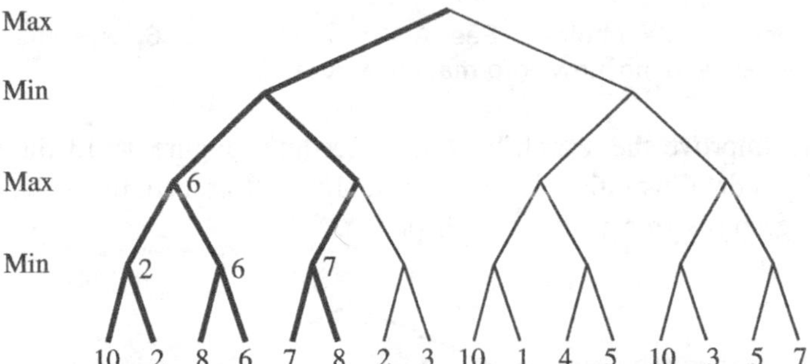

Figure 12.16 A has seen six outcomes. Knows B will not let A earn 7, so does not need to see rest.

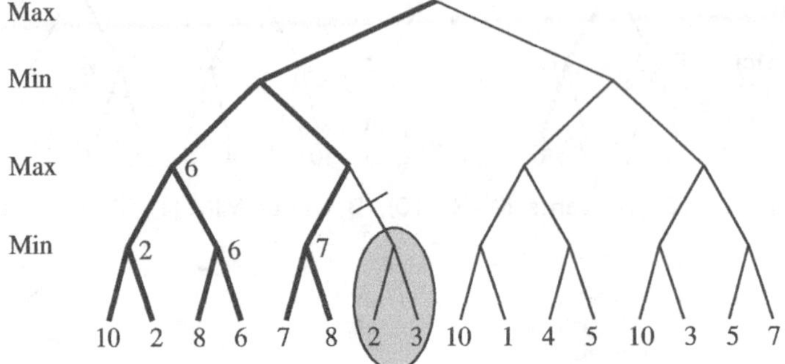

Figure 12.17 We prune the search. B will select 6 rather than give A a chance to earn 7. A has no idea what is down that path.

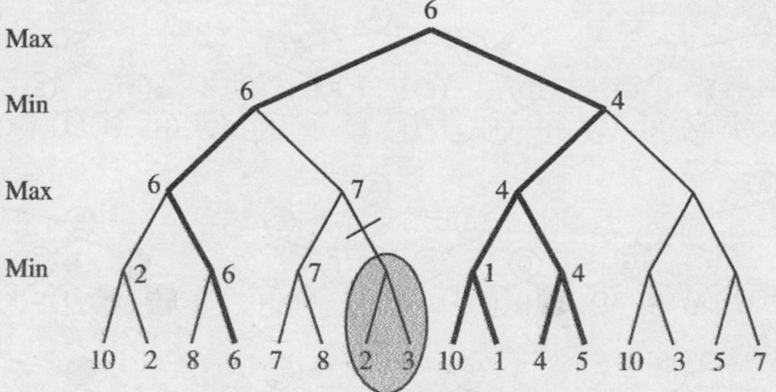

Figure 12.18 A looks at left branch of right tree. Knows B could give him a 4. He prefers a 6.

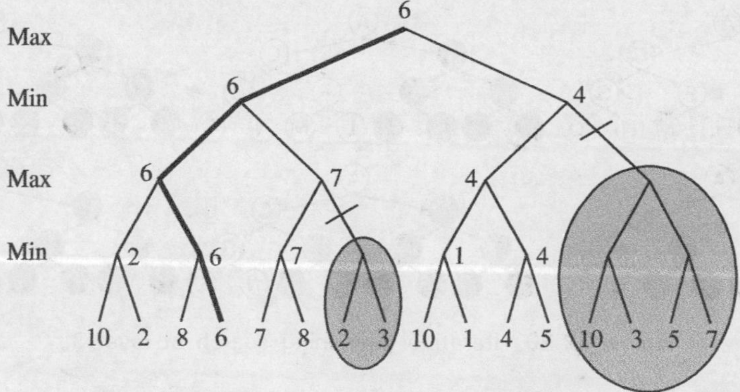

Figure 12.19 A would rather prefer a 6 than 4. Prunes choice has no idea what is there.

12.4 Iterative Deepening

Development of tree search in many games has time constraints. Identification of how long the search to a fixed depth will take is hard to estimate due to pruning. Time complexity is considerable at any point in the algorithm that we would ideally like to provide. Evaluation of choice with increasing depths is one of the solutions. The progressive increase in the bound, Iterative Deepening Search (IDS) performed as a depth limited search. The levels are l = 1, l = 2, l = 3 etc. until a solution is reached (l = *d*). If a solution is found at the finite depth, then branching factor is completed. The shortest solution is obtained if an optimal path is found. The time complexity is $O(bd)$, where b is the branching factor and d is the depth of shallowest solution. The space complexity is $O(bd)$.

There are some overheads of repeated computation in IDS. Iterative deepening search potentially explores non-solution branch of the search tree in d times approximately. When a tree is considered and its bottom layer contains most of the nodes then the search tree with an approximately constant branching factor is required. For example, in a binary tree, there are 512 nodes at level 9 and 1024 nodes at level 10. For example, iterative deepening search at level 3 is given in Figure 12.20.

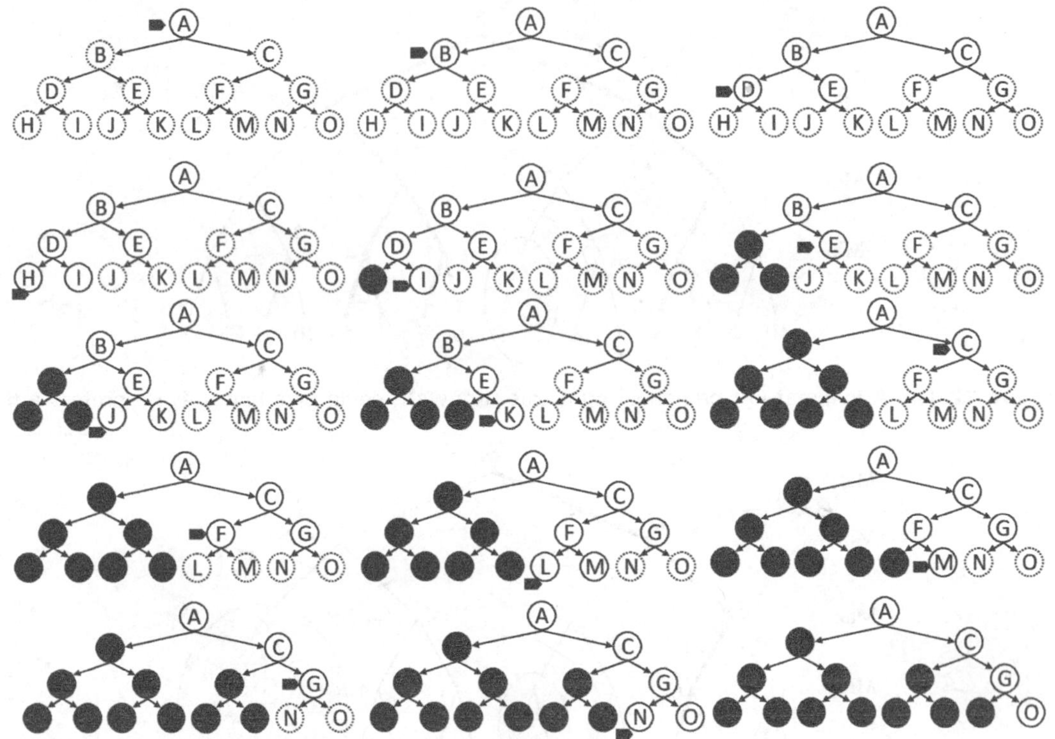

Figure 12.20 Iterative deepening search at level 3.

12.4.1 Iterative Deepening A*

Iterative Deepening A* (IDA*) algorithm uses bound on the cost of path as opposed to the depth. This algorithm is like DFS based on increasing values of total cost. Rather than increasing the cost, total cost becomes $f = g + h$. At each step, DFS cuts off a branch when its total cost exceeds a given threshold, which is set as h(start) initially. During the next iteration, threshold is the lowest cost of all values that exceeded the current threshold. The steps involved in the IDA* are given as follows:

1. Set THRESHOLD to be f(start state).
2. Depth-first search, does not explore any nodes whose f value is greater than THRESHOLD.
3. If no solution is found, increase THRESHOLD and go back to step 2.

IDA* algorithm defines a function for node x in the search space as $l(x) = g(x) + h(x)$ where $g(x)$ and $h(x)$ are the costs of getting to the node and heuristic estimate respectively. Based on heuristic, the solution found by IDA* is optimal. Like A* algorithm, IDA* expands the same number of nodes. Algorithm 20 gives the steps involved in IDA*. In a failed search, step-cost bound is incremented to that node, beaten the prior-cost bound by the least amount. IDA* gives an optimal path by setting cost cut-off. Between two cut-off, there is a gap which is the minimum value that exceeds previous cut-off.

For example, consider the graph in Figure 12.21(a). Step by step search path is given in Figure 12.22 and the final step is given in Figure 12.21(b). Suppose B is the source node, iteratively the trees expanded with the values f_{new} using the queue (Q) and stops when the goal (node D) is reached.

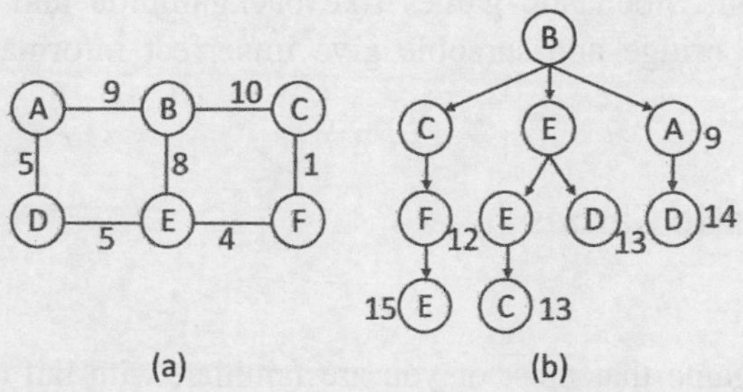

Figure 12.21 (a) Example graph (b) Final search path IDA* search from node B.

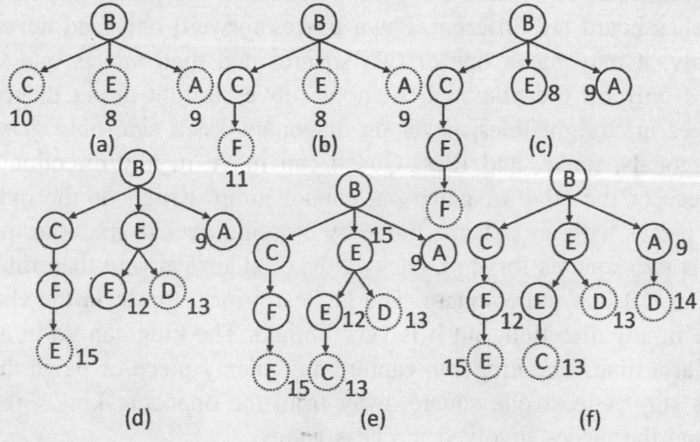

Figure 12.22 Steps in iterative deepening A* search from node B.

Algorithm 20 Iterative Deepening A*

Q ←path only containing root
f_{bound} ← natural number
f_{new} ← ∞
while Q is not empty and goal is not reached do
  Remove first path from Q
  Create paths to children
  Reject paths with loops
  Add paths with f(path) ≤ f_{bound} to front of Q (depth-first)
  f_{new} ←minimum({f_{new}} ∪ {f(P) |P is rejected path}
if goal reached **then** success **else** report f_{new}

12.5 Case Studies

'A game that is easy enough for a toddler to learn, but can be complex enough to stump a genius.' The games like chess, checkers, Go, Othello and, connect-4 are deterministic and give perfect information. Stochastic games like backgammon and monopoly give perfect information, but poker, bridge and scrabble give imperfect information, when the game is over.

12.5.1 Deterministic Games

Chess

Chess is an intelligent game that most of you are familiar, with but did you know that chess has been played for over 5,000 years? The chessboard is 8 × 8 tiles. The pieces in a chessboard are king, queen, bishop, Knight, Rook, and Pawn. This is also a two-player game which closely resembles a checkerboard, represents the battlefield which the pieces fight on. The moves of each piece in a chessboard are different. Pawn moves forward only and never backwards. On its first move only, it may move one or two squares and then moves one square at a time, capturing squares only on the diagonal. Bishops move straight along diagonals and do not jump. Rook moves in straight lines, never on diagonals. Each side only gets one queen who moves along diagonals, ranks, and files. Queen can move in any direction until encounters another chess piece or the edge of game but cannot jump. Knight is the only piece that can jump over other pieces with an L-shape move by crossing over to the other piece. Always the knight moves first two squares forward, backwards, or sideways, and then finishes the L-shape by moving right or left one more square. The highest priority piece in the chess, i.e. king can move one square in any direction, but it is very limited. The king can go in any direction, but only one square at a time. The king can capture any enemy piece or pawn that is undefended and must always stay at least one square away from the opposing king. Figure 12.23 shows the chessboard and the pieces involved in chess game.

Figure 12.23 Chessboard and pieces in the game. King, queen, bishop, knight, rook and pawn.

Initially Minimax procedure was used for the best possible move, assuming that the opponent has chosen the best move. Iterative deepening algorithm was also used for the development of chess game. We heard about the victory of deep blue super computer against the world chess champion Gary Kasparov in 1997. There was a massive parallel architecture with special purpose hardware for chess. The software is learned from many chess masters and grown by different evaluation functions for different positions. It uses heavily loaded endgame databases.

Checkers

Checkers (Draughts) is a two-player boardgame that involves diagonal moves of uniform game pieces and compulsory captures by jumping over opponent pieces. The board size may vary based on countries (example, 10 × 10 and even 12 × 12 board). In American checkers, pieces can move only forward and in Chinese checkers 2 to 6 players can play the game. Figure 12.24 shows checkers gameboard.

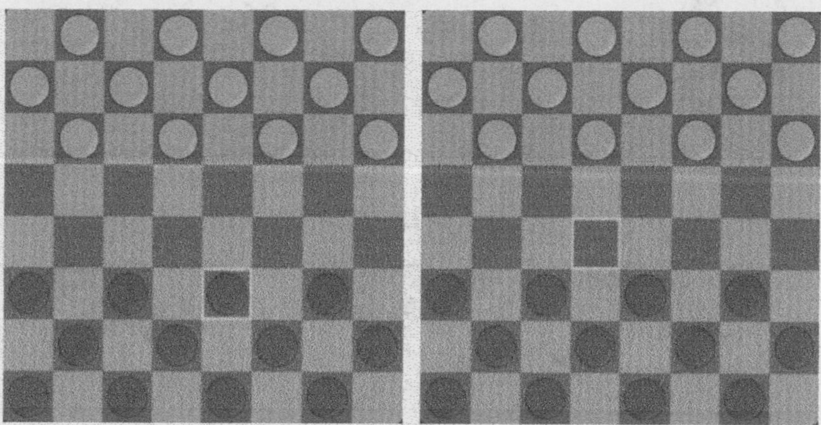

Figure 12.24 Checkersboard and pieces in the game with moves.

Go Game

Go game originated in ancient China 2,500 years ago as a two-players game with a 19 × 19 board. Beginners play on small 9 or 13-line boards. Figure 12.25 shows the traditional Go gameboard with its pieces. Using stones pieces and their turn by place, a stone is passed. The goal is to make more territory than the opponent and the game ends when both players pass. The game ends when all the walls defining territory is taken, and there are no plays left to move. There are two basic rules: capture rule and ko rule. In capture rule, stones that have no liberties, means captured a stone and removed from board. By ko rule, a player is not allowed to make a move that returns the game to previous position. The two rules are given in Figure 12.26. The winning chance is based on the position of stones in the board (4× 4) with position as mentioned in Figure 12.27.

Figure 12.25 Go game board and stones.

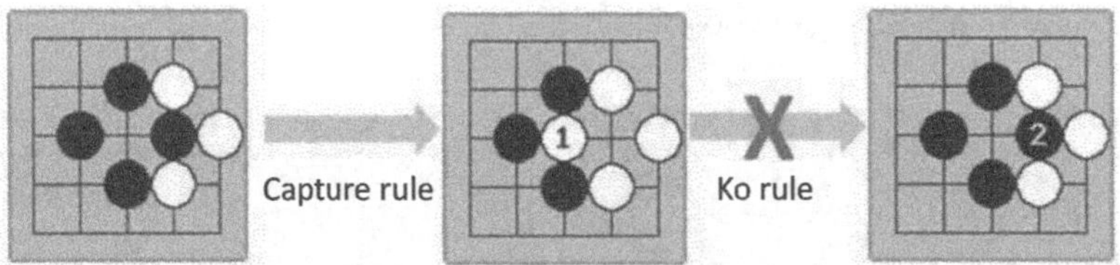

Figure 12.26 (a) Capture rule (b) Ko rule.

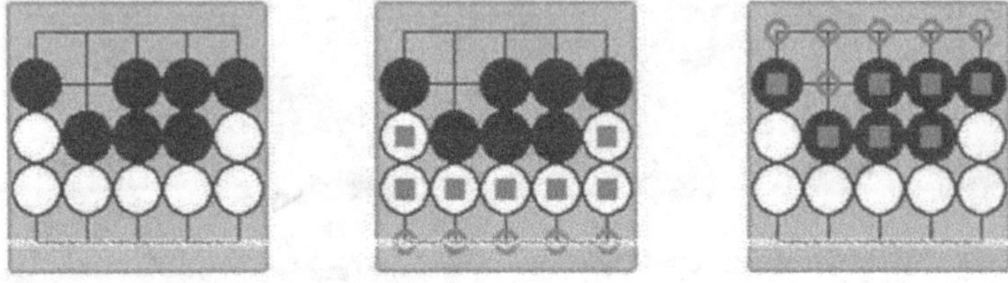

Figure 12.27 (a) Final position for board size 4 × 4 (b) White score: 12 (c) Black score: 13.

When compared to chess, Go game is easier to learn, harder to master but has unlimited possibilities. It is very hard to develop an intelligent program for Go game because possible configuration of board is extremely high $\sim 10^{700}$. It is impossible to use brute force exhaustive search for such a game. In chess game ($b \approx 35$, $d \approx 80$) but Go ($b \approx 250$, $d \approx 150$). The branching factor and value function are main challenges. In a chess game after the first two moves, there are 400 possible next moves but in Go, there are close to 130,000. AlphaGo is a computer software developed by DeepMind Technologies. AlphaGo works by combining two machine learning techniques—Monte Carlo Tree Search (MCTS) and Deep learning. Using the advantage of MCTS over Minimax-search approaches, improves the existing computer Go programs. Evaluate the board positions and the selection of next move will be decided by deep learning method. AlphaGo is built as a hardware that uses 40 search threads, 48 CPUs, and 8 GPUs.

Connect-4

It is a two player game with 7 × 6 rectangular board placed vertically. There are 21 red, 21 yellow pieces and players alternately by drop a piece into one of the seven columns and the piece falls down to the lowest unoccupied square. A player wins if he connects four pieces vertically, horizontally or diagonally. If the board is filled (42 pieces played) and no player has aligned four pieces, the game ends in a draw. Figure 12.28 gives the board and pieces in connect-4 game. There are some utility functions of this game. The odd square is a square belonging to an odd row (1,3,5) and even square is a square belonging to an even row (2,4,6). Threat is a group of three pieces of the same colour which has the fourth square empty and the square, below the empty square, is also empty. There are odd and even threats. The odd threat is a threat in which the empty square is odd and even threat is a threat in which the empty square is even. First moves assumed by red which has an odd threat and black cannot connect four pieces anywhere else red will win. A yellow has an even threat and black cannot connect four pieces anywhere else yellow will win. Placing a piece at the centre is an advantage of the game.

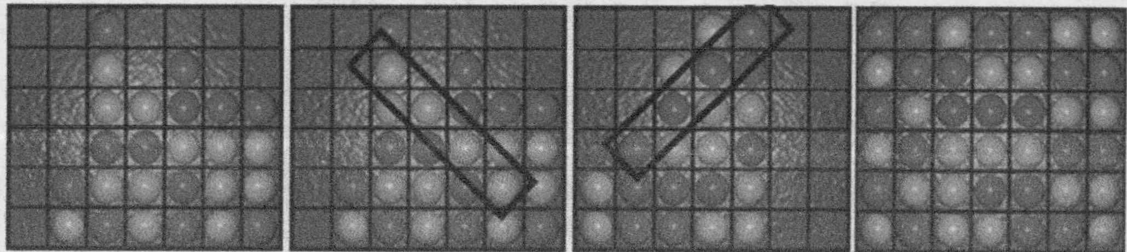

Figure 12.28 (a) Connect-4 gameboard and pieces (b) Win for yellow (c) Win for red (d) Draw.

12.5.2 Stochastic Games

Backgammon

Same as chess and Go, Backgammon is one of the oldest games in existence, probably about 5,000 years old and may have well originated in Mesopotamia. The gameboard is given in Figure 12.29. This also is a two-player game on a board consisting of twenty-four narrow triangles called points. The triangles alternate in colour and are grouped into four quadrants of six triangles each. The players' homeboard and outboard are divided into quadrants which are separated from each other by a ridge down the centre of board called the bar. Board is arranged by numbering the point for either player starting in that player's homeboard. The 24th point is the outermost point which is also the opponent's one point. Each player has fifteen checkers of his own colour. At the initial arrangements on the board, two on each player's twenty-four point, five on each player's thirteen point, three on each player's eight point, and five on each player's six point. Objective of the game is to move all your checkers into your own homeboard and then bear them off.

The game starts by two players throwing a single die to determine the player to go first. If equal numbers come up, then both players roll again until they roll different numbers. Movement of the die is decided through the roll of dice. The game has a high branching factor: 21 possible dice outcomes, ~ 20 different moves per dice role on average.

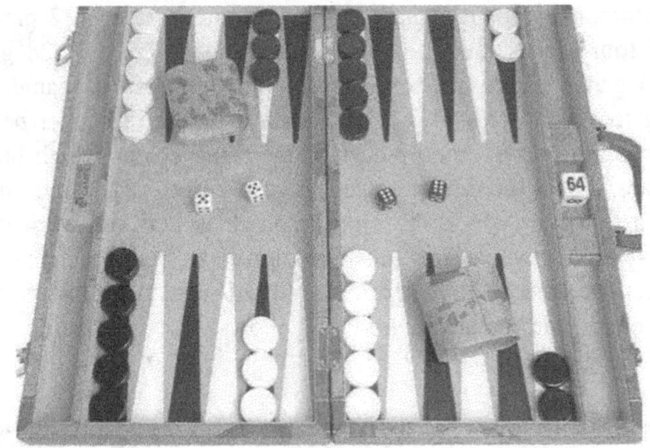

Figure 12.29 Backgammon gameboard.

Poker

This game is recognised as a challenging problem in artificial intelligence, since 1992. It is a multi-players gambling game. There is some hidden information like other player's cards. There exists uncertainty about future events, a deceptive strategies needed in a good player. For a player, given 5 cards and the goal is to obtain the best hand. Possibilities are no pair, one pair, two pair, straight, flush, full house four of a kind, straight flush and royal flush. One pair means two cards of the same face and two pair means two sets of two cards of the same face. Three of a kind is tree cards of the same face and four of a kind is four cards of the same face. The straight means all five cards sequentially ace is in either high or low. The flush state gives all five cards of the same suit and full house gives three of a kind of one face and a pair of another face. In straight flush both a straight and a flush. In royal flush, a straight flush is open, that is, cards 10, J, K, Q, A. Three different stages are given in Figure 12.30.

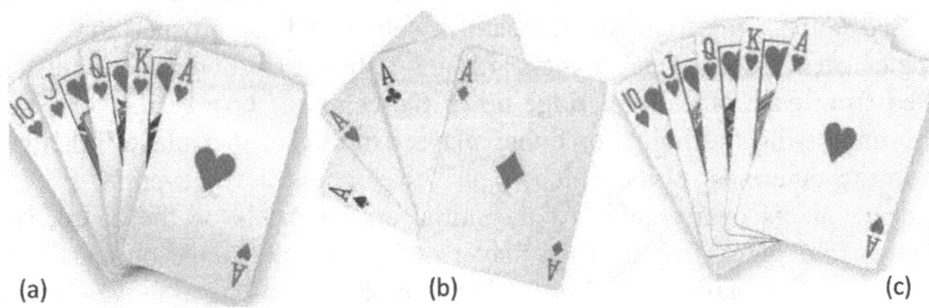

Figure 12.30 (a) Royal flush (b) Four of a kind (c) Straight flush.

12.6 Exercises

1. Examine the following versions of the NIM-game:
Consider a heap with three and a heap with two matches. Two players, MAX and MIN, drag alternate one or more matches from one of the two heaps. MAX moves first. The player who has to move next wins and no more matches are available (vice versa: the player looses who has to take the last matches).
   (a) Illustrate the whole search space of this game.
   (b) Evaluate the 'win-situation' for MAX with +1 and MIN with –1. For all nodes, compute the values of parent nodes.
   (c) Can you identify a perfect strategy to win the game? If so, describe this strategy. If not, why?
   (d) Assume, MAX can only project the results of two moves. Specify a function that leads to the same situation of the game. Name disadvantages of this function.
   (e) Assume the following situation; MAX is allowed to move next, the projection of the next four moves results in the following situation (Figure 12.31):

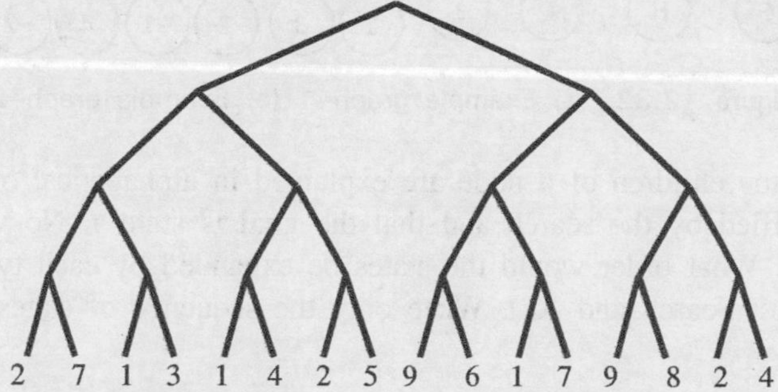

Figure 12.31 Example graph–NIM game.

   (a) Identify the parent value for each node. Which move does MAX select next? Which leaf node does MAX expect to reach?
   (b) Apply the *alpha-beta* strategy to this tree. In which order are the values evaluated? Which parts of the tree can be deleted?

2. Tic-Tac-Toe is a game for two players. The board is a square of 3 × 3 fields. Each player is assigned a type of token (O or *X*). Initially the board is empty. The players play in turn and place a token on an empty Field. A player wins, if he has first aligned three of his tokens either in a row, a column, or on one of the two diagonals. The game ends when a player wins or when there are no more empty fields.
   (a) Draw the search tree for Tic-Tac-Toe up to level 2. Take into account the symmetric game states, that is, those states that can be transformed into each other by rotation and mirroring.

(b) Give an estimation on the number of possible different Tic-Tac-Toe games.

(c) Develop an evaluation function for the Tic-Tac-Toe game.

(d) Indicate the values of your evaluation function for each node of the search tree at level 2. Use these values to compute the values of nodes on level 1 and 0 using the Minimax algorithm.

(e) Indicate all nodes in the search tree that would not have been considered when using *alpha-beta* pruning.

3. Consider the tree shown in Figure 12.32(a). Numbers on the arcs are the arc lengths; the numbers near states *B*, *C*, and *D* are the heuristic estimates; all other states have a heuristic estimate of 0.

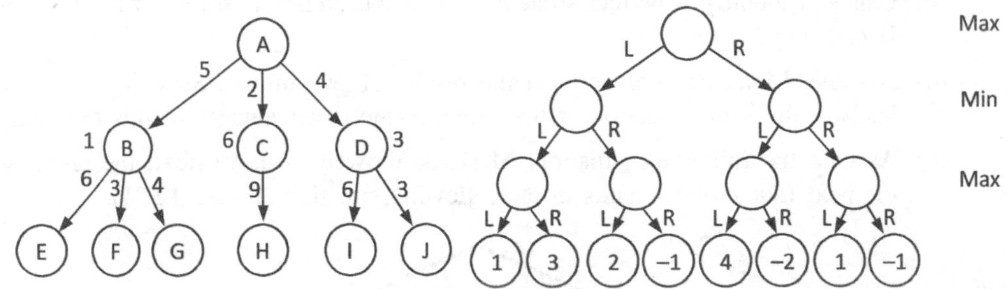

Figure 12.32 (a) Example graph–1 (b) Example graph–2.

Assume that the children of a node are explained in alphabetical order when no other order is specified by the search and that the goal is state *J*. No visited or expanded lists are used. What order would the states be expanded by each type of search (DFS, BFS, Best-First Search and A*). Write only the sequence of states expanded by each search.

4. Consider the game tree shown in Figure 12.32(b). Assume the top node is MAX node. The labels on arcs are the moves. The numbers in bottom layer are the values of different outcomes of the game to MAX player.

(a) What is the value of the game to the MAX player?

(b) What first move should the MAX player make?

(c) Assume that the MAX player makes that move, what is the best next move for the MIN player? Assume that this is the entire game tree.

5. In Figure 12.33(a) game tree, are there any *alpha-beta* cut-off?

(a) Considering the nodes from left to right, which nodes are cut-off? Circle the nodes that are not examined and label them with L.

(b) Considering the nodes from right to left, which nodes are cut-off? Circle the nodes that are examined and label them with *R*.

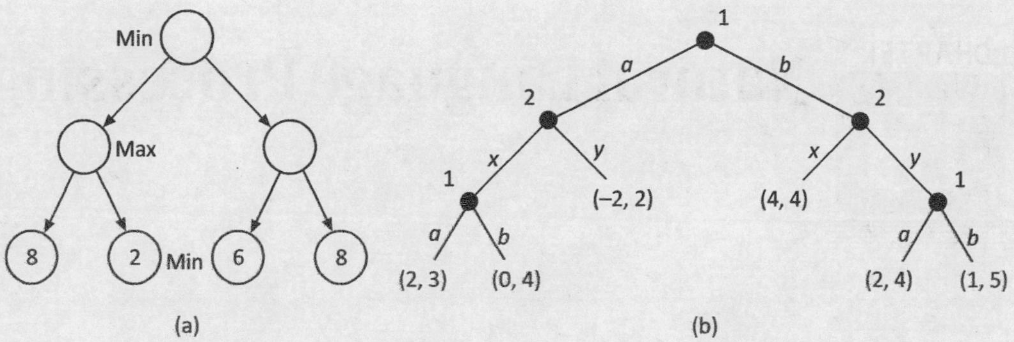

Figure 12.33 Example graph (a) Game tree–1 (b) Game tree–2.

6. Let G be a two-player zero-sum game that has a Nash equilibrium. Show that if player 1's payoffs are increased in such a way that the resulting game G_0 is also a zero-sum game then G_0 has no Nash equilibrium in which player 1 gets a lower payoff than he got in the Nash equilibria of G.
7. Consider the two player extensive form game defined by the following game tree (Figure 12.33(b)). Specify the induced strategic game. Determine all Nash equilibria and identify all non-credible threats.
8. If both min and max play a perfect game, who will win? Explain your answer.
9. The game starts with a single stack of 7 tokens. At each move a player selects one stack and divides it into two non-empty, non-equal stacks. A player who is unable to move loses the game. Draw the complete search tree.
10. If a terminal state in the search tree developed above is a win for min, a utility function of zero is assigned to that state. A utility function of 1 is assigned to a state if max wins the game. Apply the minimax algorithm to the search tree to assign utility functions to all states in the search tree.

CHAPTER 13

Natural Language Processing

Language is the most valuable single possession of human race. It is passed from generation to generation through cultural transmission. The society, culture, literature, art forms and everything that human civilisation possesses, is because of language. Speech and writing are the two main media through which language exist. Speech is the original medium for which human language evolved. When we become literate in a language, we learn to convert a written message(text) to spoken message(speech) and vice versa. We learn to read (text to speech conversion), understand and write (speech to text conversion). The artificial techniques for converting text to speech and speech to text comes under the discipline of speech and language processing.

Language is the means by which human beings communicate with each other and with the society. Many other creatures are also able to communicate with other members of their species. But there are certain unique properties for human language, including displacement, arbitrariness, productivity, cultural transmission, and duality. These features make the human language creative and provide infinite possibilities of expression.

The most remarkable feature of language is the way in which a child acquires first language within the age of 3 or 4 years. By that time almost all children, are able to form, the structure of the language being spoken, in the community where they live. Unlike other skills which each of us learns to varying degrees, every person masters their first language and generally can speak in that language with considerable proficiency, without making grammatical mistakes. According to Noam Chomsky, there is a language faculty that is, there is some part of the mind-brain, which is dedicated to the knowledge of and use of language. It is a kind of language organ, roughly analogous to the visual system.

The first language is not actually learned, it is acquired. Language acquisition is more like a growth process with the right simulation around, similar to the development of visual system. During language acquisition, the structure and grammar of the language grows inside the mind of a child.

John : "Open the side wide doors, Maya"

Maya : "I am sorry, John. I am afraid I cannot do that"

John : "What is the problem?"

Maya : "I think you know what the problem is just as well as I do"

John : "I do not know what you are talking about"

Maya : "I know that you and Salim were planning to disconnect me, and I am afraid that is something I cannot allow to happen"

The above conversation is in natural sentences which may have emotions, intelligence and plans. General speech needs language understanding and language generation capabilities. There is politeness, self-awareness and belief ascription required in natural languages. Politeness means emotional intelligence and self-awareness means a model of self, including goals and plans. The belief ascription means modelling others; reasoning about their goals and plans.

Natural language is the communication medium of human beings. There are many script oriented and scriptless languages used as a communication medium in different geographical locations. These languages propagate to exchange the idea. One of the challenging tasks in artificial intelligence is understanding of human natural languages. Natural Language Processing (NLP) is the area of artificial intelligence that extracts meaning from human language to make decisions based on the information. For example, 'google translator' is a natural language processing application.

Natural language processing is the process of computerised analysis of a human readable language input and conversion of this input into a useful form of representation. Primary concern of the NLP is getting computers to perform useful and interesting tasks with human languages and secondary concern is helping to get a better understanding of human language. The input/output of an NLP system can be a written text or speech.

In natural language processing there are mainly two components—natural language understanding and natural language generation. Natural language understanding pipes some spoken/typed sentence and working out what it means. Natural language generation spool some formal representation of what you want to say and working out a way to express it in a human language such as English. The field of natural language processing has a primary concern of computer performance that is useful and interesting task with human languages. The secondary concern of natural language processing is to facilitate a better understanding of human languages. The output of natural language processing is either speech or text. Natural language understanding and natural language generation are interconnected, as shown in Figure 13.1.

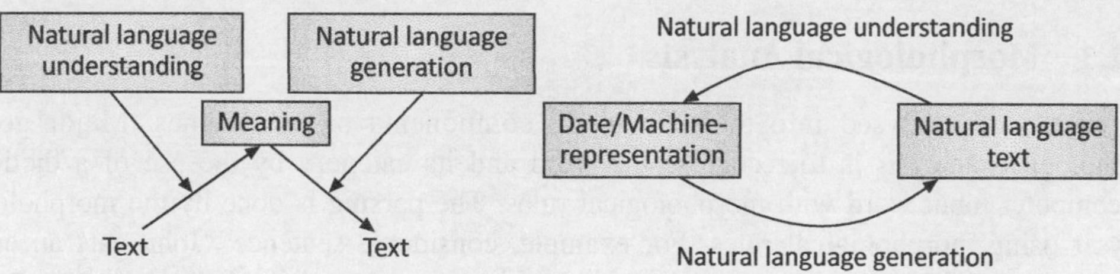

Figure 13.1 Relationship between natural language understanding and natural language generation.

13.1 Aspects of Language Processing

For a natural language there are some aspects such as phonology, morphology, syntax, semantics, pragmatics and discourse. Phonology is concerned with how words are related to the sounds that realise them. Morphology is concerned with how words are constructed from more basic meaning units called morphemes. A morpheme is the primitive unit of meaning in a language. Morphological analysis is also called as lexical analysis. Syntax is concerned with how different words can be put together to form correct sentences and determines what structural role of each word plays in the sentence and what phrases are subparts of other phrases. Syntax of a sentence deals with structure, phrase or grammatical structure of the sentence. Semantics is concerned with what words mean and how these meaning combine in sentences to form sentence meaning. The semantics studied context-independent meaning of a sentence. Pragmatics of a sentence deals with how sentences are used in different situations and how their use affects the interpretation of sentence. Discourse of sentence gives how the immediately preceding sentences affect the interpretation of next sentence. That is, relationship of the sentences. For example, interpreting pronouns and interpreting the temporal aspects of the information. Steps for the understanding of natural sentence is given in Figure 13.2. The detailed way of natural language processing is given in following sections.

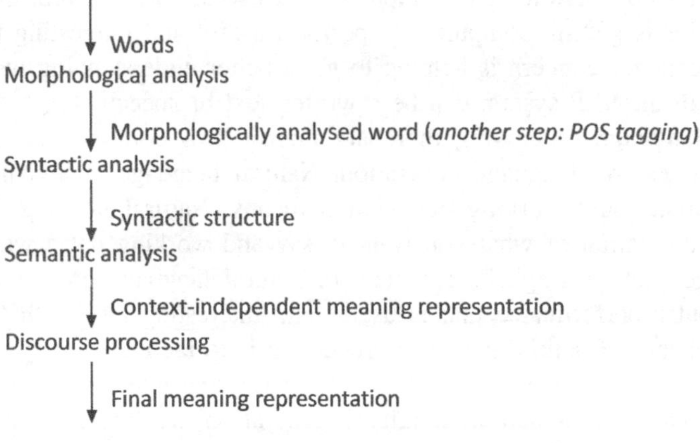

Figure 13.2 Natural language processing.

13.1.1 Morphological Analysis

The words are analysed into their linguistic components or morphemes. Major goal of morphological analysis is to recognise the word and its category by the use of a dictionary and computer input word with morphological rules. The parsing is done by the morphological analyser using morphological rules. For example, consider a sentence, 'John eats an apple'. The proper morphological rules are applied on this sentence and if an error occurs then we have to understand that the sentence is wrongly used. Morphological analysis is relatively simple for English language but for some languages such as Turkish, it is more difficult. Finite-state transducers are commonly used tools for morphological analysis.

POS tagging

Part-Of-Speech (POS) is used to describe the category of a word from a major word group or its subgroups. POS has open classes like noun, verb, adjective, adverb and closed classes like prepositions, determiners, conjunctions, pronouns, participles. Consider a sentence, 'Duck is delicious for dinner'. A morphological analyser cannot take a decision for some words (for example, duck is a verb or noun?). The POS tagger may make that decision by looking at the surrounding words. POS tagging in English is being discussed here.

Part-of-speech annotates each word in a sentence with a part-of-speech marker with lowest level of syntactic analysis. It is useful for subsequent syntactic parsing and word sense disambiguation. The Brown corpus used a large set of 87 POS tags but today's NLP system uses the Penn treebank set containing 45 tags. This is a reduced form of Brown set. The British National Corpus (BNC) has 61 tags. We have to understand the POS tags with examples. This is for understanding the structure of a sentence for language generation. The following texts give POS tags with examples:

Part-of-speech noun represents a person, place or thing. For example, Singular—NN (for example, dog, ball), Plural—NNS (for example, dogs, balls), Proper—NNP, NNPS (for example, John, Anand), Personal pronoun—PRP (for example, I, you, he, she, it) and Wh-pronoun—WP (for example, who, what).

Part-of-speech verb are actions and processes. For example, base infinitive—VB (for example, eat), Past tense—VBD (for example, ate), Gerund—VBG (for example, eating), Past participle—VBN (for example, eaten), Non 3rd person singular present tense—VBP (for example, eat), 3rd person singular present tense—VBZ (for example, eats), Modal—MD (for example, should, can) and To—TO (for example, to eat).

Part-of-speech adjective are modifying nouns like Basic—JJ (for example, red, tall), Comparative—JJR (for example, redder, taller) and Superlative—JJS (for example, reddest, tallest). The adverb is modifying verbs like basic—RB (for example, quickly), comparative—RBR (for example, quicker) and superlative—RBS (for example, quickest). Preposition—IN (for example, on, in, by, to, with). Determiner are basic—DT (for example, a, an, the) and WH-determiner—WDT (for example, which, that). Coordinating Conjunction—CC (for example, and, but, or) and Particle—RP (for example, off, up).

Basic POS tagging approaches are rule-based, learning based, statistical models and neural network based models. The rule based model is human crafted rules, based on lexical and other linguistic knowledge. In learning based model trained on human annotated corpora like the Penn treebank. In statistical models, Hidden Markov Model (HMM), Maximum Entropy Markov Model (MEMM), Conditional Random Field (CRF), etc. are widely used. HMM has a promising role in natural language processing. Recurrent networks like Long Short Term Memory (LSTMs) are the type of neural networks used in POS tagging.

Lexical processing

How will we identify the meaning of individual words? A lexical processing is used to determine meanings of individual words. A lexicon can be used to lookup in a database of meanings with the identification of non-words such as punctuation marks or question marks. Lexical anlysers identify word level ambiguity (words may have several meanings) and the correct one cannot be chosen based solely on the word itself. The ambiguity is removed by POS tagging or passing on the ambiguity to other levels.

13.1.2 Syntactic Parsing

Parsing is the process of converting a flat input sentence into a hierarchical structure that corresponds to units of meaning in the sentence. In syntax parsing there are different parsing formalisms and algorithms which have two main components, i.e. a grammar and a parser. Grammar is declarative representation describing the syntactic structure of sentences in the language. A Context Free Grammar (CFG) is a 4-tuple, $G = (V, T, P, S)$. Where V is the set of non-terminal symbols or 'variables'. T is the set of terminal symbols which is disjoint from V. P is the set of production rules, each of the form $A \rightarrow \alpha$. Here A is a non-terminal and α is a string of symbols from the infinite set of strings $(V \cup T)$. S is the start symbol. According to Noam Chomsky, the famous linguist, there are four types of grammars—regular (type 3), context free (type 2), context sensitive (type 2) and unrestricted grammar (type 0). Here we are dealing with context free grammars.

A parser is an algorithm that analyses the input and outputs its structural representation (its parse) consistent with the grammar specification. Context free grammars have many of the parsing mechanisms, but complemented by some additional features that make the formalism more suitable to handle natural languages. Grammars specify possible organisation of words in sentences and allows to determine sentence structure(s). Parsing is the process of taking a string and a grammar followed by returning one or many parse tree for that string. There are top-down and bottom-up parsing techniques. In top-down parsing, start at the root of parse tree and grow toward leaves. A top-down parser starts with the root of the parse tree. The root node is labelled with the goal symbol of the grammar. In bottom-up parsing, start at the leaves and grow toward root. A bottom-up parser builds a derivation by working from the input sentence back toward the start symbol. In parsing, a sentence is corrected according to the grammar returning a parse tree representing the structure of sentence. For example, consider the following grammar rules:

Sentence → Noun_phrase, verb_pharase
Noun_phrase → Proper_noun
Noun_phrase → Determiner, noun
Verb_phrase → Verb, noun_phrase
Proper_noun → [John]
Noun → [apple]
Verb → [eats]
Determiner → [An]

This grammar can also be represented as follows:

S → Np, Vp.	Det → [A]	Det → [The].	Det → [An].
Np → Det, noun.	Noun → [Apple].		Noun → [Mango].
Np → Proper_noun.	Proper_noun → [John].		
Vp → V, Np.	Proper_noun → [Maya].		
Vp → V.	V → [Eats].	V → [Likes].	

Syntax parsing of the sentence, 'John eats an apple' is given in Figure 13.3. The top-down and bottom-up parsing are given in the figure. Here s—sentence, det—determiner, Np—noun phrase, Vp—verb phrase and V—verb. The syntax parsing using the above grammar rule is also given.

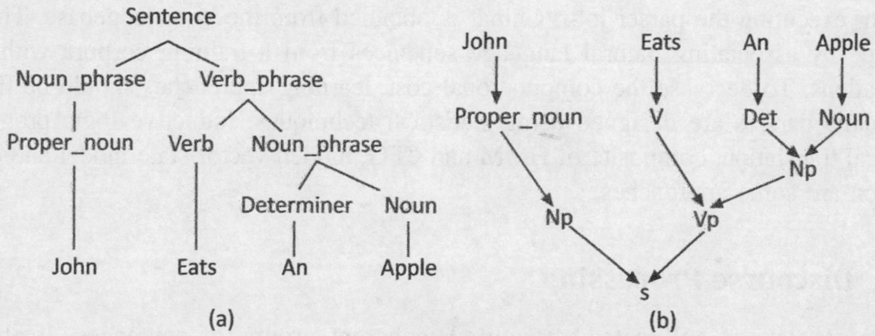

Figure 13.3 Two different representations of syntactic parsing of the sentence.

13.1.3 Semantic Parsing

This step assigns meanings to the structures created by syntactic analysis by mapping of words and structures to particular domain objects in a way consistent with our knowledge. Semantic can play an import role in selecting among competing syntactic analyses and discarding illogical analyses. For example, 'I robbed the bank'. In this sentence, the word 'bank' means either side of river or a financial institution. An appropriate decision should be taken from the sentences based on the formalisms which will be used in the meaning representation. From the syntactic structure of a sentence, partial meaning or representation is generated. For the sentence 'John eats an apple', semantic structure is generated and shown in Figure 13.4.

Due to the complexity of task semantic parsers a manual processing is carried out by learning automatically from the paired sentences with their logical form. Consider the following sentences 'Which country has the highest CO_2 emissions?'
'Which had the highest increase since last year?'
'What fraction is from the five countries with highest GDP?'

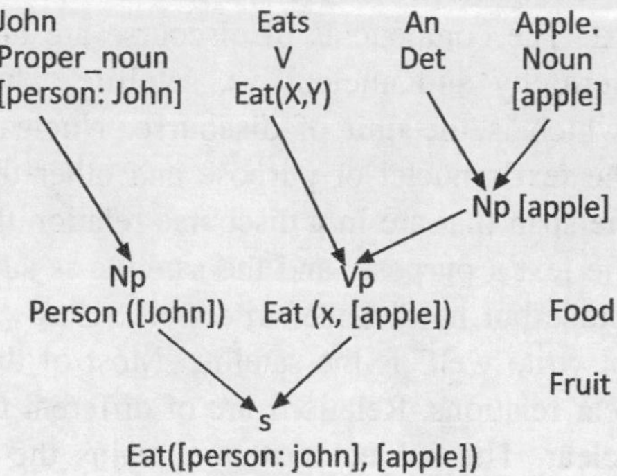

Figure 13.4 Semantic representation.

These sentences are giving answers from the knowledgebase after a semantic execution. This execution is based on the set of grammar rules. When we consider the first sentence, semantic parsing creates POS tags as follows:

$$\text{Argmax(type.country, } CO_2\text{.emission)} \rightarrow \text{China.}$$

After executing the parser job, 'China' is obtained from the knowledgebase. The parsing is achieved by associating natural language sentences from a training corpora with meaning representations. To decrease the computational cost, learning approaches should be fine-tuned. The semantic parsers are designed using statistical techniques. Inductive logic programming, hierarchical translation, composite of HMM and CFG, hidden vector state model and constraint satisfaction are some approaches.

13.1.4 Discourse Processing

Discourse consists of collocated, structured, coherent groups of sentences. It also makes continuity in unrelated sentences and deals how the text can be structured. Discourses are collection of coherent sentences with hierarchical structures. Coherence is a central theme that models the logical flow of the discourse. Discourse analysis is written, speech or sign language that uses a significant event. Consider an example, 'He hits the car with a ball'. 'It bounces back.' To understand the text:

Who/ what/ when/ where ... are involved in an event?

How to connect the semantic representations of different sentences?

What is the cause of an event and what is the consequence of an action?

Discourse is a particular unit of language and particular focus of language use. Discourse analysis studies the ways sentences and speech go together to make texts and interactions, how those texts and interactions fit into our social world. A discourse is recognised by relevance, sense and coherence of words in the sentence. It also checks whether these words are interconnected. A human property is called coherence that is used to evaluate text quality. Coherence relations are meaningful connections in the coherent discourse between its words. For example, 'Drunken John drives a car'. The result is 'He met an accident'.

The coherent texts consist of linked minimum units recursively, through verbal relations that do not show gaps or non-sequiturs. Therefore, relations are distributed throughout the text or different parts of the text. The components of discourse are units of discourse, nuclearity, relations among spans, hierarchy and nucleus vs. satellite. The text can be divided into minimum spans or units which is the unit of discourse. Nuclearity is referred as some of the span more central to the text's nuclei or purpose and other the is considered as satellite. Hierarchy or recursion is the span that are in a discourse relation that may enter new relations. The nucleus is central to the text's purpose, and the satellite is supporting to the nucleus. For example, 'John won the course but he did not write well'. Here 'John won the course' is the nucleus and 'but he did not write well' is the satellite. Most of the relations between nucleus and satellite are multi-nuclear relations. Relations are of different types such as subject matter, presentational or multi-nuclear. The subject matter explains the part of the subject or core theme of the text. Some subject matter relations are elaboration, evaluation, interpretation, solution hood, condition, unconditional, means, cause, result, otherwise, purpose and unless. For example, 'If you play well tomorrow, you will get a century'. Here 'you will get a century' is the nucleus conditionally connected to the satellite 'if you play well tomorrow'. The presentational relation makes achieve some effect on the reader. Some presentational relations are antithesis, background, justify, motivation, preparation, restatement, concession, enablement, evidence and summary. For example, 'Playing games is very interesting, but I do not like to play chess.' Here, 'playing games is very interesting' is the nucleus connected by the satellite through antithesis. Most of the relations between nucleus and satellite are multi-nuclear implemented through conjunction, disjunction, joint, list, multi-nuclear restatement, contrast, and sequence. For example, 'John is good batsman and he is good narrator too'. Here 'John is a good batsman' is the nucleus and 'he is good narrator too' is the satellite which is connected through a join 'and'.

13.1.5 Pragmatic Analysis

Pragmatic analysis deals with using and understanding sentences in different situations and how the interpretation of sentence is affected. It is a practical usage of language, means in practice like 'Do you have time?' or 'How do you do?.'

For example, consider the following conversation between two people"

P: I have a 12 year old daughter

Q: Well that is right

P: I also have a dog

Q: Oh, I m sorry

Can you understand the meaning of this exchange? It is very hard to understand what they are saying, unless you know that P. This conversation may be discussing an apartment rental case with which Q is not accepting pets. This is dependent sentence with respect to something need to understand.

13.2 Natural Language Understanding

To understand something is to transform it from one representation into another. This second representation has been chosen to correspond to a set of available actions that could be performed and where the mapping has been designed so that for each event, an appropriate action could be performed. The understanding of a sentence depends upon the context or surrounding from the framed sentence. To reproduce the same meaning with respect to the same context will be difficult for the computer.

Example, consider the sentence, 'I need to go to New Delhi as soon as possible'.

The airline officer will arrange a first fight to New Delhi, whereas relative of the person will understand that there is an emergency for his trip to New Delhi.

There are four major factors that contribute to the difficulty of an understanding problem.

1. Complexity of the target representation into which the matching is being done. Consider a sentence like, 'I want to read all about the last parliament election'

This can be translated as

Search key word = Election $\wedge$ parliament

This representation will not give unique meaning when retrieval is done, becoming the key words 'election' and 'parliament' may be taken as if some one is contesting for member of parliament post in the previous general elections. That is because we are not giving the information or knowledge of the word described by the sentence. To represent the word or knowledge in which the sentence is related is a complex problem.

2. The type of mapping: one-to-one, many-to-one, one-to-many, many-to-many

   One-to-one mapping is the simplest of all representations. That is, each different statement maps to a single target representation, which is different from that other arising representation. A very few languages are totally one to one. In such a language, one to one mapping may lead to hard understanding of sentence. Any representation using mathematical formula can be considered to be one-to-one.

   For example, many-to-one representation as follows:

   'Tell me all about the last parliament election'.

   'I would like to see all the stories on last parliament election'.

   'I am interested in the last parliament election'.

   The above sentences are giving same meaning with search keywords 'election' and 'parliament'.

3. Level of information on the components of source representation.

   In many languages, changing a single word can alter not only a single node in the trace of representation (structure) but rather the entire structure. The structures of words in the source representation are very important, since a very little change in the source representation of word will alter the entire structure of the tree in target representation.

4. Noise in the input

   Understanding is the process of interpreting an input and assigning its meaning. The noise of an input in a sentence creates problems in understanding of natural language. For example, '*** is very good'. What is the meaning of '***'? Is it a name or a place or an item? It is considered as noise.

13.2.1 Understanding of Sentence

Words in a sentence need to be known as per the human understanding. Put those words together to form a structure that represents the meaning of entire sentence.

This is not always applicable. For example, the word 'diamond' may have the following three meanings given in the dictionary:

1. A geometrical shape with four equal sides
2. A base ball field
3. An extremely hard and valuable stone

In order to distinguish with reference to a context, one can attach semantic marker like physical object, or abstract object. But this increases the knowledgebase enormously. The following analysis has to be performed in order to understand or interpret the sentences in English:

Syntactic analysis: The words are transformed into structures, which show how the words are related to each other. Some word sequence may be rejected if we violate the rules of language.

Semantic analysis: The structures created by syntax analyser are assigned meanings. In those words, a mapping is made between the syntactic structure and objects in the task domain. Structure for which no mapping is possible may be rejected.

Pragmatic analysis: The structure representing what was said is reinterpreted to determine what was actually meant. Do you know what time it should be interpreted as a request?

Understanding a single sentence usually means to assign meaning to the undirected words, and then to assign a structure to the entire sentence. But to understand a set of sentence, either a piece of text or a portion of dialogue requires the relationships among sentences. The following are the relationships between multiple sentences:

1. Identical objects

   Consider the sentence, 'Anand had a red balloon'. 'John wanted it'.

   The word 'it' should be identified as referring to the red balloon and is considered as identical objects. Such reference is called anaphoric reference.

2. Parts of entities

   Consider the sentence, 'She opened the book she just bought'. The little page was Tempest'.

   The phrase 'the title page' should be recognised as being part of the book that was just bought and is the part of entities.

3. Parts of action

   Consider the sentence, 'John went on a business trip to Bombay'. 'He left on an early morning flight'.

   Taking a 'flight' should be recognised as the part of going on a trip.

4. Object involved in action

   Consider the sentence, 'My house was broken into last week'. 'They took the TV and the stereo'.

   The pronoun 'they' should be recognised as referring to the burglars who broke into the house. That is, the objects which are involved in the action.

5. Elements to sets

   Consider the sentence, 'The decals we have in stock are stars, the moon and a flag'. 'I will take two flags'.

   The 'flags' in the second sentence should be understood to be some of the flags mentioned in the first sentence. Note that, to understand the second sentence at all requires that we use the context of first sentence to establish the words. The element sets are stars, moons and flags.

6. Names of individuals

   Consider the sentence, 'Salim went to the movie'.

   'Salim' should be understood to be some person named Salim who is an individual. Although there are many, the speaker had a particular one in mind and discourse context should tell us which are.

7. Causal chains

   Consider the sentence, 'There was a big rain storm yesterday'. 'The schools were closed today'.

   The 'storm' should be recognised as the reason due to which the schools were closed. These sentences are chained to be understood.

8. Planning sequence

   Consider the sentence, 'Maya wanted a new car'. 'She decided to get a job'.

   Maya's sudden interest in a job should be recognised as arising out of her desire for a new car, and thus for the money to buy one.

9. Illocutionary force

   Consider the sentence, 'It sure is cold in here'.

   In many circumstances, this sentence should be recognised as having, its intended effect, that the hearer should do something like closing the windows or turning on thermostat. The action related to cold is an illocutionary force.

10. Implicit presupposition

    Consider the query, 'Did Raji fail CS101?'

    The speaker's presuppositions, including the fact that CS101 is a valid course, that Raji is a student, and that Raji took CS101, should be recognised so that if any of them is not satisfied, the speaker can be informed. The implicit presupposition is related to the course CS101.

13.2.2 Focus in Understanding

Focus on the relevant part of database of available knowledge. Use that knowledge to reside ambiguities and make connections among things that were said.

'Attach the pump to the platform'.

Focus: 'The bolts are in the small plastic bag'.

Understand the two sentences, the items referred to in the second sentence is related to attach the pump. This can be done if in understanding the first sentence; the required both are brought in focus. To do this, we need a way of representing the knowledge about attaching, so that whenever attaching is mentioned, associated concepts such as both could be easily accessed. A good way of representing it is to use a semantic net, in which we can represent focus space along with the nodes and arcs.

13.3 Natural Language Generation

According to Reiter and Dale in 2000, Natural Language Generation (NLG) is the subfield of artificial intelligence and computational linguistics, which focussed on computer systems that can produce understandable texts in English or other human languages. Natural language generation is just reversion of natural language understanding. Machine representation comprises of some form of computerised data like database of daily temperature values in a city, ontology or a collection of fairy tales.

Major challenge in the natural language generation is the decision of word and formation of sentence like what to say or how to say. The output is a well-formed sentence with good orthography, morphology, syntax, reference, word choice, pragmatics, etc. For example, generating a weather report from a contextual values and sensors.

Major components in the natural language generation include content determination, document structuring, aggregation, lexicalisation expression generation and linguistic realisation. In content determination what content to express. To make an articulate text, structuring of the document by structuring the information in the text. In aggregation a combination of units in document plan tree is obtained. In lexicalisation analysed type of words are used. In the expression generation stage, noun phrase or pronoun is selected. Add some mark-ups done during the structure realisation. The voice of speech, aspect and tense of sentence, etc. are carried out in linguistic realisation step. Two different pipelined architectures are given in Figure 13.5. The tasks included in the natural language generation are document planning, microplanning and surface realization.

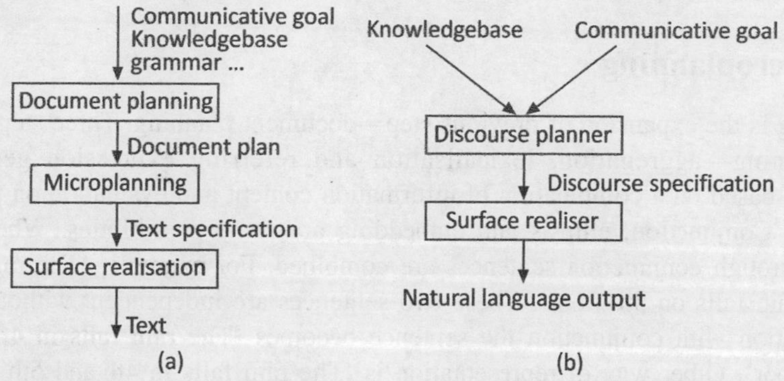

Figure 13.5 Pipelined architectures of natural language generation.

13.3.1 Document Planning

We had discussed nucleus and satellite in the previous sections. In document planning step, the content determination is done by checking importance of the sequence. It is like a domain expert system. The sentence is broadly divided into nucleus and satellite. Initially, the idea is communicated through these two. For example, weather summary report contains sentences like 'This January is very dry, The humidity is very low'. The document planning is visualised through a tree structure as given in Figure 13.6.

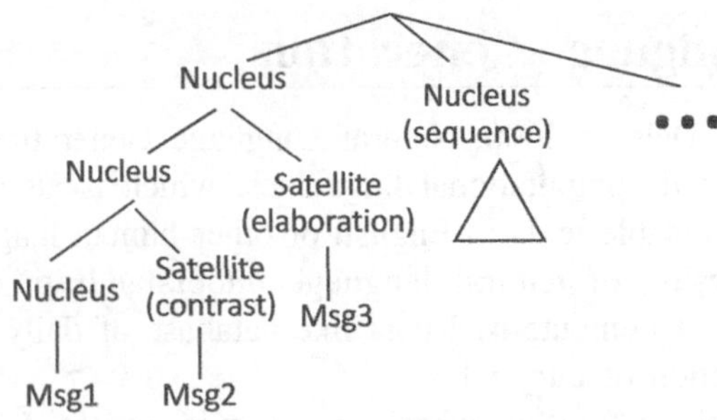

Figure 13.6 Document planning tree.

Document planning subprocesses are content determination and document structuring. In the first step, decide what information to mention in the text. Content decision is affected based on communicating goal which means the purpose of text and reader. It also decides level, size and how unusually the information is communicated. The contents can be determined through some techniques like schemas, which is a predefined template specifically the information container. Another approach is corpus based approach. The next step is document structuring, which decides the order and grouping of sentences in a generated text. Consider the following sequences:

'John went to the shop', 'John bought an apple'.

'John bought an apple', 'John went to the shop'.

Sentence structuring is important because of the difference between two given sequences.

13.3.2 Microplanning

Microplanning is the expansion of previous step—document planning. Three steps are involved in microplanning—aggregation, lexicalisation and referring expression generation. The aggregation is based on a combination of information content and evaluation on possible forms of realisation. Conjunction, ellipsis and embedding are some possibilities. When aggregation is not done through conjunction sentences are combined. For example, 'The rain falls on 4th floor'. 'The rain falls on 5th floor'. These two sentences are independent without aggregation. After aggregation with conjunction the sentence becomes 'The rain falls in 4th and the rain falls in 5th floor'. Other way of representation is 'The rain falls in 4th and 5th floor'.

The second way of aggregation is through embedding. Consider two sentences. 'June had a rainfall of 30 mm in Kerala'. 'June normally is the wettest month in Kerala'. When aggregation through embedding, sentence becomes 'June which normally is the wettest month in Kerala, had a rainfall of 20 mm this year'. Another way of representation is 'Although June is the wettest month in Kerala, this June had a rainfall of only 20 mm'. The aggregation rule can be represented in Figure 13.7. When we consider aggregation conformation to genre conventions and rules, take account of pragmatic goals. For example, making sentences shorter for poor readers. It is also important in the property structure of the sentence while applying aggregation rules. For example, aggregate only messages that are siblings in the document plan tree.

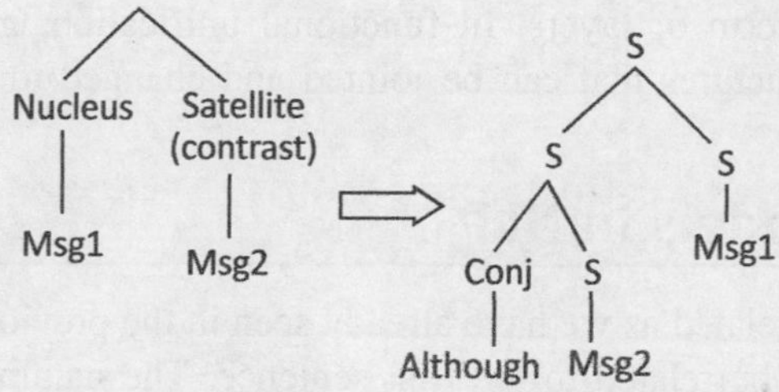

Figure 13.7 Aggregation rules.

Lexicalisation is the process of choosing words to communicate the information in messages by considering different lexicalisations such as user knowledge and preference, consistency with previous usage and pragmatics. For example, 'It is a light rain', 'It is a small amount of rain'. Lexicalisation determines the particular words to be used to express domain concepts and relations. In the above example, 'light' and 'small amount' are related to 'rain'.

How do we identify specific objects and entities where referring expression generation is concerned? Referring expression generation is concerned with how we describe domain entities in such a way that the hearer will know what we are talking about. It can be resolved by initial introduction of an object or a subsequence references to an already mentioned object. In initial reference, a few options like use of full name (for example, John Prakash Joseph). The aggregation is related to a salient entity (for example, 'the tall guy came to my office yesterday'). A pronoun, proper names and definite description can be a choice in the reference of expression. This will reduce the ambiguity in sentences. For example, 'John wants a red pen'. 'He brought it from the Premier bookshop'.

13.3.3 Surface Realisation

In the surface realisation, we have to convert text specification into actual text. In choosing reference expression, considering genre conventions and the context. For example, Jun 2019, June, this year. There is using morphology rules for word formation. For example, look + ed = looked. There is using a syntax rules for sentence formation after the morphological rule formation. It is further entered into structural realisation and linguistic realisation. A function word, selection of correct inflection, order selection of words in a sentence and giving punctuations in the sentence come under surface realisation of generated text. The punctuation, typographic issues, font size column width, etc. are orthographic concerns when we consider the surface realisation. The choices of font and layout are import in the look and feel of a sentence.

Simple surface realisation methods are canned text system and template based system. For a given input matched directly to a pre-made sentence but limited flexibility. This is commonly used system such as alert message, error message or warning message. The template system offers pre-made sentences with fill in the blank words that are filled in by the input. It is better than canned texts that can be used where the forms and letters are used. More advanced approaches use systemic grammar and functional unification grammar. When we consider sentences as a collection of function, rules can be mapped functions to grammatical forms. It is a linguistic form of layers. In functional unification, grammar sentences can be represented as feature structures that can be jointed and changed to produce sentences.

13.4 Markov Models in NLP

The sentences are highly related as we have already seen in the previous sections. That is, words in a second sentence can be related to the first sentence. The natural language processing has the stochastic property. Mathematical models are involving such interconnection of sentences. Markov models are very useful in this interconnection.

A stochastic process is a random function appearing as a result of a random experiment. It usually models the evolution of a random system in time. Consider an example of the prediction of a person's state of mind based on his feelings in the previous days. Though the human state of mind can change at any instant of time, for simplicity, assume a state will remain stable for one day. Further, assume that the state of mind of a person can be either happy, or sad or anger.

Probability that state of mind is happy, provided the three previous days' states were happy, sad, anger is represented as follows:

$$P(w_4 = \text{happy}|w_3 = \text{happy};\ w_2 = \text{sad};\ w_1 = \text{anger})$$

In general, the probability of state of mind on nth day based on $n - 1$ previous days,

$$P(w_n|w_{n-1},\ w_{n-2},\ \ldots,\ w_1)$$

When n is large, it becomes complex to collect statistics of past history. This is better solved by an assumption known as Markov assumption:

$$P(w_n|w_{n-1},\ w_{n-2},\ \ldots,\ w_1) = P(w_n|w_{n-1})$$

This is first order Markov assumption, which says that the probability of observation at nth instant of time depends on observation at $(n - 1)$th instant of time. A second order Markov assumption takes into account the observations at $(n - 1)$th and $(n - 2)$th instant of time. The states of mind and its changes can be represented using a finite automation as shown in Figure 13.8(a). The probability that process moves from one state to another is called transition probability. Assume that on analysis of previous states of mind, it is found that a person remains in the happy state with a probability of 0.7, whereas process moves to sad and angry with the probabilities of 0.15 and 0.25 respectively. Table 13.1 shows all possible state transitions and their probabilities. Simple predictions are possible by assuming the model following Markov process.

TABLE 13.1 HMM: Transition probabilities

	Happy	*Sad*	*Angry*
Happy	0.7	0.15	0.25
Sad	0.4	0.3	0.3
Angry	0.2	0.3	0.5

Given that a person is in happy mood today, the probability that he is in the angry state by tomorrow and again happy mood in the day after tomorrow is given by:

$$P(w_3 = \text{happy}, w_2 = \text{angry}|w_1 = \text{happy})$$
$$= P(w_3 = \text{happy}|w_2 = \text{angry}, w_1 = \text{happy}) \times P(w_2 = \text{angry}|w_1 = \text{happy})$$
$$= P(w_3 = \text{happy}|w_2 = \text{angry}) \times P(w_2 = \text{angry}|w_1 = \text{happy})$$
$$= 0.2 \times 0.25$$
$$= 0:05$$

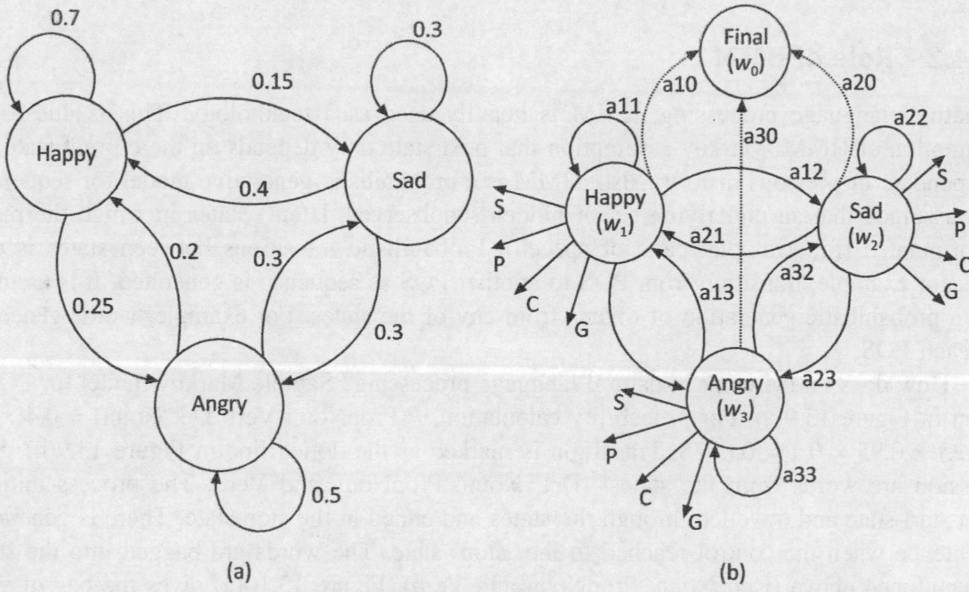

Figure 13.8 (a) An example of a Markov model with three states (b) An HMM with three states and four visible symbols.

13.4.1 Hidden Markov Models

Hidden Markov Model (HMM) is a statistical model using pattern recognition applications. In HMM, the system being modelled consists of a set of hidden states, visible states and undergoes a Markov process. Hidden states are non-observable states. In a normal Markov process, the states are directly visible to the observer. In HMM, on consumption of each input symbol, a state may transit to another state or remain in the same state and emit a visible symbol with a probability. Consider our previous example of human states. Let us assume that the visible symbols from human states are Smile(S), Play(P), Gloomy(G) and Cry(C). A person in a given mood (state) produces any one of these symbols with a probability. Revised model by the inclusion of a final state(w_0) is depicted in Figure 13.8(b). A final state or accepting state is one in which once a person enters, he cannot come back to any of the previous states.

Transition probability (a_{ij}) is the probability of transition from a state ω_i at (t = 1)th instant of time to a state ω_j at ith instant of time. The sum of transition probabilites from any state is $\Sigma a_{ij} = 1; \forall i$. Emission probability($b_{jk}$) is probability of emitting a visible symbol v_k from a state w_j, and $\Sigma_k b_{jk} = 1; \forall j$, where k is a number of visible symbols emitted from a state.

Consider an observation sequence $O = o1o2...o_T$ with states $S = \{s_0, s_1, s_2, \ldots s_N, s_F\}$. Where, s_0 and s_F are initial and final states respectively. There are N other states starting from s_1 to s_N. In natural language processing, these observations are words/letters with initial state, $q_1 = s_0$.

13.4.2 Role of HMM

In natural language processing, HMM is heavily used as a technology. This is due to the assumption of HMM. Markov assumption that next state only depends on the current state and independent of previous history. Also, HMM is a probabilistic generative model for sequences. It is assumed that an underlying set of hidden (unobserved, latent) states in which the model is formulated (for example, parts of speech). Probabilistic transitions between states is over time, for example, transition from POS to another POS as sequence is generated. It is assumed that a probabilistic generation of tokens from any of the states. For example, words generated for each POS.

How does HMM work in natural language processing? Sample Markov model for POS is given in Figure 13.9(a). For probability calculation, P(PropNoun Verb Det Noun) = 0.4 × 0.8 × 0.25 × 0.95 × 0.1 = 0.0076. Transition is marked as the dotted line in Figure 13.9(b). Here emission are words from the states—Det, Noun, PropNoun and Verb. The process initiated from start state and travelled through the states and ended in the stop state. There is generating a sentence when the control reached to the 'stop' state. The words are bagged into the states as mentioned above (Det, Noun, PropNoun and Verb). Figure 13.10(a) gives the bag of word HMM model. The sample words are only represented in the bags. Sample HMM generation for the sentence, 'John gave the apple' is represented in Figure 13.10(b).

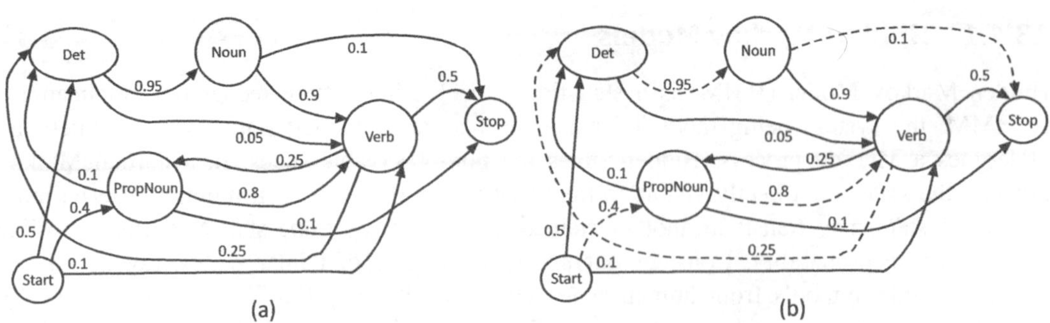

Figure 13.9 (a) Sample HMM for POS (b) Path calculation based on the state transition.

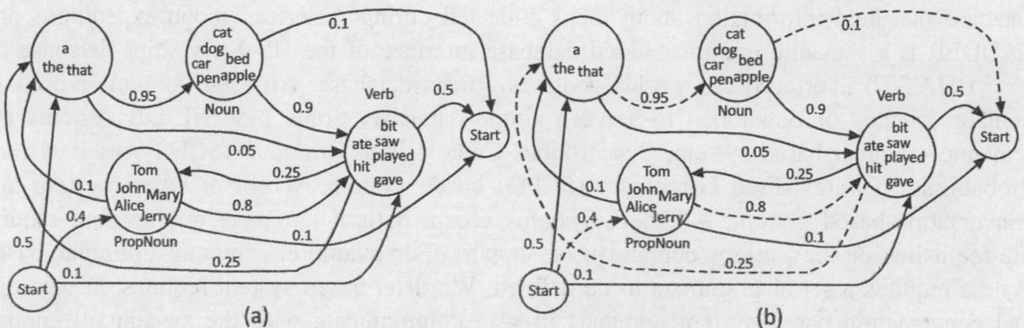

Figure 13.10 (a) HMM with bag of words (b) Path calculation based on the states.

'John gave the apple'

Start → PropNoun → Verb → Det → Noun → Stop

The probability can be calculated as $0.4 \times 0.8 \times 0.25 \times 0.95 \times 0.1 = 0.0076$. Here the emission probability is taken as 1. The travelling path is represented as a dotted line in Figure 13.10(b).

13.5 Intelligent Natural Language Query Processor

Databases are gaining importance as almost all organisations, companies and universities use it to store huge amount of data. However, only those individuals who are familiar with the query languages and methods can directly access this data. It is evident that a regular user is unaware of query languages but familiar with natural languages. It is very easy for a user in natural language queries. One of the significant areas in this direction is an application of Natural Language Interface for Databases (NLIDB). It allows user to put their queries in any form of natural language.

An Intelligent Data Base (IDBS) system is equipped with a database management system that is capable of managing large quantities of data to which various aspects of reasoning can be applied to infer additional data and information. This includes a symbolic frame system with knowledge representation techniques and inference techniques. It also includes an intelligent user interface and a database management system. The user interface is extended beyond the traditional querying method by providing natural language facilities. Intelligent user interfaces help users to make requests and receive the replies. IDBS manages information in a natural way making it easy for users to store, retrieve and use the information. IDBS provides use of knowledge and inference, enabling users to make decisions with this information. Non-expert users prefer to query databases in natural language rather than directly formulating an SQL query, which is very difficult. Intelligent interface for database systems provides a flexible querying mechanism for users. The research on Natural Language Interface for Database System (NLIDB) is considered as an important step towards the development of IDBS.

NLIDB is the key area of interest of database researches and early NLIDB systems had many roles in interface based query processing. LUNAR was developed as an interface to the database that held information about rocks collected during American moon expeditions and LADDER is a semantic grammar based database interface of the US Navy ships database.

CHAT-80 interfaced the worlds geography facts database. All these systems are either domain oriented or developed to serve a single database. Some new NLIDB systems are semantic grammar based system, Synchronous Context Free Grammar (SCFG) based system, Probabilistic Context-Free Grammars (PCFG) based system, WordNet based system and conversation-based system. All these systems accept natural language queries, but require pre-requisites or they might demand user support. For example, semantic grammar based system requires a set of grammars to be defined. WordNet based system requires an ontology and conversation based system demands user to communicate with the system till enough information for query formulation is collected. All these systems fail to formulate query accurately and hence provide incorrect results.

In order to overcome the flaws in earlier systems, a NLIDB system is implemented using shallow parsing and machine learning model which makes predictions only based on known facts. Machine learning methods such as maximum entropy model can be used to extend its performance. In this study, we used the power of maximum entropy model to improve the performance in prediction. A query translation algorithm, which utilises the linguistic dependencies and meta-data to form possible SELECT, FROM and WHERE clauses. This method increases accuracy of queries formed and eliminates formation of multiple trees or grammars for the same request. In certain situations, algorithm fails to identify the dependencies. In such instances, our model predicts the corresponding dependencies based on training set provided. Hence, this system has intelligence to predict SQL queries from natural queries.

13.5.1 Intelligent NLIDB System

It is proposed an intelligent system with the following functional components:

User request: The users query to the database is formulated in English language and is given as input to the system.

Natural language processing: The input provided in natural language must be processed to identify the attributes and predicates required to formulate the database query. Natural language processing techniques such as tokenisation, Part-Of-Speech (POS) tagging, chunking and named entity recognition are performed by this module. The resulting tagged chunks are processed to identify the attributes and predicates such as table names, column names and selection criteria of any input as natural language query.

Query formulation: The query is formulated with the help of some pre-defined query frames/structures and with the help of attribute and predicates identified using natural language processing techniques. Identification of predicates is performed based on a set of rules provided by query translation algorithm. A model for query prediction is included to predict queries in case the formulator lacks sufficient information to generate queries.

The work makes use of rules that are manually formed based on the analysis of 880 queries from the Geoquery database. Geoquery contains a small database of information about

United States geography with 800 facts (http://www.cs.utexas.edu/users/ml/geo.html). The database mainly contains the following information:

- States: their capitals, populations, areas, population densities, major cities, rivers and the bordering states.
- Cities: their populations and the states they are in.
- Rivers: their lengths and the states through which they flow.
- Mountains: their heights and the state they are in.

The work can be extendible for other databases without changing the architecture of this model.

13.5.2 Natural Language Processor

The input given by a user needs to process to identify the attributes and predicates which is required to generate the SQL query. This process is done by a natural language processor that consists of the following operations:

Tokenisation: It is the process of splitting an input natural language query into set of tokens. The input query is split at white-space to form candidate tokens. A candidate token is identified as a token if its length is less than two units or if it consists only of alphanumeric characters. Otherwise, the candidate token is examined at each character position for a possible split into more than one token. This is performed by building up a set of predicates related to possible split. Features used to generate this set of predicates include characters before and after the split, whether characters on each side are numbers or alphabets. This set of predicates is then evaluated against maximum entropy model.

POS tagging: POS tagging marks up a word into corresponding lexical category (verb, noun, adjective, adverb and so on). The tagger makes use of PENN tree bank POS tag set developed by the University of Pennsylvania for NLP related research work. It is a standard tag set accepted around the world. The tagger can be able to use a maximum entropy model trained over the PENN tag set. Tagger uses WordNet as the underlying dictionary.

Maximum entropy based learning is used for predicting tags of words. The feature selected is in the format (*a*, *b*), where *a* is the possible tag, *b* is the current word and previous two tags. Tag prediction for each word is made by considering the history *h* (sequence of tags assigned to all previous words of the sentence). Each pair (*a*, *b*) has a probability $p(a, b)$ and a tag is selected for *a* such that it maximises the entropy $H(p)$. It is computed using the Shannon's entropy equation as follows:

$$H(p) = -\Sigma_{x \in A \times B}\, p(x) \log p(x) \tag{13.1}$$

where $x \in (a, b)$; $a \in A$ and $b \in B$

A denotes the set of all tags and *B* denotes the set of words and their previous word tags. The current system makes use of twelve features to tag a word: the word, all suffixes and prefixes of the word, special character if any, capitals, regular expressions, previous two words and their tags and two words next to current word. Using these features probability of occurrence of all possible tags are computed and the tag with maximum probability is fixed as words tag.

Chunking: In SQL, the entities such as names, locations, data and so on always become part of a query. Since these entities are often distributed over a piece of text, analysing tokens one by one will not always correct to identify. This can be solved by chunking and thereby identifying the beginning and end of such sequences.

Chunking or partial parsing assigns a practical syntactic structure to a sentence and generates a flatter structure than full parsing. Chunks are non-overlapping regions of text and every chunk has a head followed by words and their respective tags. Chunker looks through the sequence of POS tags of the tagged text and identifies Noun Phrase (NP) and Verb Phrase (VP). In the context of query formulation, NPs will have all possible constraints. Chunking of 800 natural language queries from the GEOQUERY corpus in our GIS database leads to rules identification. These rules can be used to instantiate the attributes of database within a template. Based on these rules, the natural language query can be classified into the corresponding SQL query template. The rules are as follows:

- All NNP tagged words in the NP chunk are values stored in database.
- Whenever a PP chunk is followed by an NP chunk that contains a word tagged NNP, that word is an attribute value for the WHERE clause.
- The NN and NNS tagged words are either table or column name itself or their synonyms.
- The JJ and JJS tagged words give the conditions to be used in the WHERE clause.
- All questions starting with 'what' have the expected entity in the nearest NP chunk.

For example, consider the natural language query 'Give me the largest city in California'. It is chunked as [VP Give/VB] [NP me/PRP] [NP the/DT largest/JJS city/NN] [PP in/IN] [NP California/NNP].

Here, the NNP tagged word California is an attribute value and the PP chunk preceding it shows that California is an attribute value to be used in the WHERE clause. The NN tagged word gives the column name 'city' in this example. The JJS tagged word gives the condition to be checked.

Name identifier: Name Identifier (NI) is used to identify names, locations and dates specified in user query by the use of regular expressions. Name identifier extracts information from POS tagger and chunker. It uses only NPs identified by chunker as names, locations and dates will always be present in NPs alone. In NPs, POS tags of tokens are analysed and nouns that match the defined regular expressions will be either names or locations. In order to identify dates, words with tag Cardinal Number(CD) are selected and matched with defined regular expressions. In the previous example, name identifier results in,

List all details of students who joined on < date > 12 – 03 – 2013 <= date >

Query translator: Query translator converts chunked output of given input natural language query into SQL query. Every chunk generated is analysed by the query translator to recognise the attribute lists and predicate lists. A set of commonly used query formats are predefined and they are selected depending on the number of connectors, constraints as well as the number of table names identified. Algorithm used by the query translator proceeds by searching of table names, column names or their synonyms in the NP chunks. Column name synonyms are natural language alternatives for column names that belong to the database under use. A maximum entropy model is used to identify the column name synonyms in a given text. Connectors like AND, OR and NOT and conditions like greater than, less than and more than are found by searching for CC, JJR and JJS tagged words. The SQL queries are selected based on these gathered information. The procedure used by query translator is described in Algorithm 21.

There are situations in which the query translator algorithm fails to directly map the chunk values to corresponding attribute or predicate lists. For this instance, the chunked output of natural language input is given to the prediction model, which finds the highest probable SQL query that corresponds to the given input. Prediction model is a supervised maximum entropy model created using training set with different combination of values, its chunk type, and generated SQL query. The model outputs, SQL query with highest probability given the word and its chunk tag using maximum entropy. The system gives either a template based system that produces 100% correct result to the user query or the prediction module that will give the highest probable result that matches the user query.

Algorithm 21 Query translator algorithm

```
repeat
  if (Chunk contains NN or NNS tagged word) then
    if (The word is a column name or its synonym) then
      if (A value succeed it) then
        A constraint is identified
      else
        A column to be retrieved is identified
    else if (The word is a table name or its synonym) then
    The table from which data to be retrieved is identified
  if (Chunk contains NNP tagged word) then
    if (Chunk preceding current chunk is a PP chunk) then
      The attribute value for WHERE clause is identified
    if (There exists RB tagged words) then
      Check for any negative condition
    if POS tag JJR/JJS exist then
        "<" = ">" relations exist
    if POS tag CC exist then
      Predicates "AND" / "OR" exist.
until every chunked values are parsed
if number of identified tables >= 2 then
  Add condition joining the tables
Select appropriate SQL query template
Execute query and display result
```

13.5.3 Query Processing Outputs

NLIDB is preferred by users due to the advantage of non-mastering of SQL query languages. Considering all conveniences provided by the NLIDBs, it is not well accepted among the users due to the inability of the systems to process user query efficiently. Ambiguity and context of input are always a challenge. The objective of this model is to overcome these challenges. The model provides several natural language processing techniques to the NLIDB users. The techniques include sentence splitting, tokenisation, POS tagging, chunking and name recognition.

POS tagging is most important component as it is responsible for identifying the word categories correctly. Tagging is the first step towards formulating the query. Dates, numeric ids and so on can be identified after tagging as their tag always be CD (cardinal number). A name entity recogniser can be used to identify other values like name, location, etc. In the system, POS tagging is built in maximum entropy classification model. We have prepared 82 natural language queries with 500 words for the selected database (GIS database, Supply-store database and University database). The maximum entropy prediction model classified 483 tags from the 500 words correctly. The studies state that the tagger gives an accuracy of 96.6%.

For semantic grammar based system, a set of grammars needs to be defined. Context Free Grammar (CFG) based system demands input to be represented as in Codds tuple calculus expressions and Wordnet based system requires input to be converted to an intermediary format. When compared to conversation based system, our model does not bother the user with questions. It formulates a query without construction of a knowledge tree and hence no situation arises where the system cannot resolve a question. This model selects required data with the help of tags and Algorithm 21. The model is useful for any database with retrieving the SQL query from a natural language query by mapping the database.

Query translator gives 100% accuracy for a template based query processing from a natural language query by matching the attributes and constraints. Query translator is used the maximum entropy prediction model for non matching attributes and constants with the existing templates. Accuracy of prediction model depends on the input query and the system provides highest probable query. The highest probability is determined by the Shannon's entropy equation.

Consider the natural language query, 'Give the country name where city Trivandrum located'. This sentence is passed to the NLP for tokenisation. These tokens are tagged by POS. The tagged tokens are passed through the chunker to get different chunks in the query. These chunks are analysed to identify the predicates, and fill in the predefined query formats. The processing of this natural language query with respect to different module is given in Figure 13.11.

For example, consider an input in natural language queries: 'Give the country name where city Trivandrum located' and 'Give the country name where Trivandrum located'. Both queries give the same result. In the first query, it is easy to map the value 'Trivandrum' as city name attribute in the geography table and can select the appropriate SQL query template using the query translator algorithm. But in the latter query, one is not able to directly map the value 'Trivandrum' as city name attribute. In this situation, we need to calculate the probability of an attribute given 'Trivandrum' as the word and its chunk tag 'NNP'. Consequently the attribute name having highest probability is taken for the generation of SQL query.

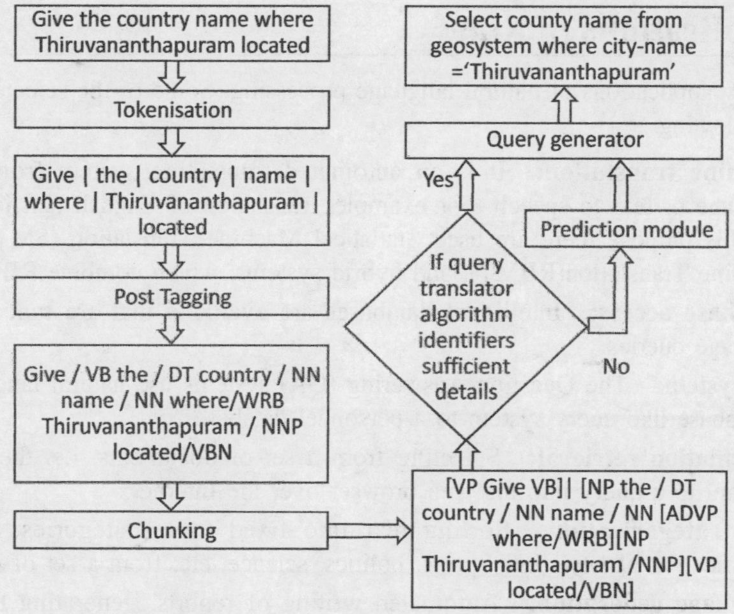

Figure 13.11 Processing of natural language query to SQL.

Complicated natural language queries given by the user are not completely affected by query translator with maximum entropy. Consider the query: 'Give the state name and count of rivers in each state of country India'

POS tagged output: Give/VB the/DT state/NN name/NN and/CC count/NN of/IN rivers/NNS in/IN each/DT states/NNS of/IN country/NN India/NNP?/

Chunked output: [VP give/VB] [NP the/DT state/NN name/NN and/CC count/NN] [PP of/IN]

[NP rivers/NNS] [PP in/IN] [NP each/DT states/NNS] [PP of/IN] [NP country/ NN India/NNP] ?/

As per the query translation algorithm, the state name and count have tag NN, so these need to be displayed in output. But count is an aggregate function, as per the query template we need to use group by statement to find the column names for the group by statements. As state_name to display, one of the groups by column name is state_name and other is river_name for displaying the count of rivers. Then rivers and states have NN tag, this method sets these tags as table names. The country tag NN is followed by NNP tag India, it is set as constraint for WHERE class. These details are passed through query generator template and the resultant query is as follows:

```
SELECT state_name, count(*) FROM rivers r, states s
WHERE r.state_name=s.state_name
AND s.country_name='India'
GROUP BY river_name, state_name;
```

13.6 Applications of NLP

There are many applications of natural language processing. Some of the selected applications include the following:

1. **Machine translation:** It is an automated translation system from language to language or text to speech. For example, translation of English text into Hindi text. For this purpose there are used Statistical Machine Translation (SMT), Rule-Based Machine Translation(RBMT), and hybrid systems, which combine RBMT and SMT.
2. **Database access:** Intelligent databases are available that are tuned with natural language queries.
3. **QA system:** The Question-Answering (QA) systems use natural language to query a database like query system to a personnel database.
4. **Information retrieval:** Selecting from a set of documents, i.e. the ones that are relevant to a query into the web browser over the Internet.
5. **Text categorisation:** Sorting text into fixed topic categories. For example, classification of news into sports, politics, science, etc. from a set of news tracts.
6. **Language generation:** Automated writing of reports, generating texts based on analysis of retail sales data, producing textual weather forecasts from weather data, summarising electronic medical records and even producing jokes.
7. **Extracting data from text:** Converting unstructured text into structure data from a raw text. This is useful for data mining.
8. **Speech recognition:** NLP system developed in neural networks for home automation, mobile telephony, virtual assistance, hands-free computing, video games, etc.
9. **Healthcare industry:** A technology that improves care delivery, disease diagnosis and bringing costs down while healthcare organisations are going through a growing adoption of electronic health records.
10. **Part-of-speech tagging:** Parsing, text-to-speech conversion, information extraction, etc. are some areas of NLP. The system automatically answers the questions.
11. **Spelling and grammar checkers:** At present, most of the text editors are provided with this facility. Neural net based text editors users check if their text contains spelling mistakes.
12. **Automated tutoring systems:** The machine interacts with a student for the automated learning. For example, an automated mathematics tutoring system for a student for calculus teaching.

13.7 Case Studies

13.7.1 Indian Languages

The languages of India may be grouped by major language families. The largest of these in terms of speakers is the Indo-European family, predominantly represented in its Indo-Aryan branch (accounting for some 700 million speakers), but also including minority languages such as Persian, Portuguese or French, and English as lingua franca[1]. The second largest is the Dravidian family, accounting for some 200 million speakers. Minor linguistic families include the Austro-Asiatic and Tibeto-Burman families (with some 10 and 6 million speakers, respectively). Kashmiri, considered a Dardic language, has some 4.6 million speakers in India. There is also a language isolate, the Nihali language.

13.7.2 Two-letter Word

The word is a sequence of characters. The sequence is important when a word is meaningful. For example, 'qc' may have less chance to occur than 'is'. We can say that the probability of occurrence of the second word is greater than the first. How can we identify such sequence of occurrences in word? We have hidden Markov model, which tells the probability distribution over a finite set of states. hidden Markov model is a probabilistic process over a finite set, $\{S_1, S_2, ..., S_n\}$, usually called states. Each state-transition generates a character from the alphabet of the process. There are transition and emission probabilities in an HMM. These are values, when state transition occurs and generate values at the state. If this probability depends on the previous state then it is order 1(first order) Markov model and its memory size is one. The words are considered as first order Markov model because the pronunciation depends on the combination of letters. General architecture of HMM for two letter word is given in Figure 13.12.

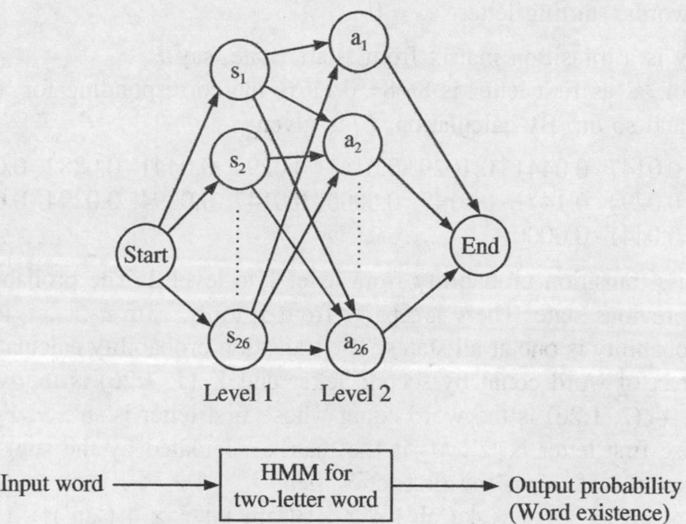

Figure 13.12 HMM for two letter word.

1. A lingua franca (from Italian, literally meaning Frankish language, see etymology under Sabir and Italian below) is a language systematically used to communicate between persons not sharing a mother tongue, in particular when it is a third language, distinct from both person's mother tongues.

We would like to generate an HMM for two-letter word in English language. The probabilities of occurrence of letters are different. For example, the probability of occurrence of 'as' is more than 'pr'. How will you form an HMM for this problem? Here the length of the word is two. So, there is a transition from one letter to other, if letters are taken as states. A 'start' state and 'stop' state is required. The transition probability from 'start' state and towards 'end' state is one. See, there is nothing emitted by these states. This model is shown in the figure. The output at s_1 is 'a', s_2 is 'b' etc. Also state a_1 gives an output 'a', a_2 gives 'b' etc.

The letters at first position are considered as level 1 and letters at second position as level 2. How will you calculate the transition probability from 'start' state, level 1 state to level 2 state? Initial training is required for calculating the probabilities. We use a training set with number of two-letter words. The efficiency depends on number of two-letter words in the training set. Consider the training set contains 68 two letter words and is given as follows:

'ag', 'ah', 'am', 'an', 'as', 'at', 'aw', 'ax', 'be', 'bo', 'by', 'co', 'de', 'do', 'di', 'ed', 'eh', 'em', 'en', 'er', 'et', 'ex', 'fu', 'gi', 'go', 'ha', 'he', 'hi', 'id', 'if', 'in', 'iq', 'is', 'it', 'ja', 'je', 'ku', 'lo', 'me', 'my', 'no', 'nu', 'of', 'oh', 'oi', 'op', 'ok', 'on', 'or', 'ow', 'ox', 'oy', 'pi', 're', 'se', 'so', 'ta', 'to', 'uh', 'um', 'up', 'us', 'vu', 'we', 'xi', 'ya', 'ye', 'yo'.

The number of words, starting letter 'a' = 8

The number of words, starting letter 'b' = 3

⋮

The number of words, starting letter 'z' = 0

Initial probability is a transition matrix from 'start' state, say P_1

The probability of 'a' as first letter is 8/68= 0.1176, the corresponding for 'b' is 3/68 = 00.0441 and so on. By calculation, P_1 is given

0.1176 0.0441 0.0147 0.0441 0.1029 0.0147 0.0294 0.0441 0.0882 0.0294 0.0147
0.0147 0.0294 0.0294 0.1471 0.0147 0.0000 0.0147 0.0294 0.0294 0.0588 0.0147
0.0147 0.0147 0.0441 0.0000

Next, discuss transition probability from level 1 to level 2. The probability of next state depends on the previous state. There are paths from s_1 to a_1, s_1 to a_2,..., s_2 to a_1, s_2 to s_2, etc. The emission probability is one at all states. The transition probability calculation is as follows.

T_2 is a matrix of word count by second letter and Y_2 (1, 1:26) is the word count whose first letter is 'a', Y_2 (2, 1:26) is the word count whose first letter is 'b',..., Y_2 (26, 1:26) is the word count whose first letter is 'z'. M_2 is the matrix calculated by the sum of Y_2 (1, 1:26) × T_2 (1:26). For example, if 'a' as the first letter then:

$M_2(1)$ = 1 × 1 ('ag') + 1 × 4 ('ah') + 1 × 3('am') + 1 × 4 ('an') + 1 × 3 ('as') + 1 × 3 ('at') + 1 × 2 ('aw') + 1 × 3 ('ay') = 23 (see the training set).

Here 'g' occurs one time as a second letter in whole training set, 'h' occurs 4 times etc. Similarly we can calculate the number of word count for other letters, as the first letter. The probability calculation is as follows:

$$P_2(i, 1:26) = \frac{T_2(1:26) \times Y_2(i, 1:26)}{M_2(i)}; \quad \text{for } i = 1,\ldots, 26.$$

For example, the probability matrix for 'a' as the starter

$$P_2(s_1, 1 : 26) = 0.0000\ 0.0000\ 0.0000\ 0.0000\ 0.0000\ 0.0000\ 0.0435\ 0.1739\ 0.0000\ 0.0000\ 0.0000\ 0.0000\ 0.1304\ 0.1739\ 0.0000\ 0.0000\ 0.0000\ 0.0000\ 0.1304\ 0.1304\ 0.0000\ 0.0000\ 0.0870\ 0.1304\ 0.0000\ 0.0000$$

Probability matrix for 'e' as the starter

$$P_2(s_5, 1 : 26) = 0.0000\ 0.0000\ 0.0000\ 0.0952\ 0.0000\ 0.0000\ 0.0000\ 0.1905\ 0.0000\ 0.0000\ 0.0000\ 0.0000\ 0.1429\ 0.1905\ 0.0000\ 0.0000\ 0.0000\ 0.0952\ 0.0000\ 0.1429\ 0.0000\ 0.0000\ 0.0000\ 0.1429\ 0.0000\ 0.0000$$

For all letters we got a vector size 1 × 26. The resulting transition probability is a matrix of size 26 × 26 at level 2. Output probability = product of probabilities of transition and emission. Some of the examples are given in Figure 13.13.

Example 1 Input word: 'as'

$$\text{Output probability} = 0.1776 \times 0.1304 \times 1 = 00.0153$$

Example 2 Input word: 'kw'

$$\text{Output probability} = 00.0147 \times 0 \times 1 = 0.$$

If any of the word count is zero in T_2 then denominator becomes zero. The division by zero problems can be solved by considering the count as a small number. So, the probability becomes a small number, i.e. nearly zero. Table 13.2 shows different two letter words and existence of probability (rounded in four figures). The comment is made by the comparison of each word in the table.

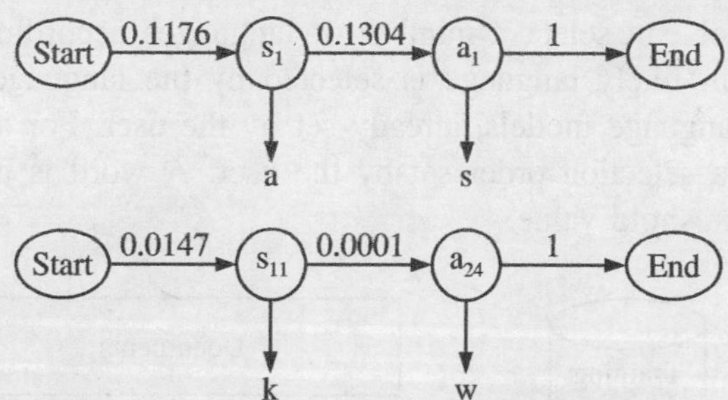

Figure 13.13 (a) Example 1 (b) Example 2.

TABLE 13.2 Sample cases of probability calculations

Word	*Probability*	*Comment*	*Word*	*Probability*	*Comment*
ao	00.0005	Low probable	re	00.0138	High probable
me	00.0211	High probable	hi	00.0136	High probable
pe	00.0002	Low probable	is	00.0170	High probable
ox	00.0150	High probable	it	00.0170	High probable
kw	00.0001	Very low probable	st	00.0001	Very low probable
we	00.0138	High probable	no	00.0195	High probable
ee	00.0004	Low probable	in	00.0227	High probable
ja	00.0087	Relatively probable	fe	00.0003	Low probable

13.7.3 Linguistic Colouring

The Oxford dictionary contains over one million words. A common man may not be familiar with more than 3–10% of the words in it but give him two words—'obnoxious' and 'xzjwwxz', he will immediately say that the first may be an English word whereas the second is not. How does he do this classification? There may be a hidden Markov model of English words working in his brain which checks out for the English word.

It is very common in non-English speaking countries to use vernacular language words in email messages, chat room communication and SMS over mobile phones. This text undergoes natural language processing, the oblivious preliminary requirement would be to separate out the transliterated words before parsing can be attempted. Also, in case of plain reading, some readers may not be comfortable with usages like above which will reduce their reading speed. Readability can be considerably improved if messages with mixture of different languages are shown in different colours. This is the concept of 'linguistic colouring' and is useful, most of the time, for unstructured text documents.

For the above problem, we use a first order hidden Markov model (HMM) for identifying such words. Hidden Markov model describes a probability distribution over a set of possible sequences. Starting from some initial state with an initial probability, a sequence of states is generated by moving from one state to another according to the state transition probabilities until a final state is reached, creating an observable sequence of symbols as each state emits a symbol when it is visited. The designed model gives an output probability for every word. Based on the probability, we can identify which language it is. Figure 13.14(a) shows the general architecture of word recognition by HMM. A word is split into letters and sent to all language models. A user can select a number of languages according to the trained HMM available. The maximum likely language is selected by the language selector based on the value returned by the language models, already set by the user. For an equal returned value, language selector uses a selection order set by the user. A word is treated as a noise, if all models give a bolero-threshold value.

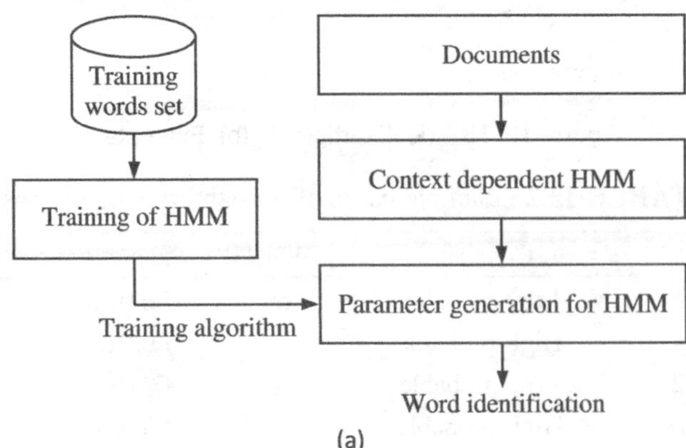

Figure 13.14 (a) Architecture of HMM based word identifier.

Each language model possesses a word identifier. The language selection is based on the largest probability returned among these models. Using the identified information, the word is coloured appropriately, thus presenting a possibility of adding this feature in standard text editors. Three languages are chosen in this study : English, Hindi and Malayalam. The language selection model is shown in Figure 13.14(b).

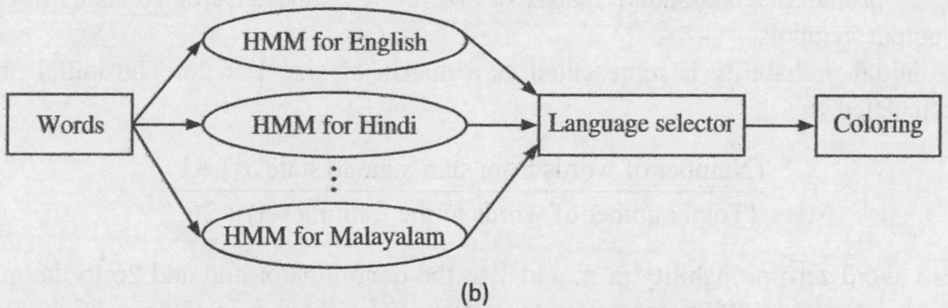

Figure 13.14 (b) Language selection.

HMM for word identification

A bag of words with a maximum of length 15 was used for training the HMM. A training set containing 4000 English words for HMM for English, 3000 words each for Hindi and Malayalam HMM language models. All the observed symbols are English letters. In this section, English HMM language model is discussed. The same principles are applicable for other languages. The HMM for word identification is shown in Figure 13.15. The elements of the HMM contain an initial probability vector, transition probability matrix and emission probability matrix. There are 15 levels of states and a pair of letter transition is described through two levels of states. At each level, there are 26 states marked 1 through 26. The levels are connected as left-to-right model. All the states except the 'start' state are connected to the 'stop' state. All words are passed by the HMM from 'start' state to the 'stop' state. If the given word length is greater than 15, the model takes only first 15 letters of that word. Figure 13.15 is the HMM word model architecture.

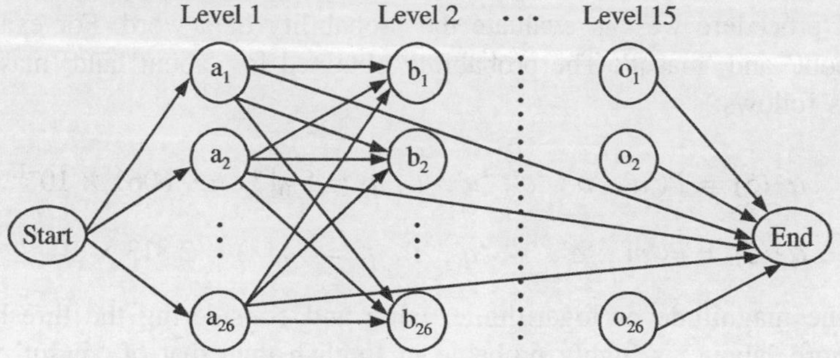

Figure 13.15 Word identification model.

The model HMM can be defined as follows:

$$M = <\pi, A, B>$$

with initial probability π, transition probability A and emission probability B. In this model, transition probabilities and emission probabilities are A_i and B_i respectively. Where $i = 1, 2, \ldots, 15$. Each probability is a square matrix of size 26 × 26, because of 26 states in each level with 26 output symbols.

The initial probability is represented as a matrix of size 1 × 26. The initial probability can be calculated as

$$\frac{(\text{Number of words from start state to state } a_0) + 1}{(\text{Total number of words in the training set}) + 26} \tag{13.2}$$

In order to avoid zero probability in π, add 1 to the denominator and add 26 to the numerator. For example, there are 300 words starting at first state, then π_1 is (300 + 1)/(4000 + 26) = 0.07476.

The transition probability is a 26 × 26 matrix for each level. $A^1_{1,1}, A^1_{1,2}, \ldots, A^1_{1,26}, A^1_{2,1}, A^1_{2,1}, \ldots, A^1_{26,26}$ are the elements in the first level.

$$A^1_{1,1} = \frac{(\text{Number of transitions from state } a_1 \text{ to state } a_2) + 1}{(\text{Total number of training words}) + 26} \tag{13.3}$$

Similarly we can calculate all the elements in $A_1, A_2, \ldots, A_{15}$. There are 26 hidden outcomes in each state. So, the emission matrices denoted as $B_1, B_2, \ldots, B_{15}$. At first level, $B_{1,1}, B_{1,2}, \ldots, B_{1,26}$ are obtained.

$$B_{1,1} = \frac{(\text{Number of emission symbols, } a) + 1}{(\text{Total number emissions at level 1}) + 26} \tag{13.4}$$

Similarly we may get all the elements in B.

HMM evaluation

Using forward procedure we can evaluate the probability of a word. For example, consider two words ‘about’ and ‘masqt’. The probability obtained for ‘about’ and ‘masqt’ by forward procedure is as follows:

$$\alpha^{\frac{1}{5}}(5) = P(\text{‘}a\text{’}, \text{‘}b\text{’}, \text{‘}o\text{’}, \text{‘}u\text{’}, \text{‘}t\text{’}, q_5 = S_{20}|\lambda) = 9.6061 \times 10^{-15}$$

$$\alpha^{\frac{2}{5}}(5) = P(\text{‘}m\text{’}, \text{‘}a\text{’}, \text{‘}s\text{’}, \text{‘}q\text{’}, \text{‘}t\text{’}, q_5 = S_{20}|\lambda) = 2.813 \times 10^{-25}$$

Taking the magnitude or logarithmic value and considering the threshold fixed, the existence of word ’about’ is highly probable in English than that of ‘masqt’. So, ‘masqt’ is considered as a noise word. Transition diagram is shown in Figure 13.16.

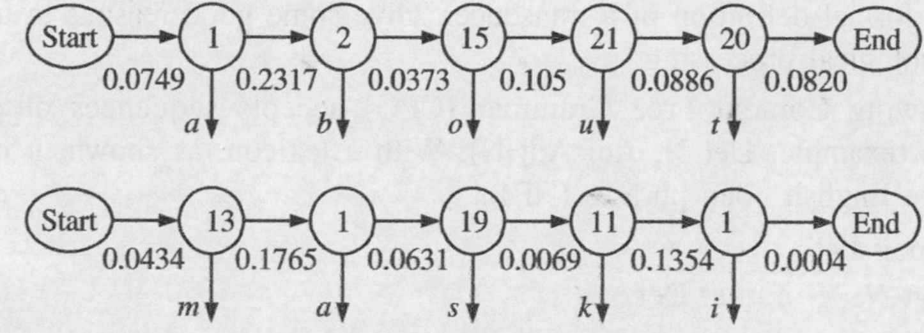

Figure 13.16 (a) Transition for 'about' (b) Transition for 'masqt'.

In the 'linguistic colouring' model, the above two words are given to all language models. The highest probability given by a model is selected and given the corresponding colour. If the entire models give an under threshold, then it is coloured as a noise word. The threshold value depends on the length of the word given to the HMM. Table 13.3 shows a result of this model, with three languages (English, Hindi and Malayalam).

13.8 Exercises

1. What is a finite state transducer and what is it used for in computational linguistics? How does it differ from a finite state automaton?
2. Give an equation for finding the most probable sequence of part-of-speech (POS) tags that could be utilised by a stochastic POS tagger.

TABLE 13.3 Sample cases of probability calculations

Word	*Value for English*	*Value for Hindi*	*Value for Malayalam*	*Language Selected*
kappada	12.2456	1.2301	5.5058	Hindi
was	3.5058	12.5020	15.1643	English
message	3.2685	9.2090	13.2237	English
chayakkada	20.5833	6.6744	0.6150	Malayalam
land	2.5204	4.7698	4.9692	English
manushayan	7.7762	2.3422	0.6571	Malayalam
pithaji	20.3362	1.0018	2.0147	Hindi
bride	2.0880	7.8650	9.8399	English
kwsp	10.4705	18.6788	24.4752	High Noise
Qt	11.4726	17.3456	17.4465	High Noise
ISO	5.6630	19.0870	18.2900	Low Noise
Manmohan	9.6825	5.7605	4.3695	Low Noise
makkan	12.2010	0.3403	2.5414	Hindi
probability	4.6105	28.9066	31.2666	English
tiger	2.1234	5.3201	7.2097	English
bagh	3.5550	2.3451	12.7225	Hindi
kaduva	17.6277	4.5655	0.4804	Malayalam

3. Provide a formal definition of a transducer. Give some good reasons to use such a tool for morphological processing.
4. The following Context Free Grammar (CFG) accepts sequences of part-of-speech categories (example, Det N, Adj Adj N). With a lexicon, as shown, it can be used to parse some English noun phrases (NPs).

   Start symbol: NP

$NP \rightarrow Det\ N$	a, the: Det
$NP \rightarrow N$	dog, dogs, house, houses, model, models: N
$N \rightarrow Adj\ N$	brown, red, model: Adj
$N \rightarrow N\ PP$	in, under: P
$PP \rightarrow P$ NP	

   (a) Give a Non-deterministic Finite-state Automaton (NFA) which accepts the same sequences of part-of-speech categories as CFG. Explain the notation that you use.
   (b) Give two examples of over generation that can be demonstrated with the lexicon shown, and explain how the CFG and NFA (and, if necessary, part-of-speech categories and lexicon) could be modified to prevent them.
   (c) The CFG does not accept noun-noun compounds (example, the dog house, house dogs). Indicate how you could modify the original CFG and NFA to allow for them.
   (d) Hand-constructed Finite State Automata (FSA) have sometimes been used for part-of-speech tagging. Outline the possible practical and theoretical advantages and disadvantages of such an approach when compared to stochastic tagging using hidden Markov models.
5. Briefly describe the specific objectives of the morphological module in the general perspective of automated natural language processing.
6. What are the different types of morphologies that can be considered? Briefly describe the main differences between them.
7. Explain the difference between inflectional and derivational morphology. Illustrate your explanation with concrete examples in English or Hindi.
8. Given a sequence of observations O and a model with a set of parameters, λ, what is the probability that this observation was generated by this model: $P(O|\lambda)$?
9. Two players, A and B, play the game of matching money. For each time n, each player has money and must secretly turn the money into heads or tails. The players then reveal their choices simultaneously. If the moneys match (both heads or both tails), player A wins the money. If moneys do not match (one heads and one tails), player B wins the money. Suppose players have between them a total of 5 moneys. If at any time one player has all the money, to keep the game going, he gives one back to the other player and the game continues. (a) Show that this game can be formulated as a Markov chain. (b) Is the chain regular (irreducible + aperiodic?) (c) If player A starts with 3 moneys and player B with 2, what is the probability that A will lose his moneys first?
10. What is a finite state transducer (FST), and what is it used for in computational linguistics? How does it differ from a finite state automaton?
11. Give an equation for finding the most probable sequence of POS that could be utilised by a stochastic POS tagger. You should assume a bigram model.

CHAPTER 14

Planning

How will we prepare a sequence of action? How will we prepare a good knowledgebase? How is an artificial intelligence system useful? The answer is planning. It is a task of coming up with a sequence of actions that will achieve the desired goal. A classical planning environment contains fully observable, deterministic, finite, static and discrete elements. That is, operators, current state and a goal state are the basic elements. A static environment change happens only when the agent acts, and discrete environment is in time, action or objects. A plan is complete only if every precondition is achieved. Precondition is achieved when it effects an earlier step and there is no possibility in the intervening steps that are undergone. For example, we have discussed the Block World problem in Chapter 3. Here domain object is a set of blocks and a table. The states are blocks that stacked on each other and the table, and we want to change their positions. Examples of some discrete actions are PickUp(block), PutDown(Block), Unstack(Block, Block), Stack(Block, Block), etc.

In artificial intelligence based planning, we are combining actions in the real-world search and logical representation of objects and relations. We have to use many search techniques in a solution space like depth first search, best first search, A* algorithm, etc. We can define planning as a search problem that requires to find an efficient sequence of actions that transform a system from given starting state to the goal state. As we have discussed, it is given an initial state, a goal state and some finite set of conditions (pre-conditions and post-conditions) with cost. Output of the planning gives a sequence of actions that meet some criteria. Driven actions are matched with current system state. The sequence can transform initial state to goal state. Cost of the actions must be below a specific value. Applications of planning in real-world are planning in daily life, game world and workflow management. Many planner softwares have been developed by the use of artificial intelligence in different business areas. Figure 14.1 refers to the possible application area that uses artificial intelligence based planning software. Some planning domains that use artificial intelligence are Web management agents, Robot planning, Manufacturing planning, Image processing management, Logistics transportation, Crisis management, Bank risk management, Blocksworld, Puzzles and Artificial domains.

Figure 14.1 Application area that uses artificial intelligence based planners.

14.1 Planning Problem

The real-world problems are solved with the help of planning. So, planning problems are important in the field of artificial intelligence. Planning is the task of coming up with a sequence of actions that will achieve a goal. Use of problem-solving agent in a search space and logical planning agent are used to solve complex or large scale problems. In search spaces, it is difficult to find a good heuristic function and hard to take advantage of problem decomposition. It is rare to find a perfect decomposable problem when dealing with real-world. Partial-order planner is based on the assumption that most real-world problems are nearly decomposable. But it should be done with care because, working on some subgoal may undo another subgoal. Planning agents have some similarities with problem-solving agent. That is, constructs plans to achieve goals and then executes the algorithm or procedure. Major difference between a planning and problem-solving agents is their representation of goals, states, actions and use of explicit, logical representations. In problem-solving, data structures are used as states while in planning, logical sentences are used as states.

A planning problem is a 4-tuple, (P, A, S, G), where P is a set of predicates, A is the set of operators (actions), S is the initial state and G is goal(s). In a planning system, it is required to set actions, states, goals and plans. For example, a robot control system defines the following representations:

- **Actions:** Defining preconditions and its effects that generate successor state descriptions.
- **States:** Current situations are described through data structures.
- **Goals:** Final achievement.
- **Plans:** A sequence of actions that leads to a solution.

Problem-solving and planning offers solutions to a problems. Planning offers better solution than problem-solving because of the representations and methods used. States, goals, and actions are decomposed into sentences. These sentences are represented in first-order logics. The search process is performed in plan space, but state space is used in problem-solving. To reduce the complexity of the planning problem, subgoals can be planned. There are some typical assumptions for a planning problem as follows:

- Each action is indivisible.
- Concurrent actions are not permitted.
- There is no uncertainty in the deterministic actions, hence results are completely determined.
- Planning agent is the only cause of change in the world.
- Agent in the planner has complete knowledge of the state of the world.
- Closed world is known to be true in the world included in the state description.
- Anything not listed is false.

14.1.1 STRIPS

STRIPS is a language with conjunction of positive, ground, function-free literals. STRIPS stands for Stanford Research Institute Problem Solver, developed by Fikes and Nilsson in 1971 as a theorem proving system. This represents a conjunction of ground literals like at(Home) $\wedge \neg$have(Milk) $\wedge \neg$have(bananas), etc. Goals are conjunctions of literals, but may have variables which are assumed to be existentially quantified like at(x) $\wedge \neg$have(Milk) $\wedge \neg$have(bananas), etc. Actions are preconditions that must hold before execution and the effects after execution. One of the assumptions is that any literal not mentioned is assumed to be false. STRIP planning system is a tuple $T = (L, O, I, G)$, where L is a finite set of ground literals, the conditions. O is a finite set of operators, I is the initial state, a subset of P and G is the goal state, a subset of P.

Operator O has the following attributes:

LC is a set of ground literals, defines the precondition of operator. D is a set of ground literals, defines the conditions that will be removed after the operation is executed. A is a set of ground literals, defines the conditions that will be added after the operation is executed. Finally, C is the cost of operation.

The original STRIPS system uses a goal stack to control its search. The system has a database and a goal stack, and it focusses attention on solving the top goal, which may involve solving subgoals, that pushed onto the stack, etc. STRIPS maintains two additional data structures, i.e. state list and goal stack. State list is a true predicate and goal stack is a push-down stack of goals to be solved, with current goal on the top of stack. If current goal is not satisfied by present state, examine add lists of operators, push operator and preconditions list on stack. When a current goal is satisfied, POP it from stack. When an operator is on top stack, record the application of that operator on the plan sequence and using the operators add and delete lists to update the current state. STRIPS procedure is given in Algorithm 22.

Algorithm 22 STRIPS algorithm

```
STRIPS (initial-state, goals)
state = initial-state; plan = empty; stack = empty
push goals on stack
repeat until stack is empty
  if top of stack is goal that matches state, then pop stack
  elseif top of stack is a conjunctive goal g, then
    select an ordering for the sub goals of g, and push them on stack
  elseif top of stack is a simple goal sg, then
    choose an operator o whose add-list matches goal sg
    replace goal sg with operator o
    push the preconditions of o on the stack
  elseif top of stack is an operator o, then
    state = apply (o, state)
    plan = [plan; o]
```

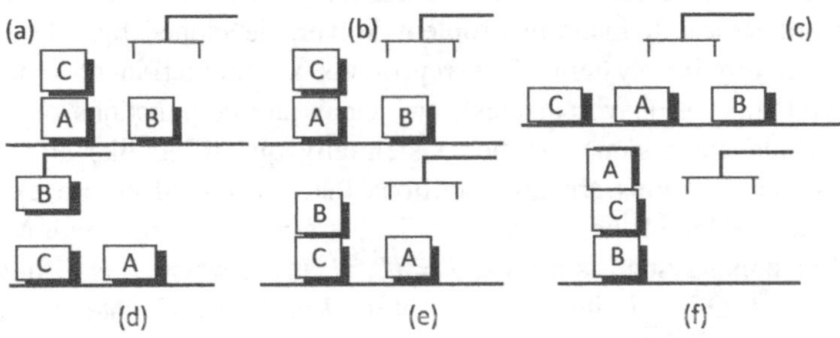

Figure 14.2 Block world example step by step.

Example: Consider Block world example with start state in Figure 14.2(a) and the goal state is given in Figure 14.2(f). The operations are given in Figure 14.3 for the steps described in Figure 14.2(a) through Figure 14.2(f). The steps are processed based on the list of actions by predicates. List of predicates is as follows:

empty: the gripper is not holding a block

holding(B): the gripper is holding block B

on(B_1, B_2): block B_1 is on the top of block B_2

ontable(B): block B is on the table

clear(B): block B has no blocks on the top of it and is not being held by the gripper

List of actions are listed in Table 14.1.

TABLE 14.1 List of actions in block world example

Action	*Preconditions*	*Add List*	*Delete List*
unstack(B_1, B_2)	empty and clear(B_1) on(B_1, B_2)	holding(B_1), clear(B_2)	empty, on(B_1, B_2), clear(B_1)
pickup(B)	empty and clear(B) and ontable(B)	holding (B)	empty, ontable(B), clear (B)
stack(B_1, B_2)	holding(B_1) and clear(B_2)	empty, on(B_1, B_2), clear(B_2)	clear(B_2), holding(B_1)
putdown(B)	holding(B)	empty, ontable(B), clear(B)	holding(B)

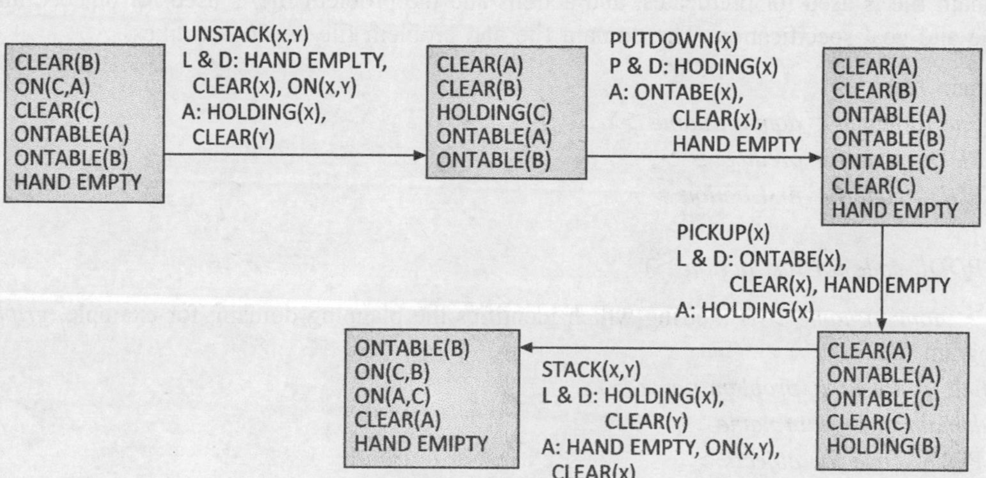

Figure 14.3 Step by step illustration of block world example.

14.1.2 ADL

Action Description Language (ADL) addresses expressive limitations of STRIPS which adds features not in STRIPS such as negative literals, quantified variables, conditional effects, equality, etc. The expensive power of STRIPS was observed by Pednault so as to suggest an improvement. It is allowed conditional operators and proportional fragments that is a deviation from STRIPS. The conditions in the environment is the sense of a planning language. It is also noted that for leading to a desired goal, automatically generate a chain of actions that are performed. Set of preconditions must be satisfied to get effects bringing environmental changes. This environment is described by means of certain predicates. ADL allows negative literals and disjunction but STRIPS allows positive literals and conjunctions only. ADL scheme consists of optional parameters, action name and four optional groups of clauses labelled Precond, Add, Delete and Update. Action execution formulae are listed in the Precond group. The value TRUE is inserted for an empty set. Add and Delete operations are specified in the Precond. ADL located between STRIPS and the Situation Calculus. ADL is strictly more brief than STRIPS.

14.1.3 PDDL

Planning Domain Definition Language (PDDL) is a standard encoding language for classical planning tasks. The components of a PDDL planning task are as follows:

- Objects: Things in the world that are interested.
- Predicates: properties of objects that we are interested in which can be true or false.
- Initial state: state of the world that we start in.
- Goal specification: things that we want to be true.
- Actions/Operators: ways of changing state of the world.

Planning task in PDDL is specified in two files, i.e. domain file and problem file. The domain file is used for predicates, and actions and the problem file is used for objects, initial state and goal specification. The domain file and problem file are look a likes.

domain file:
(define (domain < domain name >)
< PDDL code for predicates >
< PDDL code for first action >
[...]
< PDDL code for last action >)

< domain name > is a string which identifies the planning domain, for example, *gripper*.
program file:
define (problem < problem name >)
(: *domain < domain name >*)
< PDDL *code for objects* >
< PDDL *code for initial state* >
< PDDL code for goal specification >)

< problem name > is a string that identifies the planning task, for example, gripper-four-balls.

< domain name > must match the domain name in the corresponding domain file.

Different variations or extension of PDDL are PDDL+, NDDL (New Domain Definition Language), MAPL (Multi-Agent Planning Language), OPT (Ontology with Polymorphic Types), PPDDL (Probabilistic PDDL), APPL (Abstract Plan Preparation Language), RDDL (Relational Dynamic influence Diagram Language) and MA-PDDL (Multi-Agent PDDL).

14.2 Planning with State Space Search

All planning procedures are search procedures that have different search spaces. These are of two types, i.e. state-space planning and plan-space planning. In a state-space planning, each node represents a state of the world and that plan is a path through the space. In plan-space planning, each node is a set of partially-instantiated operators with some constraints. It imposes added constraints, until we will get a plan. The components of state-space planning are forward state space search (progression), backward state space search (regression), lifting and block stacking.

14.2.1 Forward State Space Search

In a progression procedure, there is initial state of the planning problem, applicable to the current state (actions preconditions are satisfied), goal of planning and step action. The output plan is a sequence of STRIP operators. Forward search is a complete and sound procedure (Algorithm 23) that returns any of its non-deterministic traces. When it reaches the final stage, solution is generated. In forward search, large branching factors are involved like applicable actions that do not progress towards a goal. Is this large branching factor bad? Yes, and this is due to deterministic implementations, which can waste time by trying lots of irrelevant actions. To improve forward procedure, a good heuristic function and/or pruning procedure is required. The process flow of forward search is given in Figure 14.4(a).

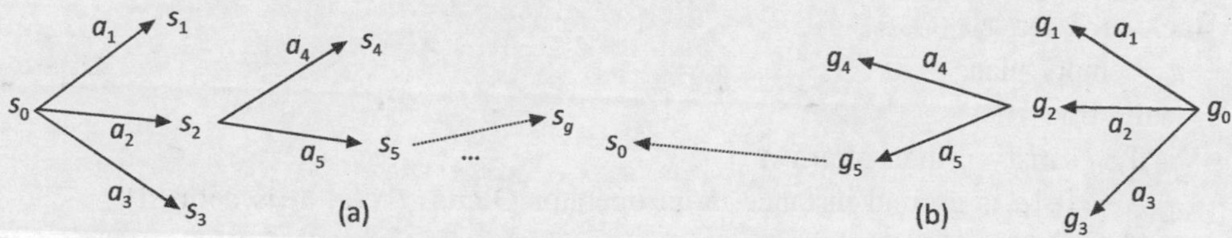

Figure 14.4 (a) Forward search flow path (b) Backward search flow path.

Algorithm 23 Forward search

Forward-search (O, s_0, g)
  $s = s_0$
  π = empty plan
  while true
    if s satisfy g then return π
    E = $\{a \mid a$ is ground instance in an operator O and precond(a) is true in $s\}$
    if $E = \phi$ then return failure
    non-deterministically choose an action $a \in E$
    $s = \gamma(s, a)$
    $\pi = \pi \cdot a$

14.2.2 Backward State Space Search

We have started at the initial state and transitions are computed in the forward search and have reached a new state, $\gamma(s, a)$. In backward search, start from a goal and compute inverse state transitions and reach to the set of subgoals, $\gamma^{-1}(g, a)$. The backward search procedure is given in Algorithm 24. A large branching factor is produced by backward search. Figure 14.4(b) gives how branching is done by backward search procedure. For example, a search graph is given in Figure 14.5(a) and the backward search tree is given in Figure 14.5(b).

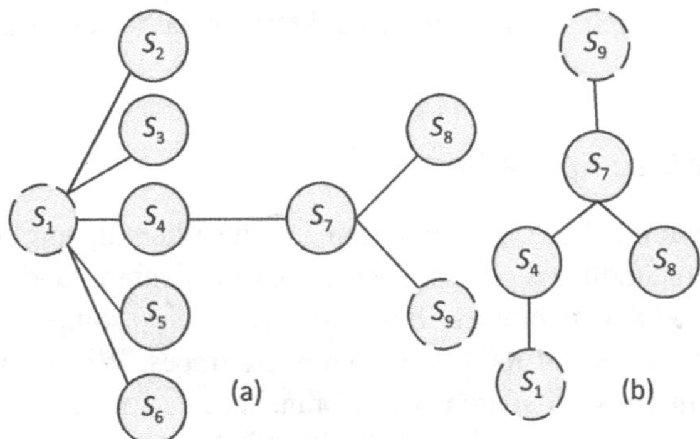

Figure 14.5 Darkened dotted circles are start (S_1) and goal (S_9) nodes (a) Example graph (b) Backward search tree.

Algorithm 24 Backward search

Backward-search (O, s_0, g)
  π = empty plan
  while true
    if s_0 satisfy g then return π
    A = $\{a \mid a$ is ground instance in an operator O and γ^{-1} (g, a) is defined$\}$
    if A = ϕ then return failure
    non-deterministically choose an action $a \in$ A
    $s = \gamma^{-1}(g, a)$
    $\pi = \pi.a$

14.2.3 Lifting

How can we reduce the branching factor of the backward search? Partially instantiating the operators we can reduce the branching factor of backward search. This is called lifting. The lifting procedure is given in Algorithm 25. It is more complicated than backward search but it keeps track of the substitutions that were performed. Example for before and after lifting backward procedure is given in Figure 14.6(a) and Figure 14.6(b) respectively.

Algorithm 25 Lifted search

Lifted backward-search (O, s_0, g)
  π = empty plan
  while true
    if s_0 satisfy g then return π
    A = $\{(0, \theta) \mid 0$ is a standardisation of an operator O θ is an mgu for
      an atom of g and atom of effect$^+$(o) and $\gamma^{-1}(\theta(g), \theta(o))$ is defined$\}$
    if A = ϕ then return failure
    non-deterministically choose a pair (o, θ) $\in$ A
    π = the concatenation of $\theta(o)$ and $\theta(\pi)$
    $s = \gamma^{-1}(\theta(g), \theta(\pi))$

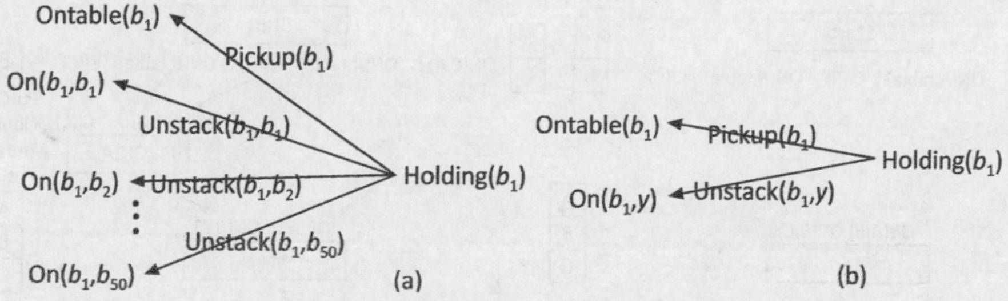

Figure 14.6 (a) Before lifting the backward procedure (b) After lifting the backward procedure.

14.2.4 Sussman Anomaly

Sussman anomaly is a problem in the block world first described by Gerald Sussman in his PhD thesis under the supervision of Mervin Minsky in 1970s. This anomaly shows limitations of non-interleaved planning method. People do planning by considering different subgoals, before considering this anomaly. This anomaly shows naively pursuing one subgoal *X* after satisfying other subgoal *Y* which may not work, because steps required to complete *X* might undo things for subgoal *Y*.

Consider blocks labelled *A*, *B* and *C* resting on a table. There is only one block that moves at a time. The agent must stack the blocks such that *A* is atop *B*, which in turn is atop *C*. That is, final state required is ON(*A*, *B*) or ON(*B*, *C*). That is,

Initial state: ON(*C*, *A*)∧ONtable(*A*)∧ONtable(*B*)∧Clear(*C*)

Goal state: ON(*A*, *B*)∧ON(*B*, *C*).

Figure 14.7(a) gives initial state, Figure 14.7(b) gives goal state. For the subgoal ON(*A*, *B*), trying to put on top of *C* would cause ON(*A*, *B*) to be undone. An alternative goal is given in Figure 14.7(c). It is trying to focus on subgoal ON(*B*, *C*). That is, trying to put *A* on top of *B* cannot done without undoing ON(*B*, *C*). Step by step solution to the Sussman anomaly is given in Figure 14.8(a), Figure 14.8(b) and Figure 14.9(a). The final solution is given in Figure 14.9(b). In Sussman anomaly, the term clobbers means a potentially intervening step that destroys the condition achieved by a causal link.

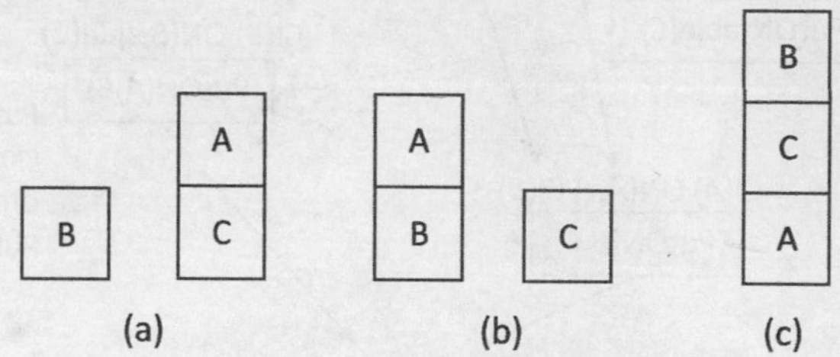

Figure 14.7 Cl:clear (a) Initial state (b) Goal state 1 (c) Goal state 2.

Start
ON(C,A) ON(A,table) Cl(B) ON(B,table) Cl(C)
B
A
C
OB(A,B) ON(B,C)
Stop
A
B
C
(a)
Start
ON(C,A) ON(A,table) Cl(B) ON(B,table) Cl(C)
B
A
C
Cl(B) ON(B,Z) Cl(C)
PutON(A,B)
PutON(A,B) clobbers Cl(B) orderafter PutON(B,C)
OB(A,B) ON(B,C)
Stop
A
B
C
(b)

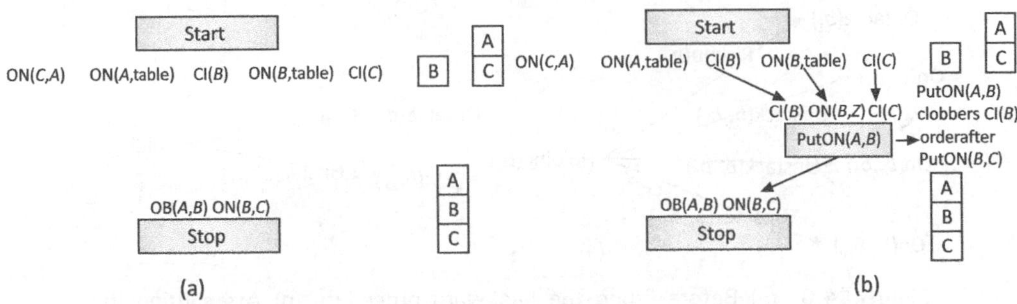

Figure 14.8 Anomaly illustrates the need for interleaved plans (a) Step 1 (b) Step 2.

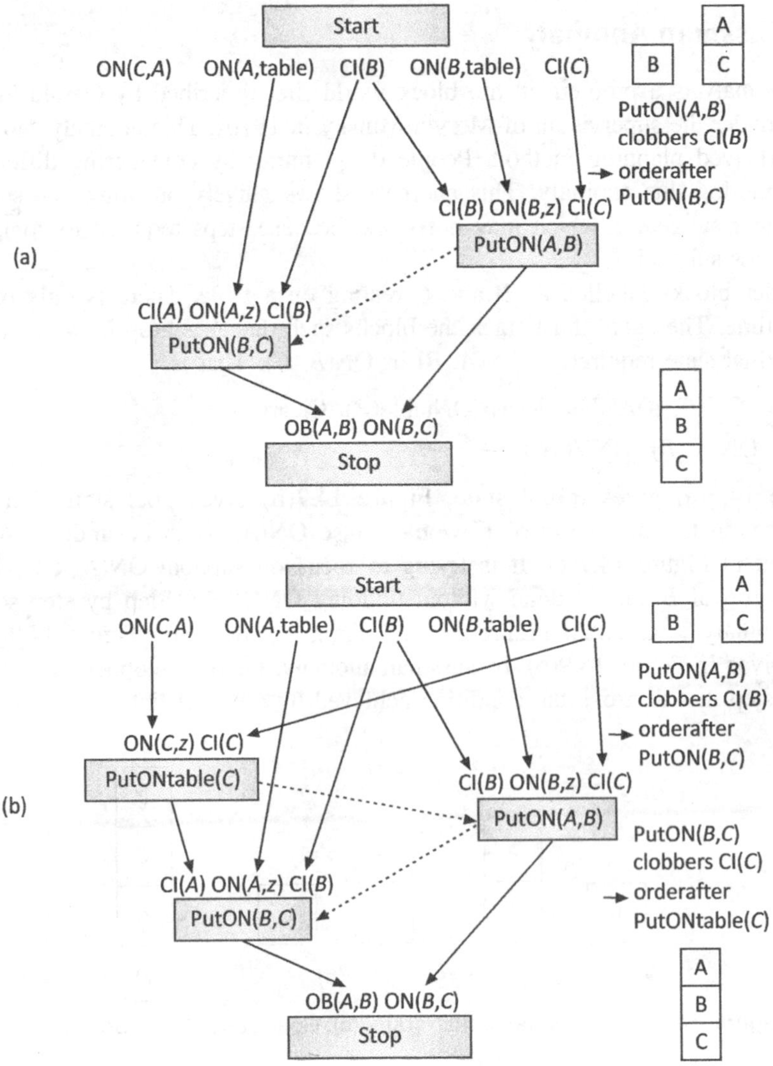

Start
ON(C,A) ON(A,table) Cl(B) ON(B,table) Cl(C)
B
A
C
PutON(A,B) clobbers Cl(B) orderafter PutON(B,C)
Cl(B) ON(B,z) Cl(C)
PutON(A,B)
(a)
Cl(A) ON(A,z) Cl(B)
PutON(B,C)
OB(A,B) ON(B,C)
Stop
A
B
C
Start
ON(C,A) ON(A,table) Cl(B) ON(B,table) Cl(C)
B
A
C
PutON(A,B) clobbers Cl(B) orderafter PutON(B,C)
ON(C,z) Cl(C)
PutONtable(C)
(b)
Cl(B) ON(B,z) Cl(C)
PutON(A,B)
PutON(B,C) clobbers Cl(C) orderafter PutONtable(C)
Cl(A) ON(A,z) Cl(B)
PutON(B,C)
OB(A,B) ON(B,C)
Stop
A
B
C

Figure 14.9 Need to reorder plan steps dynamically (a) Step 3 (b) Final state.

14.3 Partial Order Planning

Planning generates a sequence of actions to achieve a goal. Partial-order planning consists partially ordered set of actions with sequence constraints existing on these actions. In partial order planning, a partially ordered collection of steps are start step, finish step, causal links and temporal ordering. Initial step with its effect and goal step with its preconditions are described in start and finish step respectively. Causal link gives the outcome of one step to precondition another step and temporal ordering performs pairs of steps. Open condition is precondition of a step not yet causally linked. A complete plan has achieved every precondition. For example, suppose we would like to paint the ceiling of a room. The partial planning is given in Figure 14.10. Planning problems are operated on a large search space. The action is associated with system states restricted on the action sequence. There may not be a valid solution, so an optimisation is required.

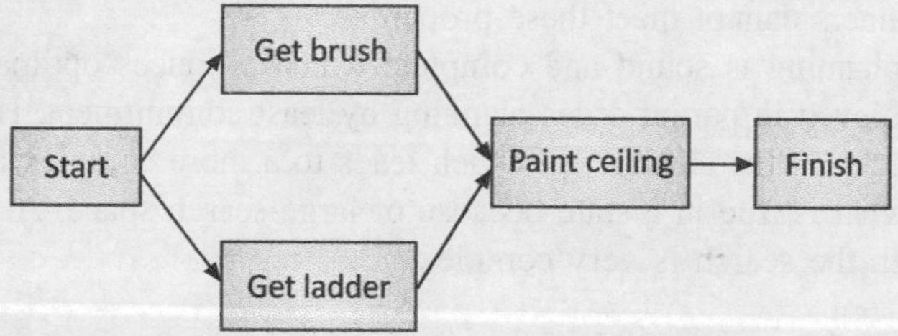

Figure 14.10 Partial order planning example.

Least commitment is one of the basic ideas behind the partial order planning. It makes only those choices that are appropriate to solve the present part of the problem, is a basic idea of least commitment. The choices of least commitment are orderings, bindings and actions. In orderings, only sequential actions are taken. Unified conditions are achieved with leaving variables is the binding actions as we had discussed earlier. Partial order planning technologies are total ordered, instantiated and consistent plans. In total order plans there exists sufficient orderings O such that all actions in A are ordered with respect to each other. Partial-order plan consists partially ordered set of actions that are sequence constraints existing on these actions. Partial order plan generation algorithm can be applied to transform partial order plan to total order plan. Total-order plan consists totally ordered set of actions. For example, in a partial order plan 'paint the celling of a room' in Figure 14.10 can be represented in total order plan as shown in Figure 14.11. In a fully instantiated plan sufficient constraints in 'B' such that all variables are constrained to be equal to some constant. In a consistent plan, there are no contradictions in 'O' or 'B'. For a complete plan, every precondition p of every action a_i in 'A' is achieved.

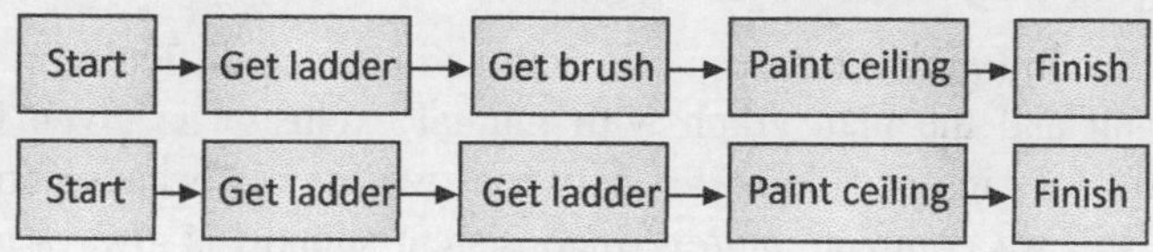

Figure 14.11 Total order planning example.

The partial order planner maintains partial order constraints on operators in the plan. For example, there are two situations, say S_1 and S_2. S_1 must come before S_2, but not necessarily immediately before it. A total order planner maintains a partial solution as a totally ordered list of steps found so far. STRIPS is an example.

Casting the planning problem as search problem is the simple way to build a planner. This can be represented as a graph where nodes are represented as states of the world and arcs connect worlds that can be reached by executing a single action. There are two types of planners, i.e. progression planner and regression planner. The progression planner searches forward from the initial situation to the goal situation. The regression planner searches backward from the goal situation to the initial situation.

When we build a plan, do not make an ordering choice. If it is essential, ordering can be done based on the request. The choice of ordering should be as general as possible like, $S_1 < S_2$ vs. $S_1 \rightarrow S_2$. To avoid wastage of time, reduce maximum backtracking process. While designing a partial order planner these properties are required to be considered. Most of the time situational planners cannot meet these properties.

Partial order planning is sound and complete which produces optimal solutions. Shorter search times are achieved in partial order planning by least commitment. Higher per node cost is one of the problems in this technology which leads to a more complex algorithm. It is very hard to determine what is true in a state because of large search space. A concurrent action is allowed even though the search is very complex.

14.4 Planning Graphs

We can represent a planning action through graphical structure. Planning graphs give an efficient way to create a representation of a planning problem. For a better heuristic estimate and direct construction of plans we have to use planning graphs. A planning graph consists of sequence levels that correspond to time-steps in the plan. The initial state is represented by level 0 followed by other levels containing a set of literals and a set of actions. For time step, literals are true and actions connected with their preconditions are also true. The graph representation works only for propositional planning. To build a planing graph, the following steps are required:

1. Start from S_0
2. $i = 0$
3. Find all the actions of A_{i+1} applicable in S_i given *mutexes*
4. Compute *mutexes* between the actions of A_{i+1}
5. Compute literals reachable in $S_i + 1$
6. Compute *mutexes* in S_{i+1}
7. If S_{i+1} 6 = S_i, then $i = i + 1$ and go to step 3

A mutual exclusion (*mutex*) between two actions indicates difficulty in parallel execution. A sample plan graph and the plan graph with mutual exclusion is given in Figure 14.12(a) and Figure 14.12(b). It is impossible *mutex* between two literals because both literals are true at this stage. How will we complete *mutex*? Inconsistent support is always required to achieve two literals in mutually exclusive (Figure 14.12(c)). For actions, there are inconsistency effects, interference and inconsistency support. Precondition of the first action that negates the other action is the interference. Precondition of first action mutually exclusive to the second action is the competing need. A sample planning graph with *mutex* relationships for actions like inconsistency effect, interference and competing need are given in Figure 14.13.

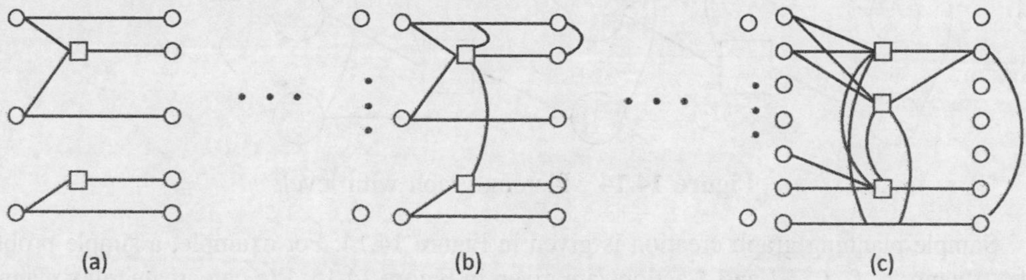

Figure 14.12 (a) Sample plan graph (b) Plan graph with mutual exclusion—curved lines (c) Mutex relationships for literals—inconsistent support.

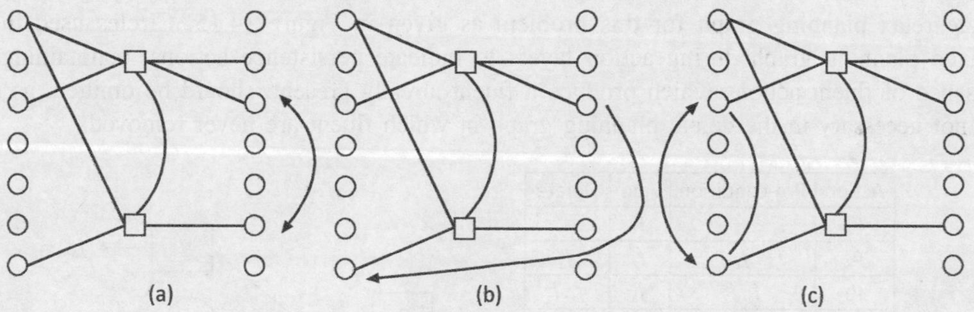

Figure 14.13 Mutex relationships for actions (a) Inconsistent effects (b) Interference (c) Competing needs.

A planning graph is given in Figure 14.14. The propositions, actions and *mutex* are given. Graph plan procedure can be listed in Algorithm 26.

Algorithm 26 Planning graph algorithm

```
function Graphplan (problem)
  returns a solution, or failure
graph ← initial-planning-graph (problem)
goals ← goals[problem]
loop do
  if goals all non-mutex in last level of graph then do
     solution ← exact-solution (graph, goals, length (graph))
  if solution
     failure then return solution
  elseif no-solution possible(graph) then return failure
graph ← expand-graph (graph, problem)
```

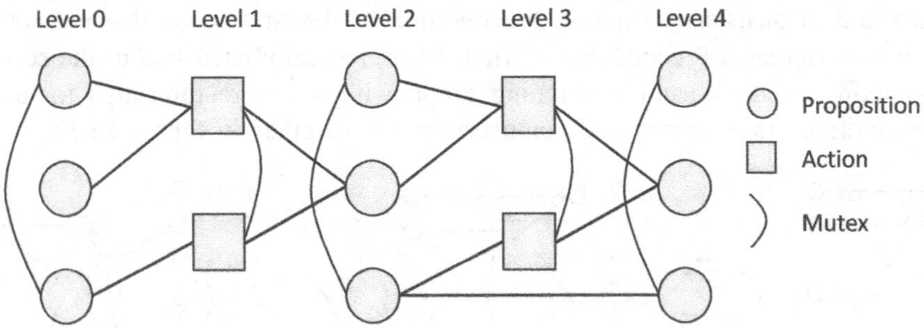

Figure 14.14 Planing graph with levels

Sample planning graph creation is given in Figure 14.14. For example, a simple problem with 4 fluent $\{f_1, f_2, f_3, f_4\}$ and 4 actions are given in Figure 14.15. We can create relax planning graph by removing delete lists of the actions. The action layer contains appropriate actions and expand graph until a layer comprises all the goals. *Mutex* is not included in the graph because delete lists were removed. Starting from null initial state and goal stat $\{f_1, f_2, f_3, f_4\}$, we can create relax planning graph for this problem as given in Figure 14.15. Circles used in the relaxed planning graph during action layers to indicate persistence actions, maintaining the presence of fluent actions which produce a fluent already present, should be omitted, as they are not necessary in the delete planning graph in which fluent are never removed.

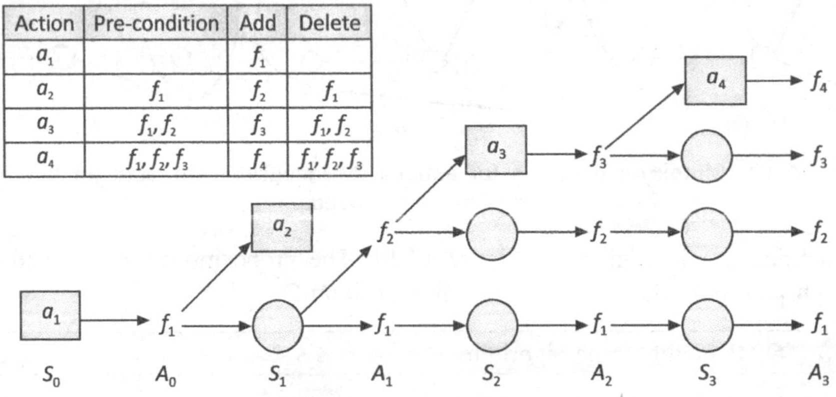

Action	Pre-condition	Add	Delete
a_1		f_1	
a_2	f_1	f_2	f_1
a_3	f_1, f_2	f_3	f_1, f_2
a_4	f_1, f_2, f_3	f_4	f_1, f_2, f_3

Figure 14.15 Planing graph example

Consider another example,

'Have coffee and drink coffee too'

Initial state: have(Coffee)

Goal: have (Coffee) ∧ Drink (Coffee)

Drink (Coffee)

Precondition: have(Coffee)

Effect: ¬have(Coffee)∧ Drink (Coffee)

Prepare(Coffee)

Precondition: ¬have(Coffee)

Effect: have(Coffee)

Figure 14.16 gives the solution in planning graph. Consider inconsistency and interference in the planning graph for the above problem. There are two options in inconsistent post conditions: one operation negates effect of the other; drink (Coffee) and no-op have (Coffee). The interference also has two options: a post condition of one operation negates a precondition of the other; drink (Coffee) and no-op have (Coffee). There are two options in competing needs: a precondition of one operation is *mutex* with a precondition of other because they are the negate of each other, like for prepare (Coffee) and drink (Coffee), or because they have inconsistent support. Two conditions in the inconsistency: each possible pair of operations that achieve the two conditions is *mutex*, have (Coffee) and drink (Coffee), are *mutex* in S_1 but not in S_2 because they can be achieved by drink (Coffee) and drink (Coffee).

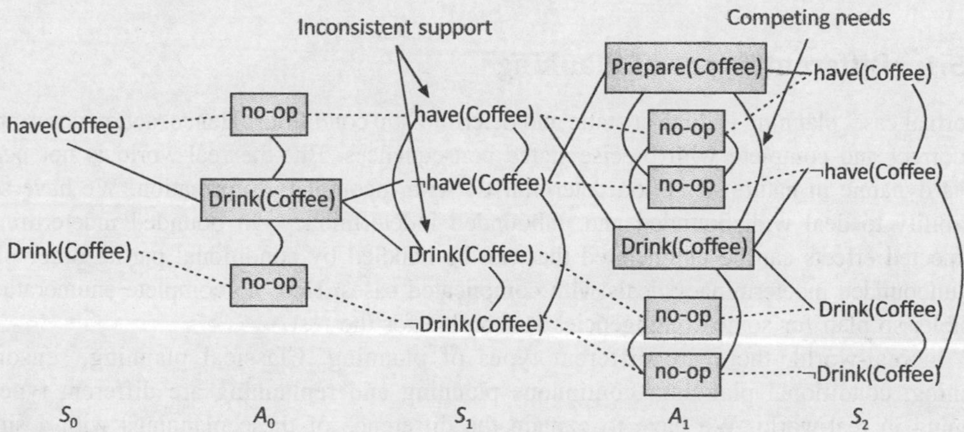

Figure 14.16 Planing graph example.

14.5 Planning in Real-World

In basic planning, there arise two questions. What to do and in which order? But in real-world planning with limited resources, questions asked are: what to do and when to do? In real-world, action occurs at certain moment of time with a beginning and end. Each action should take certain moment of time. For example, job shop scheduling problem. Job shop scheduling is the process with a set of jobs that consists a sequence of actions. Each action has a given duration and need resources. Minimise the total processing time of completion of all jobs is the constrains of job shop scheduling. Critical path method is used to determine start and end times. This means, path is a linear sequence from start to end and critical path is the path with longest total duration.

A job shop scheduling has finite set of *n* jobs, each job consists of a chain of operations in a finite set of *m* machines, one operation is handled at a time. A schedule can be introduced to these jobs and allocate these operations to a time interval of the machines that have minimal length. The operations need to process with minimum interruptions. Formally we can say that:

Job set, $J = \{J_1, J_2,.., J_n\}$

Machine set, $M = \{M_1, M_2,.., M_m,\}$

Operations, $O = \{O_1, O_2.., O_n\}$ with $O_i = \{O_{i1}, O_{i2}.., O_{im}\}$
Each operation has processing time, $\{t_{i1}, t_{i2}.., t_{im}\}$

A binary relation is defined as a procedure to the operation O. Let A induces the total ordering belonging to the same job. If $(v, w) \in A$, then v has to be performed before w. No precedence exists between operations of different jobs. A schedule is a function into the operation.

Artificial intelligence based planning helps to reduce the effective size of the search space. The area planning related work scheduling are assignment problem, transportation problem, travelling salesman problem, graph theory, flow networks, long-range planning and combinatorial optimisation.

14.5.1 Different Types of Planning

In normal case, planning is done at static and deterministic conditions. Hence, action descriptions are correct and complete with precise stated consequences. But the real-world is not perfect and is dynamic in nature. It is non deterministic with incomplete information. We have some possibility to deal with bounded and unbounded indeterminacy. In bounded indeterminacy, unexpected effects can be enumerated that can be handled by conditional partial order plans. The unbounded indeterminacy deals with complicated cases. Here no complete enumeration is possible, so plan for some contingencies and replan for the rest.

In real-world, there are different types of planning. Classical planning, sensorless planning, conditional planning, continuous planning and replanning are different types of planning in real-world. We have to explain the difference of these plannings with a simple example. Suppose we have a chair, a table and cans of different unknown colour paints. The goal is for the chair, and the table has the same colour. In classical planning, we have to check each can before painting. Sensor-less planning coerces to identify the choice of a can. The conditional plan identifies whether these paints are the same and painted with the available colour or choose both. In a continuous planning, painting can be stopped when one meets unexpected events then continues the painting process. Replanning offers to check the effect and identify the missing spot.

In conditional planning, we have to identify the number of possible conditions versus number of steps in the plan. Remember that only one set of conditions will occur at a time. In replanning, fix problems as they arise during execution. Breakable plans occur due to replanning. We need an intermediate plan between conditional planning and replanning. The most likely ones are done by conditional planning and the rest will be done by replanning.

14.5.2 Hierarchical Planning

A huge set of actions is involved in the real-world applications. For example, a detailed motor plan contains about 10^{10} actions. It is difficult to deal with such a huge number of actions. Artificial intelligence does what humans appears to do and higher level of plan abstractions are expected. Decompose abstract actions into several levels till we get actions required to be executed. The planning plays its role before and during the execution of plans. We can decompose abstract actions into several plans which are hierarchical in nature for managing complexity. In hierarchical planning, a task is reduced to small activities for the next level. The small plans are executed so as to reduce the computational cost.

Planning for 'Tour to Goa in this summer vacation', we have certain number of steps— switch on your desktop, open web browser, Open Indian Railways website (*www.irctc.co.in*), select date, class and train, book the ticket, pay the money online, etc. That is, practical problems are too complex to be solved at one level. One reason for such complexity is no structure and distinction between important and unimportant properties and there is no distinction between important and unimportant operators. For example, consider a cricket test match. The captain of team plans the order of five bowlers in two days of a test match for nearly 180 overs. The possibilities are $5^{180} = 25^{90}$.

Hierarchy of actions is made in terms of major action or minor action. The lower level activities would detail more precise steps for accomplishing the higher level tasks. For example, in the above example 'Tour to Goa', major steps are ticket booking, hotel booking, reaching Goa, local conveyance booking, staying and enjoying there and coming back. The minor steps are, take a taxi to reach station/airport, have candle light dinner on beach, take video/photos and a cruise trip. The major motivation of the hierarchical planning is to reduce the size of search space. Hierarchical planning limits the way in which an agent can select and order primitive operators, instead of having to try out a large number of possible plan orders. If we plan at the level of detailed actions it would be impossibly long. This is one of the naturals of 'intelligent' agents. Consider our cricket example, 180 overs means 15 spells (12 overs each). If there are 5 bowlers in 3 categories (2 pacer/2 spinner/1 pacer and 1 spinner), the top level possibilities are 3^{15} and total possibilities $< 3 \times 3^{15}$. Which is much less than 5^{180}.

A planner used in hierarchical planning is to identify a hierarchy of conditions, construct a plan in levels, postponing details to the next level and patch higher levels as details become visible. Such a planner is ABSTRIPS (Abstraction-Based STRIPS). ABSTRIPS is a modified version of STRIPS that incorporates hierarchical planning. ABSTRIPS provides a hierarchy of conditions that reflect the intrinsic difficulty of achieving various conditions. Main idea of ABSTRIPS is to introduce weights for each literal and consider only the most significant ones in first loop, then polish by considering literals with second highest weight.

For example, 'Travel to Goa', we can make required actions in hierarchical plan as, opening *www.irctc.co.in*(1), finding train(2), buy ticket(3), get taxi(2), reach station(3), pay-driver(1), check in(1), boarding plane(2), reach Goa(3). Here we can make hierarchical plans as:

Level 0 plan:
Tour to Goa

Level 1 plan:
buy Ticket(3), reach station(3), reach Goa(3)

Level 2 Plan:
Finding train(2), buy ticket(3), get taxi(2), reach station(3), boarding plane(2), reach Goa(3)

Level 3 Plan (final):
Opening *www.irctc.co.in* (1), finding train (2), buy ticket (3), get taxi(2), reach station(3), pay-driver(1), check in(1), boarding plane(2), reach Goa(3)

Performance of hierarchical planning comes down when the number of operator increases. The highly increased nodes in search trees for STRIPS and ABSTRIPS for a problem will reduce the performance in planning process.

14.5.3 Hierarchical Task Network

There are some drawbacks in STRIPS style planning. Compound goal and intermediate constraints are two such drawbacks. A compound goal is collected using operators similar to those employed. For example, a round-trip selection for a travel (New Delhi–Thiruvananthapuram–New Delhi). Immediate constraint is an action which is scheduled to the previous action. For example, in 'travel to Goa' trip, reach station, boarding train are immediate constraints. These problems are overcome by Hierarchical Task Network (HTN). Most of the practical planners in artificial intelligence are HTN.

A task network is a collection of tasks and constraints on those tasks. HTN contains hierarchy abstraction achieved through methods. A method is a pair (α, d), where α is the non-primitive task, and d is the task network to achieve the task α. Example of HTN is given in Figure 14.17. In our previous example, 'travel to Goa', we can represent the HTN as shown in Figure 14.18.

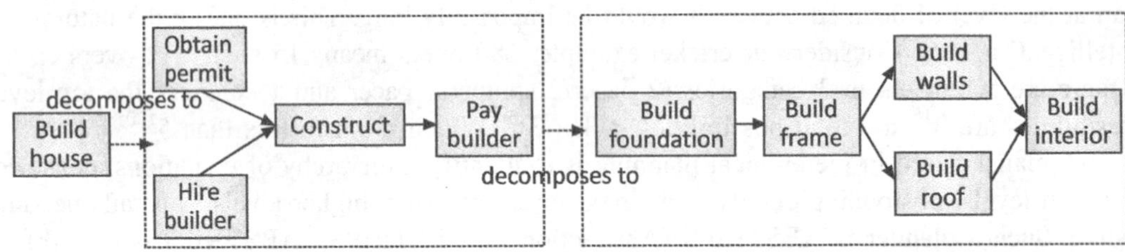

Figure 14.17 HTN example—building construction.

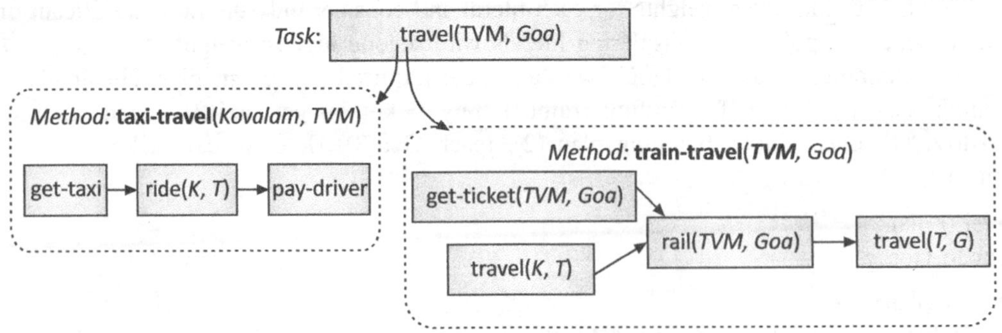

Figure 14.18 HTN example—travel to Goa. T, TVM: Thiruvananthapuram, K: Kovalam, G: Goa.

14.6 Case Studies

14.6.1 PDDL: Planning Robot Moves

Gripper task with four balls

A robot can move between two nearest rooms for taking balls, there are four numbers. Robot can pick up or drop balls with either of its two arms. At an initial stage, robot and balls are in room 1. We need to get all balls into room 2.

For this problem, the objects, predicates, initial state, goal specification and actions/operators are given below.

- **Objects:** The two rooms, four balls and two robot arms.
- **Predicates:** Is *x* a room? Is *x* a ball? Is ball *x* inside room *y*? Is robot arm *x* empty? [...]
- **Initial state:** All balls and the robot are in the first room. All robot arms are empty. [...]
- **Goal specification:** All balls must be in the second room.
- **Actions/Operators:** The robot can move between rooms, pick up a ball or drop a ball.

Gripper tasks for objects
Rooms: *rooma, roomb*
Balls: *ball1, ball2, ball3, ball4*
Robot arms: *left,right*

Corresponding in PDDL
(:objects *rooma roomb*
ball1 ball2 ball3 ball4
left right)

Gripper task in predicates
ROOM(x): true if *x* is a room.
BALL(x): true if *x* is a ball.
GRIPPER(x): true if *x* is a gripper (robot arm).
at-robby(x): true if *x* is a room and the robot is in *x*.
at-ball(x,y): true if *x* is a ball, *y* is a room, and *x* is in *y*.
free(x): true if *x* is a gripper and *x* does not hold a ball.
carry(x,y): true if *x* is a gripper, *y* is a ball, and *x* holds *y*.

Corresponding in PDDL
(:predicates *(ROOM ?x) (BALL ?x) (GRIPPER ?x)*
(at-robby ?x) (at-ball ?x ?y)
(free ?x) (carry ?x ?y))

Gripper task for initial state

ROOM(rooma) and *ROOM(roomb)* are true.

BALL(ball1), ..., BALL(ball4) are true.

GRIPPER(left), GRIPPER(right), free(left) and *free(right)* are true.

at-robby(rooma),at-ball(ball1, rooma), ..., at-ball(ball4, rooma) are true.

Everything else is false.

Corresponding in PDDL

(:init *(ROOM rooma) (ROOM roomb)*
(BALL ball1) (BALL ball2) (BALL ball3) (BALL ball4)
(GRIPPER left) (GRIPPER right) (free left) (free right)
(at-robby rooma)
(at-ball ball1 rooma) (at-ball ball2 rooma)
(at-ball ball3 rooma) (at-ball ball4 rooma))

Gripper task for goal specification

at-ball(ball1, roomb), ..., at-ball(ball4, roomb) must be true.

Everything else we do not care about.

Corresponding in PDDL

(:goal *(and (at-ball ball1 roomb)*
(at-ball ball2 roomb)
(at-ball ball3 roomb)
(at-ball ball4 roomb)))

Gripper task for Action/Operator – Movement operator

Description: The robot can move from *x* toy.

Precondition: *ROOM(x),ROOM(y)*and *at-robby(x)* are true.

Effect: *at-robby(y)* becomes true.
at-robby(x) becomes false.

Everything else does not change.

Corresponding in PDDL

(:action move :parameters *(?x ?y)*
:precondition (and (ROOM ?*x*) (ROOM ?*y*)(at-robby ?*x*))
:effect (and (at-robby ?*y*)(not (at-robby ?*x*))))

Gripper task for Action/Operator – Pick-up operator

Description: The robot can pick up *x* in *y* with *z*.

Precondition: *BALL(x),ROOM(y),GRIPPER(z),at-ball(x,y),at-robby(y)*and *free(z)*are true.

Effect: carry(*z,x*)becomes true.
*at-ball(x,y) and free(z)*become false.

Everything else does not change.

Corresponding in PDDL

```
(:action pick-up :parameters (?x ?y ?z)
        :precondition (and (BALL ?x) (ROOM ?y) (GRIPPER ?z)
                (at-ball ?x ?y) (at-robby ?y) (free ?z))
        :effect (and (carry ?z ?x)
                (not (at-ball ?x ?y)) (not (free ?z))))
```

Gripper task for Action/Operator Drop operator

Description: The robot can drop *x* in *y* from *z*.

(Preconditions and effects similar to the pick-up operator.)

Corresponding in PDDL

```
(:action drop :parameters (?x ?y ?z)
        :precondition (and (BALL ?x) (ROOM ?y) (GRIPPER ?z)
                (carry ?z ?x) (at-robby ?y))
        :effect (and (at-ball ?x ?y) (free ?z)
                (not (carry ?z ?x))))
```

TABLE 14.2 Practical artificial intelligence planners

Planner	*Reference*	*Applications*
STRIPS	Fikes 1971	Mobile robot control, etc.
HACKER	Sussman 1973	Simple program generation
NOAH	Sacerdoti 1977	Mechanical engineers apprentice supervision
NONLIN	Tate 1977	Electricity turbine overhaul, etc.
NASL	McDermott 1978	Electronic circuit design
OPM	Hayes-Roth 1979	Journey planning
ISIS-II	Fox et. al. 1981	Job shop scheduling (Turbine production)
MOLGEN	Stefik 1981	Experiment planning in molecular genetics
DEVISER	Vere 1983	Spacecraft Mission planning
FORBIN	Miller et al. 1985	Factory control
SIPE/SIPE-2	Wilkins 1988	Crisis action planning, oil spill management, etc.
SHOP/SHOP-2	Nau et al. 1999	Evacuation planning, forest fires, etc.
I-X/I-Plan	Tate et al. 2000	Emergency response, etc.

14.6.2 Piratical Planners

There are many practical artificial intelligence based planning softwares. The planners established before 2000 are listed in Table 14.2. Typically these planners have features like Hierarchical Task Network (HTN) planning, Partial Order Planning (POP), rich domain model, detailed constraint management, simulations and analyses. These planners are integrated with other systems like user interfaces, databases, spreadsheets, project management systems, etc.

14.7 Exercises

1. Describe the differences and similarities between problem-solving and planning.
2. Explain why dropping negative effects from every action schema in a planning problem results in a relaxed problem.
3. What is Sussman anomaly? How is it solved?
4. Write Tower of Hanoi domain in PDDL. Use 3 predicates: unary predicate clear, binary predicates on and smaller. One action move is sufficient.
5. Suggest a good heuristic for a forward planner for using in the robot delivery domain. How well does it work?
6. In a forward planner, we can represent a state in terms of the sequence of actions that lead to that state. How do you check that the precondition of an action is satisfied?.
7. Explain how multiple-path pruning can be incorporated into a regression planner. When can a node be pruned?
8. Prove that if a classical planning problem *P* is solvable, then there will always be an execution trace backward-search that returns a shortest-length solution for *P*.
9. Produce a planning graph for the block world problem shown in Figure 14.19(a).

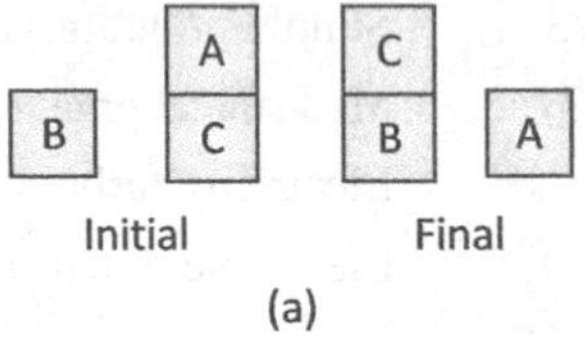

Figure 14.19 (a) Block world 1.

10. Compare and contrast planning and scheduling.
11. What are the principles of least commitment? How is it lined with partial order planning?
12. What is dynamic world planning? What is probabilistic planning?
13. Write SRTIPS codes for the block world problem shown in Figure 14.19(b).

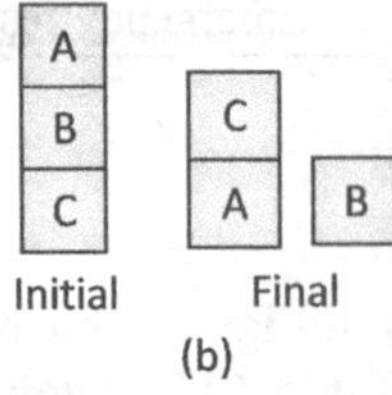

Figure 14.19 (b) Block world 2.

14. Explain how case based planning can be used in the development of planning system that is able to learn.

15. Identify the difference between STRIPS and ADL.

16. Formalise the following instance (Figure 14.20) of the bridges domain in PDDL and solve it with the same fast downward configuration, where each bridge may be traversed only once. Provide the runtime, number of expanded states and the plan that was found if the problem is solvable.

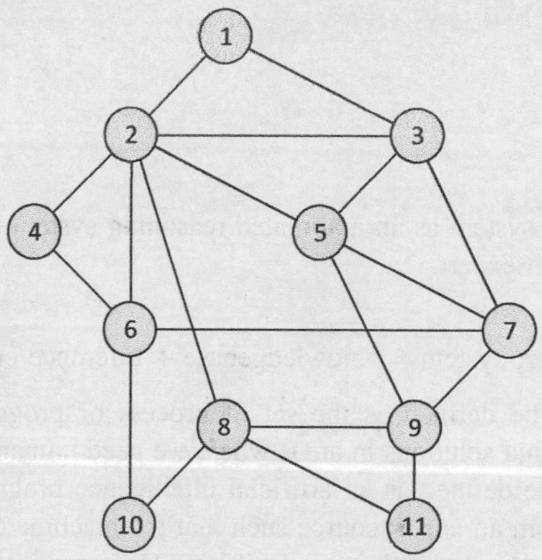

Figure 14.20 Example instance.

17. What are the components of a planning system?

18. What is non-linear planning? Where is it using?

19. Differentiate search and planning.

20. What are the differences and similarities between problem solving and planning?

21. What are the types of planner?

CHAPTER 15

Expert Systems

Rosenman defined expert system as an automated reasoning system that attempts to mimic the performance of the human expert.

Simply,

Expert system = Knowledgebase + Inference engine

Expert system can be defined as the set of process or programs which makes use of embedded knowledge to find solutions in areas where we need human expertise. In other words, expert system may also be defined as an artificial intelligence program in which the system's knowledge is obtained from an expert source such that the machine can offer intelligent advice or take an intelligent decision in solving a problem. Here artificial intelligence performs a decision-making task much faster than the humans can. Expert systems are not necessarily used to replace human experts. They can be used to make their knowledge and experience more widely available (for example, allowing non experts to work better). The areas applied to expert system is shown in Figure 15.1.

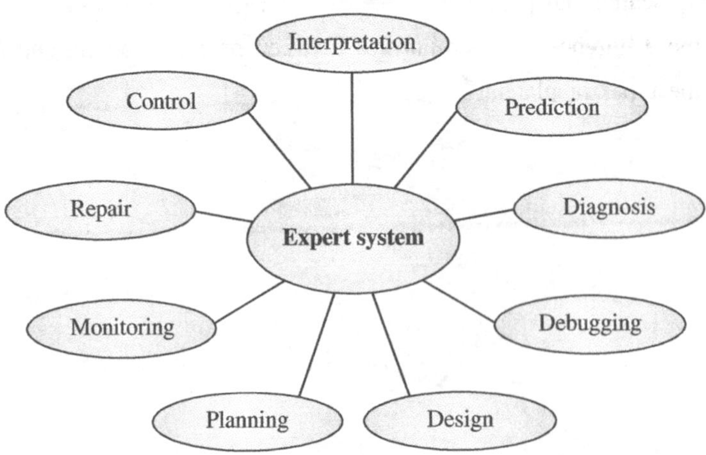

Figure 15.1 Expert system areas.

Expert systems were used in the following areas:

1. Credit granting
2. Information management and retrieval
3. Artificial intelligence and expert systems embedded in products
4. Plant layout
5. Hospitals and medical facilities
6. Help desks and assistance
7. Employee performance evaluation
8. Loan analysis
9. Virus detection
10. Repair and maintenance
11. Shipping
12. Marketing
13. Warehouse optimisation

Examples of the early and famous expert systems

DENDRAL Stanford University (1965)

Role: Analysis of chemical compounds

Method: Rule-based system

CADACEUS University of Pittsburgh (1970)

Role: Diagnosis of human internal diseases

Method: Rule-based system

MYCYSMA MIT (1971)

Role: Symbolic mathematical analysis

Method: Rule-based system

MYCIN Stanford University (1972)

Role: Diagnosis and treatment of infectious blood diseases

Method: Rule-based system

EMYCIN Stanford University (1978)

Role: Evolved from (Empty) Mycin

Method: Rule-based system, first expert system shell

XCON Digital Equipment Corporation (1980)

Role: configures VAX computer system

Method: Rule-based system. First expert system used commercially

15.1 Characteristics of Expert System

Expert systems explain their reasoning or suggested decisions and display intelligent behaviour. It can draw conclusions from complex relationships and provide portable knowledge.

1. For controlling the solution process, expert systems make use of knowledge rather than data.
2. The encoded knowledge and the control programme are maintained as separate entities. The control programme is compiled separately and this permits the refinement of knowledge at a later stage. Even if new knowledge is added the source programme needs not be recompiled as it stands as an independent unit.
3. Expert systems can explain how a particular conclusion was made, and the type of requested information that is needed during a consultation.
4. Symbolic representations are used by expert systems to display knowledge, rules or networks. The inferences are also computed symbolically which resemble closely to the natural language manipulations.
5. Expert systems often reason with meta-knowledge, that is, they reason with knowledge about themselves, and their own knowledge limits and capabilities.
6. Many a times, experts differ opinion on processes and conclusions.
7. Process will not be that easy to codify.

Although the words expertise and intelligence are related, they are not the same things. It takes time to get expertise in a particular domain but getting intelligent is easy by simply learning the facts. Intelligence allows you to use your expertise and apply the knowledge at a particular place. Thus, expertise enables to find solution quickly. Capabilities of expert systems are shown in Figure 15.2.

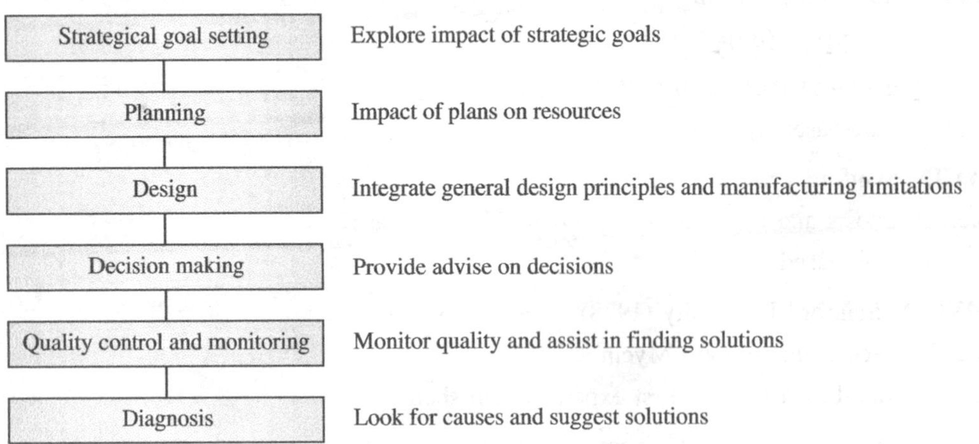

Figure 15.2 Expert system capabilities.

Use an expert system

Experts often do not really know how they do things themselves. Although the expert can perform the tasks, he does not necessarily have access to the mechanism used.

The following are the tasks performed by experts:

1. Provide a high potential pay-off or significantly reduced downside risk.
2. Catch and store irreplaceable human expertise.
3. Provide support at numerous locations at the same time or in an hostile environment which is dangerous to human health.
4. Provide expertise which are expensive or rare.
5. Produce effective and more optimised solutions faster than human experts.
6. The expertise required for training and development is provided. It also helps in sharing the wisdom of human experts with a larger mass of people.

Limitations

1. Not widely used or tested.
2. The system performed well only to relatively narrow situations/problems.
3. Cannot readily deal with mixed knowledge.
4. Probability of error is high.
5. Cannot refine own knowledgebase.
6. Difficult to maintain.
7. The cost of development is also high.
8. Raise ethical and legal concerns.
9. While having large number of rules, the consequence of adding new rules will be difficult to assess.
10. Experts hardly codified into rules.
11. Experts often lack access to their own analysis mechanisms.
12. Validation/Verification of large systems is very difficult.
13. Track record does not seem to contain many successes. Relatively high-risk to implement.

15.2 Components of an Expert System

Quick summary

Components of an expert system are given in Figure 15.3. There are external and internal components for an expert system. Each of these components has a specified role.

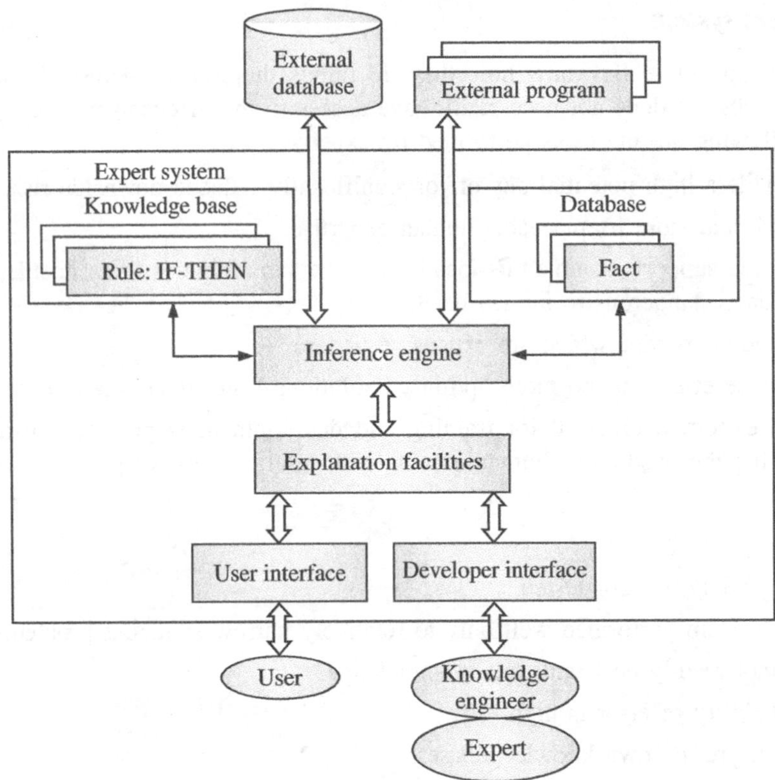

Figure 15.3 Components of an expert system.

1. Inference engine: Mines the knowledge database to find relationship and information for prediction, query analysis and suggests results in the same way as human experts do. Inference engine examines the knowledgebase to answer questions, solve problems or make decisions within the domain. Theorem provers or logic programming language, production systems (rule-based) or frame systems and semantic networks are some methods of inferencing. The inference movement is either forward chaining(normally used for analysis and interpretation, for example, DENDRAL (1971) used for finding the molecular structure of soil sample) or backward chaining (widely used for diagnosis, for example, MYCIN (1976) used for the diagnosis of infectious blood diseases).
2. Rules: A conditional statement always maps given conditions to particular actions or outcomes.
3. Explanation facility: It is that part of the expert system, that helps the user to understand how the system came to a particular conclusion.
4. Knowledge acquisition facility: It gives an efficient method of capturing and storing the components of the knowledgebase.

The heart of expert system is the knowledgebase, which contains the knowledge of a particular application domain. In a rule-based expert system this knowledge is represented in the form of **if . . . then . . .** rules.

The knowledgebase contains facts and rules about some specialised knowledge domain. The knowledgebase contains both general knowledge as well as case specific information.

The inference engine accepts user input queries and responses to questions through the I/O interface and uses this dynamic information together with the state knowledge stored in the knowledgebase. When consistent match is found, corresponding rues are placed in a conflict set. To find an appropriate and consistent match, substation may be required. The inferring process is carried out recursively in three stages (Figure 15.4).

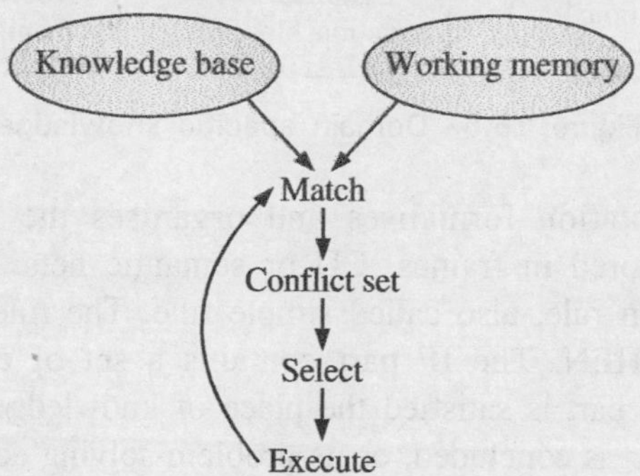

Figure 15.4 The production system inference cycle.

1. Match
2. Select
3. Execute

During the match stage, the contents of work memory are compared to fact and rules contained in the knowledgebase. When consistent matches are found, the corresponding rules are placed in conflict set. To find an appropriate and consistent match, substitutions may be required.

15.2.1 Building Blocks of Expert Systems

The knowledgebase and the reasoning (inference) engine are the two major parts of every expert system. One of the most difficult problems in the building of expert systems is capturing the knowledge from the experts. There are many knowledge elicitation techniques to aid this process, but the fundamental problem remains: things that are normally implicit, inside the experts head, must be externalised and made explicit.

Factual and heuristic knowledge are the knowledgebase of an expert system. Factual knowledge is that part of knowledge of the task domain, which is widely shared and available. For example, knowledge found in textbooks or journals. These types of knowledge are commonly agreed by others in that particular field. Domain specific knowledge inferencing is shown in Figure 15.5.

Heuristic knowledge is less stringent but more of experiential and judgemental knowledge of performance. It is seen as largely individualistic knowledge and rarely discussed. Factual knowledge can been seen as a scheme which is ‘opposite to heuristic knowledge. It is the knowledge of good practice, good judgement, and plausible reasoning in the field. It is the knowledge that underlies the ‘art of good guessing’.

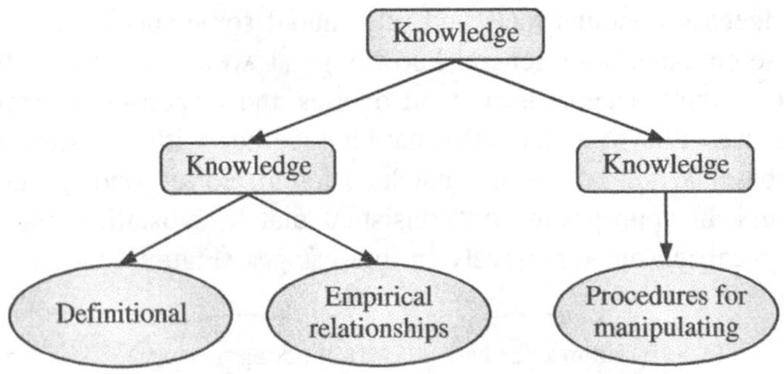

Figure 15.5 Domain specific knowledge.

Knowledge representation formalises and organises the knowledge (Figure 15.6). The knowledge can be stored in frames, CD or semantic nets, etc. Most commonly used representation is production rule, also called simple rule. The rule contains a condition part, IF and an action part, THEN. The IF part contains a set of conditions in some logical combination. When the IF part is satisfied the piece of knowledge represented by the rule is relevant and the THEN part is concluded, or its problem-solving action is taken. Such systems where the knowledge is represented in the rule form are called rule-based systems.

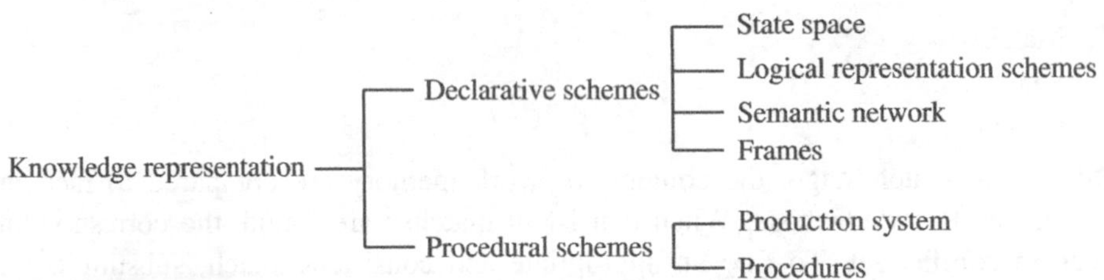

Figure 15.6 Domain knowledge representation.

Another commonly used representation is the unit (also termed as frame, schema, or list structure). Its concept basically depends more on the passive view of knowledge. The unit is a package of associated symbolic knowledge about an item to be represented. Property of the attributes and associated values are generally included in a unit.

Every task contains various attributes which are used in many relations. The property is used to specify the relations and the values of the property denote the name of other units linked to it. One unit can represent knowledge about another unit, or some units can become part of another unit.

The problem-solving model controls and organises the steps which are used for solving the problem. One common but powerful paradigm involves chaining of IF-THEN rules to form a line of reasoning (sample rules are given in Figure 15.7). Forward chaining is the method where the process of chaining starts from a set of conditions and proceeds to a basic conclusion. In some cases, the conclusion that needs to be obtained will be known, but the process of how to reach the goal remains unknown. This kind of chaining is called backward chaining. For reasoning, there should be some engines or procedures to act upon knowledge in the database.

1. IF engine_getting_petrol & engine_turns_over
   THEN problem_with_spark_plugs
2. IF NOT engine_turns_over & NOT lights_come_on
   THEN problem_with_battery
3. IF NOT engine_turns_over & lights_come_on
   THEN problem_with_starter
4. IF petrol_in_fuel_tank
   THEN engine_getting_petrol

Figure 15.7 Sample rules related to a vehicle.

The knowledgebase can be anything that one gains from experience or the things one studied in school or college. The more exposure to the experience, the more fit he becomes for the task. It helps him in more sharp design, analysis and debugging. Basically the expert system consists of a knowledgebase and the inference engine, but we can also mention about reasoning with uncertainty and explanation of the line of reasoning.

Data are never complete and so is the knowledge. In every database or a dataset there will be some values that are missing or some attributes whose value is indeterminable. These missing values can be so critical, and hence there must always be some methods for finding the missing values and unknown knowledge. For that, a rule may have associated with it a confidence factor or a weight. In the process of reasoning, there are methods for making use of such uncertain data along with uncertain knowledge. Such a process is termed as reasoning with uncertainty. A major subclass of this method of reasoning with uncertainty is called 'fuzzy logic,' and such systems are called 'fuzzy systems'.

It is human nature to believe things only after getting a valid justification to a situation or a question. Similarly as an expert system uses uncertain or heuristic knowledge, its credibility lies in the area of responding to the questions raised against it. Even then, the answer to the question can be again questioned. So most of the expert systems should have the ability to answer the questions of the form: 'Why is the answer *X* and not *Y*?' Explanations are generated by tracing the line of reasoning by the inference engine.

Knowledge is the basic ingredient in any expert system. The strength of the system resides on the quality and extent of knowledge available for a particular task. The researchers working in the area of artificial intelligence will explore more into these methodologies and add new techniques for the representation of knowledge and for the process of reasoning. But as we mentioned before, the whole power of the system resides in knowledge. Current knowledge acquisition methods are slow and tedious; a vast unexplored area is still open to researchers. The main aim will be to find methods to remove the data bottlenecks and to find new methods for knowledge represented.

Types of expert system

Based on the implementation, we can categorise the expert system as follows:

- **Rule based expert system:** Represented as a series of rules, basically if.. then .. rules.

- **Fuzzy expert system:** Fuzziness of information is caused by appearance of noise, caused by observations (especially on bounds of intervals of working parameters), incompleteness of description of states classes, fuzziness of bounds between states classes and, respectively, between classes of its diagnostic features.
- **Frame based system:** Representation of the object-oriented programming approach and using frame data structure for knowledge storage.
- **Hybrid System:** Include several knowledge representation approach combined to handle a single unit.
- **Model based system:** Structured around the model that stimulates the structure and function of the system under study.
- **Readymade (Off-the-shelf) system:** Custom-made, similar to application package such as an accounting general ledger or project management in operation management.

15.3 Expert System Development

Expert system development is involved in many steps (Figure 15.8). The expert system development is done though many experts including domain expert, knowledge engineer and programmer (Figure 15.9). Primary step is to identify the area of knowledge that needs to be addressed by the expert system. Major knowledge contributors of the expert system are domain expert (the individual or group whose expertise and knowledge is captured for use in an expert system), knowledge user (the individual or group who uses and benefits from the expert system) and knowledge engineer (someone trained or experienced in the design, development, implementation, and maintenance of an expert system). The stages of expert system development can be summarised as follows:

1. **Identification and definition of problem:** Determination of the scope of the problem and its characteristics and then identifying the experts, resources needed, and goals of the expert system. Next step is to determine whether problem can be addressed by the expert system.

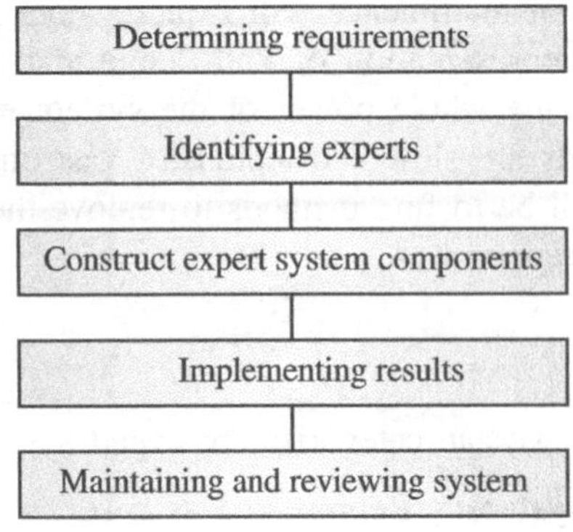

Figure 15.8 Steps in expert system development.

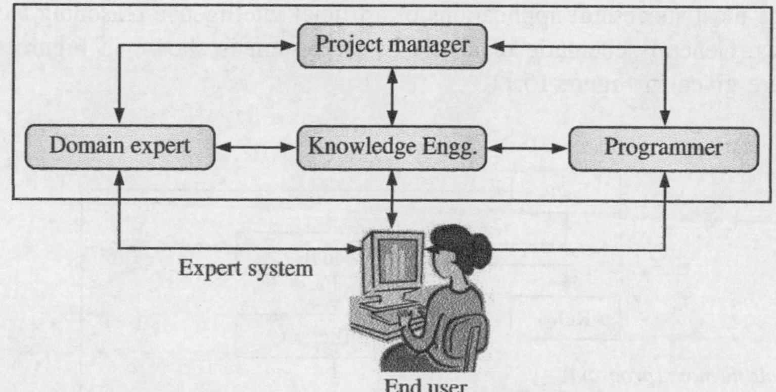

Figure 15.9 Main players of expert system development.

2. **Conceptualisation:** It is simply knowledge acquisition step which decides on possible decomposition of problem into subtasks. Study these subtasks in detail, nature of domain knowledge to identify concepts required to arrive at a solution of the problem and consider the control strategy for properly using these concepts.
3. **Formalisation:** Knowledge engineer selects suitable representation schemes for the domain knowledge and then the knowledge is organised and formalised on paper using knowledge nets or decision trees, so that domain expert can also browse through to check for inconsistency or redundancy. These inference mechanisms are selected depending on the flow of information and the underlying structure of the domain.
4. **Item implementation:** An expert system development tool of shell is selected, which has the capabilities to represent the formalisms and inference mechanisms adopted in the previous stage. Knowledgebase is coded according to the syntax of shell interpreter and the program code is developed in computer.
5. **Testing and validation:** The developed expert system is made to run on a large number of representative test cases and debugged. Validation is done with the help of domain expert.
6. **Integration:** This is an often overlooked and unattended stage of expert system development. Developed expert system is integrated with the existing systems or with the day-to-day operations of the organisation.
7. **Maintenance:** Minor modification and corrections in the programs are done if necessary. The knowledge update is also highly required.

15.3.1 Rule Based Expert Systems

In many applications, the system has access only to uncertain rules, and the user may not be able to answer questions with certainty. A rule based expert system suitable for the diagnosis of a system. MYCIN (Shortliffe in 1976) is one of the systems which uses bacterial infections.

It is one of the most successful applications of artificial intelligence reasoning technique using facts and rules. General schematic of a rule based system is shown in Figure 15.10 and its components are given in Figure 15.11.

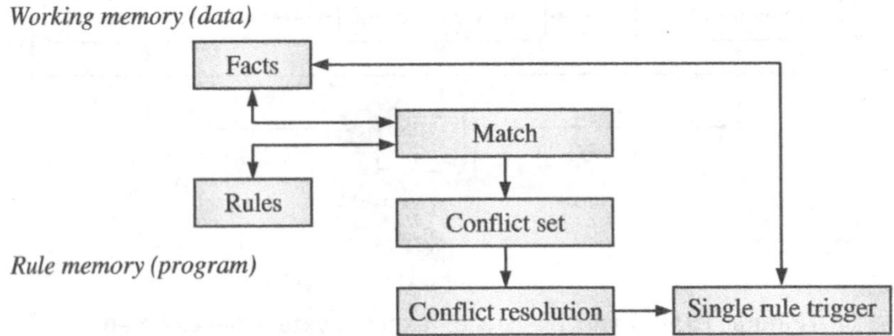

Figure 15.10 Rule based expert system.

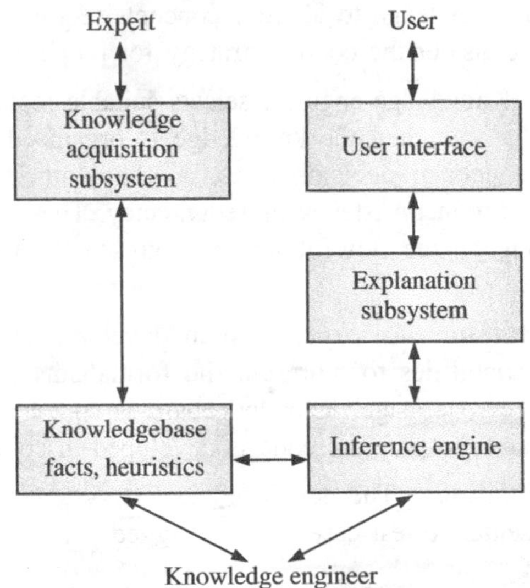

Figure 15.11 Components of a rule based expert system.

A rule based expert system's knowledgebase consists of predicate calculus facts and rules about subject at hand. The inference engine consists of all the processes that manipulate the knowledgebase to deduce information requested by the user (Figure 15.12). An inference chaining example forming a set of rules is shown in Figure 15.13. Here three rules are given. When the condition part of a rule is satisfied, the rule is said to fire and the action part is executed. The knowledgebase includes a set of facts used to match against the IF (condition) parts of rules stored in that knowledgebase. The inference engine searches fro AND/OR proof tree using either backward or forward chaining. The forward chaining (data-driven search) starts with known data (facts). Fire the rules that have an antecedent that match facts in the database, and add any resulting facts to the database. Each rule can fire only once and stop when no more rules can fire. In backward chaining (goal-driven search) system has hypothetical solutions and tries to prove it. It finds rules that consequence a match the goal and stops if it reaches to antecedents match in the facts. If not, make the antecedents the new subgoals, and repeat the cycle. This process is given in Figure 15.14.

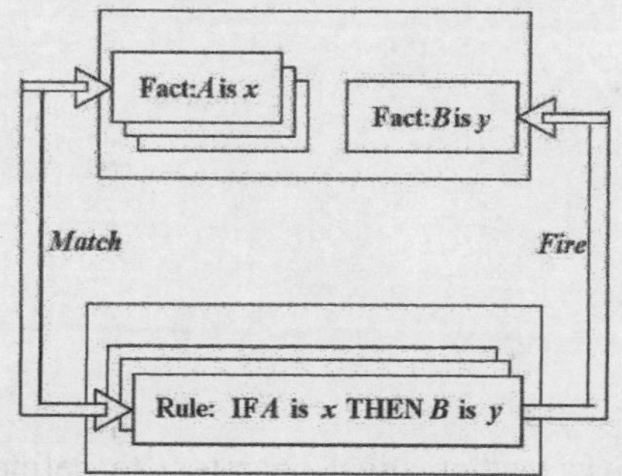

Figure 15.12 Inference engine cycles via a match-fire procedure.

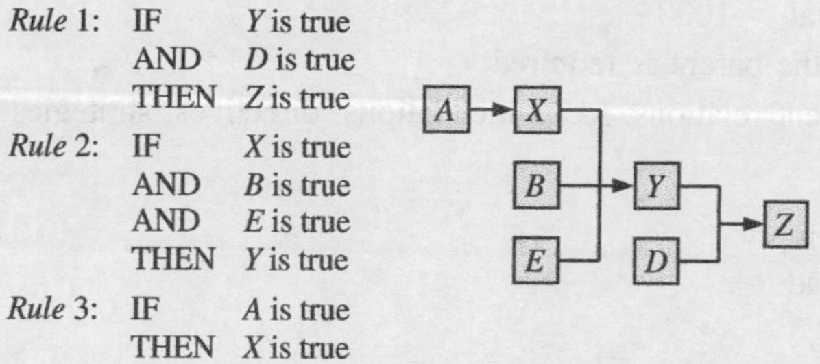

Figure 15.13 An example of inference chain.

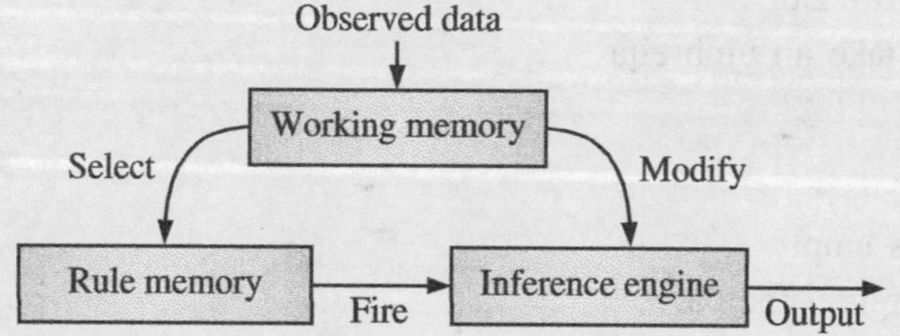

Figure 15.14 Reasoning with production rules.

The knowledge acquisition subsystem checks the growing knowledgebase for possible inconsistencies and incomplete information. Some kinds of natural language processing systems or graphical user interfaces with menus are included in the user interface. The expert system knowledge engineer of the rule based system works with an expert in the field of application in order to store a form of knowledge into the knowledgebase of the expert.

Production rules

The antecedent of a rule incorporates two parts: an object or linguistic object and its value. The object and its value are linked by an operator. The operator identifies the object and assigns the value. Operators such as *is, are, is not, are not,* are used to assign a symbolic value to a linguistic object. The production rules is as follows:

IF ⟨*antecedent*⟩
THEN ⟨*consequent*⟩

For example:

IF fish is floating
THEN fish is dead
Alternate syntax:
Fish is floating fish is dead.

Expert systems can also use mathematical operators to define an object as numerical and assign it to the numerical value. For example,

IF age of the customer < 18
AND cash withdrawal > 1000
THEN signature of the parent is required

Rules can represent relations, recommendations, directives, strategies and heuristics:
Relation

IF the fuel tank is empty
THEN the car is dead

Recommendation
IF the season is autumn
AND the sky is cloudy
AND the forecast is drizzle
THEN the advice is take an umbrella

Directive
IF the car is dead
AND the fuel tank is empty
THEN the action is refuel the car

Strategy
IF the car is dead
THEN the action is check the fuel tank;
step1 is complete

IF step1 is complete
AND the fuel tank is full
THEN the action is check the battery;
step2 is complete

Heuristic
IF the spill is liquid
AND the spill pH < 6
AND the spill smell is vinegar
THEN the spill material is acetic acid

Multiple antecedents
Multiple antecedents may be connected by conjunction (AND) and/or disjunction (OR).
IF animal is horse-shaped
AND animal has stripes
THEN animal is zebra

IF animal is hippo
OR animal is lion
THEN animal is dangerous
Multiple consequences are possible, and are connected by conjunctions. For example,

IF tsunami alarm is sounding AND date is not first Monday in month
THEN Condition is dangerous AND
Advice is move away from the ocean

Conflict resolution

What happens when two rules provide conflicting conclusions? For example,

If it has feathers then it is a bird
If it can not fly then it is not a bird
What if has feathers, but can not fly?

Major problem is that, more than one rules fire at once. Conflict resolution strategy decides which conclusions to use. Give rules priorities and to use the conclusion that has the highest priority. Apply the rule with the longest antecedent and apply the rule that was most recently added to the database. For example,

We can consider three simple rules for crossing a road as follows:

Rule 1:
IF the traffic light is green
THEN the action is go

Rule 2:
IF the traffic light is red
THEN the action is stop

Rule 3:
IF the traffic light is red
THEN the action is go

If we have two rules, Rule 2 and Rule 3, with the same IF part. Thus, both of them can be set to fire when the condition part is satisfied. These rules represent a conflict set. The inference engine must determine which rule to fire from such a set. A method for choosing a rule to fire when more than one rule can be fired in a given cycle is called conflict resolution.

In forward chaining, both rules would be fired. Rule 2 is fired first as the topmost one, and as a result, its THEN part is executed and linguistic object action obtains value stop. However, Rule 3 is also fired because the condition part of this rule matches the fact 'traffic light is red', which is still in the database. As a consequence, object action takes new value 'go'.

A subset of the rules in a knowledgebase that can fire at the same time, but have inconsistent consequence. For example,

X is dog → X is not smart

X is dog and X has breed = 'border colli' → X is smart

For conflict resolution,

1. Use rule-order as an implied priority and the first rule to provide an answer is used.
2. Assign a priority to each rule, the rule with higher priority is sustained.
3. Longest matching strategy uses the rule with most specific information (For example, If it cannot fly and has feathers then it is a bird).
4. Certainty based conflict resolution is used to measure certainty provided for data and rules. Most certain rule is sustained.
5. Highest priority rules are fired first. In simple day to day applications, by placing the rules in the appropriate order in the knowledgebase, we can establish priority.
6. Fire the most specific rule (longest matching strategy). This assumes that a specific rule processes more information than a general one (For example, border collie rule would fire before more general dogma rule).
7. Fire the rule based on the data most recently entered. This assumes a recent data more important than older data.

15.3.2 Meta Rules and Meta Knowledge

Meta rules are those which decide the conflict resolution strategies. They define the knowledge part on how the systems work. For example, if a problem is being solved by two experts Meta rules might predict that the knowledge from the expert X can be more trusted than from the expert Y. The Meta rules are considered as normal rules by the system but a high priority is given to Meta rules.

So, how can we define Meta knowledge? It is the knowledge about knowledge. It can be about the knowledge on the use and control of domain in an expert system. In the case of rule-based expert systems, meta-rules represent the meta-knowledge. The meta-rules describe the procedure for the use of task-specific rules in the expert system.

15.4 Knowledge Engineering

The art of designing and building the expert systems is called as knowledge engineering, and knowledge engineers are its practitioners. Gerald M. Weinberg in The Psychology of Programming mentioned 'Programming,' as 'loving,' because it is a single word that encompasses infinitude of activities. Knowledge engineering is the same, perhaps more so. We have defined before that knowledge engineering is the application stream of artificial intelligence which comes under the area of computer science. Then theoretically we may say that knowledge engineer is a computer scientist who designs and implements program under the stream of artificial intelligence using its concepts. Now, the face of knowledge engineering is changing, and so are knowledge engineers. We can see the evolution of knowledge engineering later.

Now, we have two different ways to build an expert system. First one is to begin working from the scratch and the other using some development software called tools or shells. Let's first look at the role of knowledge engineers before mentioning about the tools. Even though there exist various advanced knowledge engineering techniques and styles, the primary approach is the same. The knowledge engineer observes and interviews a group of experts and checks how they use their knowledge to justify the results. The engineer then translates the knowledge into a computer that has an interpretable format and creates designs an interface engine. He also designs the structure of reasoning and the way the knowledge should be accessed. He also decides how to integrate the uncertain knowledge in reasoning, and what kinds of explanation would be useful to the end user.

Next level of work involves the incorporation of inference engine and knowledge representation facilities into the program. The domain knowledge is also entered sidewise to the program. Sometimes the inference engines will not be suitable, which make feel that the knowledge adapted is not the correct one and the experts will finally decide not to take such inputs. All these results are studied and modified as the expert system gradually gains competence.

Finding new ways of knowledge representation and new techniques of reasoning always comes under the scope of artificial intelligence researchers. Finding and gathering of knowledge of a task domain comes under the roof of domain experts. Domain knowledge includes the knowledge that we get from the formal textbooks, and experiential knowledge—the expertise of the experts.

15.5 Applications of Expert System

Applications of Expert system include the area of technology, industry, commercial, medical and many. The applications find their way to most areas of knowledge in the field. Expert system applications can be clustered into the following classes:

1. **Diagnosis and troubleshooting of devices:** It comprises system that deduces faults and suggests corrective measures for a malfunctioning device or process. One of the first such areas is medical diagnosis, which uses rule based technology to build the expert system. Troubleshooting for a device (for example, some parts of a robot)

is based on proper diagnosis. This is basically achieved by experience (knowledge base).

Expert systems have found various applications in the field of power plant management and monitoring. Fault diagnosis is one of the main areas where the expert system technology excels as it mimics the problem-solving characteristics of an expert plant operator. Various expert systems have also been developed for alarm processing, plant maintenance, fault prediction operator training, evaluation, alarm processing and design. Based on methodology used there are many types of expert systems such as rule based, frame based, object oriented, fuzzy based, genetic algorithmic and neural network, hybrid, etc. Expert systems have found broad application in fault diagnosis field from their early stages because an expert system acts as a human expert itself when a fault occurs in the plant giving directions to diagnose the fault and taking corrective actions to restore the system operations as early as possible.

2. **Planning and scheduling:** To analyse a set of one or more potentially complex and interacting goals in order to determine a set of actions to achieve those goals. This will provide a detailed temporal ordering of those actions, taking into account personnel, material and other constraints. Planning is highly essential in most of the critical applications.
3. **Financial decision making:** Use of expert system shells is very important in financial decision making companies such as stock exchange, insurance companies, banks, etc. The expert system shells are blended with advisory programs which provide financial services to the user. A typical application in the financial markets is in foreign exchange trading.
4. **Knowledge publishing:** Primary objective of this method is exploring knowledge in a knowledgebase. For example, a tax advisor provides tax programs which advise the use on tax strategy, tactics, and individual tax policy.

15.6 Case Studies

First we have a look into building of a simple medical expert system. This is our first case study. Next what are the old successful expert systems? Discussion of old expert systems are considered as the second case.

15.6.1 A Simple Medical Expert System

Medical expert systems are commonly used to diagnos patients and suggest a medicine. The following rules are assigned to this system.

Rules

1. If headache then prescribe pain killer
2. If headache and sore throat and coughing than diagnose flu

3. If tired and headache then diagnose glandular fever
4. If tired and sore throat then diagnose tonsillitis
5. If tired than diagnose stress

IF patient has pain THEN prescribe pain killers (priority 10)
IF patient has chest pain
THEN treat for heart disease (priority 100)

IF patient has pain AND patient is over 60
AND patient has a history of heart conditions
THEN take to emergency room

User interaction Patient: I have a headache
System: I suggest you take a pain killer. Do you have a sore throat?
Patient: Yes, I do.
System: Have you been coughing?
Patient: No, why do you ask?
System: If you had a headache and a sore throat and had been coughing, I would have to conclude that you might have the flu. It does not sound as though you have the flu? Have you been feeling tired recently?
Patient: Yes, I have.
System: It sounds as though you may have glandular fever or tonsillitis.

15.6.2 ORTESS – Expert System Shell for Power Plants

A process power plant is a complex system which contains enormous machineries and equipments. It has both critical and real system characteristics. So, the maintenance of a power plant is an important issue. More than these, it holds a huge cost factor for its maintenance. The plant should be continuously available for power generation for a grid. In case a problem happened, the cost and time to reinstate the power plant is a crucial issue. So the better option to maintain the power plant is accommodating a facility for preventive maintenance. General architecture of the process power plant is given in Figure 15.15. Supervisory control And Data Acquisition (SCADA) is a data acquisition and control system for power plant operations. The SCADA can give the sensory status of the power plant parameters, and equipment and subsystems. The plant operator is the direct stake holder of this system. The unavailability of an expert plant engineer can also cause problems in plant operations. The ORTESS is an expert system shell which has been developed by CDAC.

ORTESS has a knowledgebase for representing rules and capability for doing diagnosis of a fault. But it cannot do a prognosis of plant fault condition based on the online SCADA log data. In order to provide monitoring and fault forecasting facility to ORTESS a new module has to be added. Based on the SCADA online log data as input the entire power plant shall be monitored and any faulty situation, if it culminates, shall be forecast. A module can be developed from this knowledge. This module can advise and guide the plant operator regarding the precautions and remedies to be taken against an irregular situation to avoid a critical alarm to be happened.

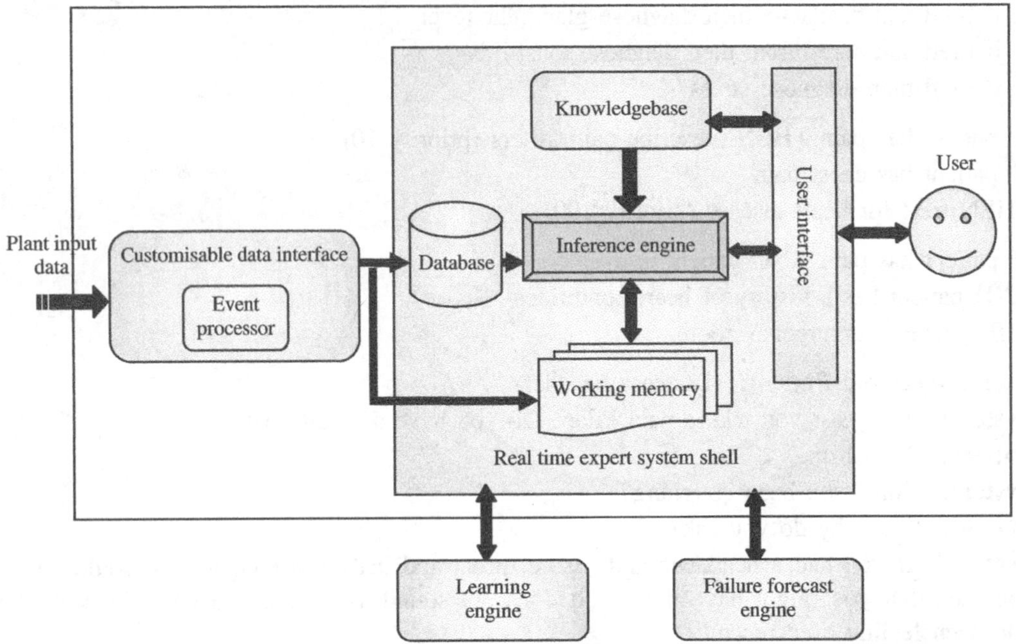

Figure 15.15 Architecture of the expert system for process plant.

Analysis of large scale of data from SCADA to formulate the knowledge on the live plant health condition and the possibility of future failures is a cumbersome task for a plant operator even though the operator is an expert. The existing ORTESS can only perform diagnosis of a 'black out' situation that already happened based on its rule base. The perspective of forecasting engine is prognosis of plant fault conditions. It is developed as a module of ORTESS to extend the functionality of the existing shell to monitor and forecast the plant events by analysing the online SCADA log data. It can monitor and forecast all critical alarm events which are already learned and stored in the rule base of ORTESS.

Main inputs of the learning engine are the knowledgebase of ORTESS and historic data. It interfaces with the ORTESS to access and update the knowledgebase while learning. The proposed system requires two types of datasets, i.e. training dataset and test dataset. The training dataset is the pre-processed historic data logged via open platform communication. The training set is a set of data to discover relationships between different alarm tags to be used as the input to the learning engine to create model. The test dataset is pre-processed historic data in the same format as the training dataset. The test set is a set of data to assess the strength and utility of the model and to test and verify that the model created is valid. The expected output is the rules created in the form of IF - THEN. The domain expert has to confirm the new rules created before updating in knowledgebase.

SCADA is a data acquisition and control system for power plant operations. The SCADA can give sensory status of power plant parameters, and equipment and subsystems. The plant operator is the direct stake holder of this system. The unavailability of an expert plant engineer can also cause problems in plant operations. The different modules for learning engine are the data pre-processing of the historic data from the SCADA system, rule mining, post processing and rule validation. The input to learning engine is the historic data from SCADA system, during pre-processing the data with quality zero are removed and the data is converted to a form that is suitable for learning engine. Learning engine processes the data and gives valid rule as output. The rules thus generated will contain rules for normal condition along with critical alarm condition. The normal condition rules are eliminated during post processing module. Finally, the rules are to be validated by the expert and update the knowledgebase. The plant people such as plant engineers, plant experts and plant operators are real stake holders of this work.

15.6.3 Successful Expert Systems

Table 15.1 shows summary of some earlier expert systems which was successfully demonstrated and tested by different experts. Most of these systems are developed at universities.

TABLE 15.1 Expert systems summary

System	*Representation*	*Inference mechanism*
MYCIN	Rules in LISP	Backward chaining
PROSPECTOR	Rules and semantic net	Forward chaining and Bayesian reasoning
DENDRAL	Rules	Graph generation and test (using forward chaining)
XCON	Rules	Forward chaining; subtasking and constraint satisfaction
ESIE	Rules	Backward chaining

R1/XCON

This is a rule based system which was developed by DEC and CMU for configuring Vax computers. R1 represents the prototype and XCON denotes the commercial application. Customer order is the primary input to the system and the corrected order with diagrams showing component layout and wiring suggestions are the output from the system. The system gives output in minutes what used to take humans' days and has a much lower error rate. This system uses forward chaining in the inference engine. The steps included for this expert system include check order for missing/mismatched pieces, layout processor cabinets, put boxes in input/output cabinets and put components in boxes, put panels in input/output cabinets, layout floor plan and finally indicate cabling of the machine.

Example rules for XCON expert system is as follows:

IF context is layout and
you are assigning power supply
THEN
add appropriate power supply

ESIE

This is a freeware expert system shell written in Pascal language that uses backward chaining inference mechanism. The conflict resolution is rule ordering (can use maximum specificity with careful rule placement). The facts stored as object/value pairs can use 100 question rules and 400 if-then rule lines.

MYCIN

MYCIN was developed at Stanford in the 1970s by Feigenbaum, Buchanan and Ted Shortliffe. This expert system was recommended therapy for blood/meningitis infections and the diagnosis normally involved growing cultures of the infecting organism within 48 hours. Major beneficiaries are doctors who can prescribe drugs to deal with immediate problems. The system was developed to explore how doctors make these rough, but important, guesses with partial information. It is also important in practice as there are many junior doctors or non-specialised doctors.

MYCIN is a goal-directed system that uses a basic backward chaining technique with 450 rules written in LISP. It performed as well as some experts and significantly better than junior doctors. MYCIN was good in a way that it could calculate dosages very precisely. MYCIN dealt well with interactions between drugs. It retrieves and displays relevant rules to offer explanation of its behaviour.

Example rule used in MYCIN,

IF

The site of the culture is blood

The gram of the organism is negative

The morphology of the organism is rod

The burn of the patient is serious

THEN

there is weakly suggestive evidence (0.4) that the identity of the organism is pseudomonas.

The certainty factor is 0.4 that gave conclusion to the given evidence.

15.7 Exercises

1. What is different between human expert and expert system? Can you say that one is more intelligent than another?
2. What are the fundamental characteristics of an expert system? What are the differences between expert systems and conventional programs?
3. What is different between intelligence system and expert system?
4. Can a computer do what an expert does?
5. What are the mechanisms provided by a typical expert system shell?
6. What is the role of heuristics in expert system?

7. What is the criterion for electing a problem for expert system?
8. Can an expert system make mistakes? Why are such mistakes evolved?
9. With a suitable example, describe the forward chaining inference process used in an expert system.
10. Describe the backward chaining inference process used in an expert system.
11. List problems for which the forward chaining inference technique is appropriate. Why is backward chaining used for diagnostic problems?
12. What is a conflict set of rules? How can we resolve a conflict? List and describe the basic conflict resolution methods.
13. List the advantages of rule based expert systems. What are their disadvantages?
14. What kind of mistakes might expert system make and why? Why is it easier to correct mistakes in expert system than in conventional programs?
15. Discuss the pros and cons of rule based expert systems.
16. How do we choose a tool to develop an expert system? What are the advantages of expert system shells?
17. List some advantages of frame based expert systems. What are the difficulties involved in developing a frame based expert system?
18. Assume that you are developing an expert system for troubleshooting a complex electronic system. What are the sources of uncertainty that can arise in this application?
19. You have to design an expert system for troubleshooting a mobile phone. You are going to have your first meeting with the domain expert. You have to plan a set of questions, answers to which will help you better understand the domain. Create a set of questions you should ask the domain expert in this meeting.
20. You have developed an expert system for a large insurance company which is supposed to advise people, based on their responses to questions, what policies they are eligible for and what their premium would be. How would you go about validating this expert system?
21. Why is MYCIN considered important in the development of expert systems and how did it lead to EMYCIN?
22. What is the main difference between conventional computer programs and production systems (rule based systems)?
23. Why a production system model was used to implement the first rule-based expert systems?
24. What types of knowledge are used by neural networks and by rule-based systems? What kinds of systems are they with respect to the type of knowledge they use?
25. Describe at least three advantages that expert systems offer organizations that would otherwise have to employ human experts.

CHAPTER 16

Robotics

Can all jobs be done by humans? Consider repair jobs in blast furnace, which is a steel making chamber using iron and carbon. The furnace is a huge machinery that may contain toxic carbon monoxide and harmful gases at the lower side of the furnace. So, maintenance involving humans at lower level is very difficult. How can we overcome this? An intelligent robot makes this job easy. We have designed a specified intelligent machine for this purpose. Similarly, many jobs are done by robots at present that make the human life easy.

One of the major application areas of artificial intelligence is robotics. It was once considered as the future of artificial intelligence, but as time progressed, robotics proved to be stepping stone of futuristic artificial intelligence. Different aspects or forms of robots were in the mind of humans since the time they started building things. There are a lot of innovations in that field itself where machines mimic the human behaviour. They were built to do tasks which the humans felt difficult. They were used in environments where it was difficult for humans to work. In this chapter, we have to discuss the kinematics and dynamics of robot making and the basic mathematical foundation.

16.1 Introduction to Robotics

Robots are very powerful elements of today's industry. Japan accounts for more than half of the world's robots installed worldwide. The word 'robots' was derived from a Czech word 'robota' which means 'slave labour'. The modern industrial robots can be divided into three categories. The first classification is 'environment centric' robots which are used at places where humans find difficult to work. For examples, in a nuclear power plant or places where there are more chances of radiation, etc. the second classification is 'work centric' where a large number of tasks need to be done repetitively or works which are monotonous. This kind of works make humans feel boredom and the quality delivered by using humans will be very low. The third classification where the robots are used is 'requirement centric' process. Here number of products that need to be produced is less but the requirements of clients/models change frequently. Robots look like as shown in Figure 16.1(a).

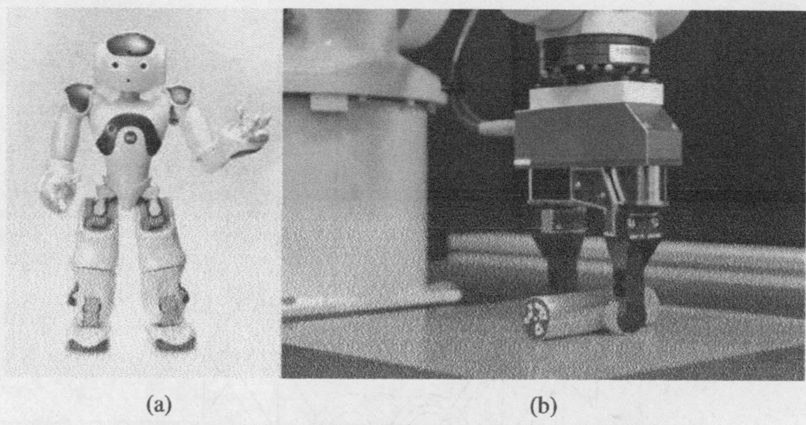

(a) (b)

Figure 16.1 Sample robot (b) Robot grasp objects.

Robots are generally classified according to the number of degree of freedom which indicates the capability of a robot. A general orientation or position of object can be achieved by a robot with 6 Degree Of Freedom (DOF). A welding or painting robots will have 5 DOF and an assembly line robot will have 4 DOF. Based on the configuration, a robot may be a cartesian, spherical or cylindrical robot.

The technology and hardware of robots are changing continuously. A typical robot consists of mechanical components, actuators, power transmission devices, sensors, an electronic controller and computers. The main mechanical component are links connected by joints. Most common industrial robots are designed as a series of links connected by motor-actuated joints that extend from a base to an end-effector, termed as 'serial manipulators.' Here the links are arranged sequentially starting from the base and ending in the end-effector with no loops. In a parallel manipulator, there is one or more loops available. Links need to be strong enough and light as possible to have more flexible movement and for high accuracy positioning. These kinds of robots are used in object alignment and for grasping objects. Figure 16.1(b) shows an object grasping robot.

The links are moved by actuators, which can be electric motors or hydraulic cylinders or any other hardware supporting such movement. The amount of rotation needed, translation, etc. are measured by the sensors at each joint. There will be tachometers on each arm which measures the angular velocity. All these controllers are implemented using microprocessors and contain circuitry for analogue to digital and vice versa converters. The controllers bear 60% of the total cost of robot assembly. Figure 16.2 shows the basic architecture of gripping robots.

Computers and the software residing inside the system basically have two tasks to perfrom. One is to control the actuators and other is the supervisory role. This does the fault prediction, diagnosis and corrective actions that need to be taken. At the actuator level, programs are fast and relatively simple but the supervisory module works at slow speed and the program definition is complex. This is a very basic thing for well-defined and simple user interface.

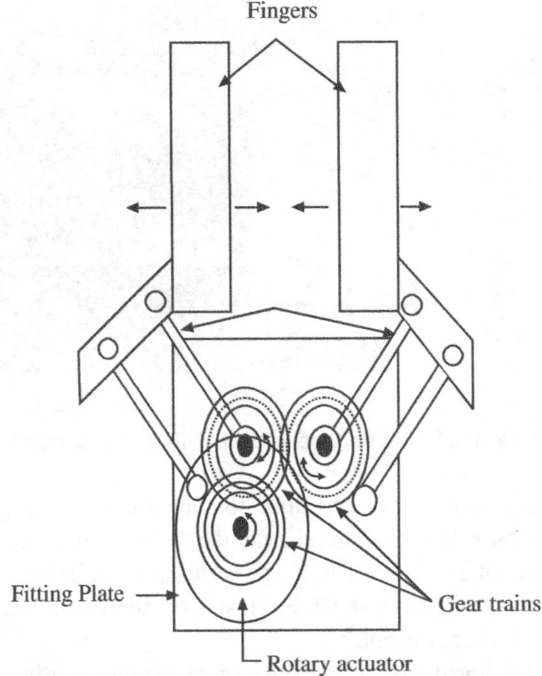

Figure 16.2 Architecture of finger griping robot.

16.2 Principles and Components of Robots

Robotics need to be seen as an interdisciplinary subject which takes input from basic physics, maths, dynamics, control, sensors, intelligence and computing. Let us see each one in detail. Kinematics deals with motion of rigid bodies in 3D space. A rigid body moving in a 3D space will have six degrees of freedom. In other words, such a body requires six different parameters to specify its coordinates. Among the six parameters, three specify the position of interest and other three specify the orientation. In dynamics the major concern is in the study of motion of links and end-effector under the action of external forces and torques from the actuators. Methodology is that first obtain the mass and inertia of the moving links and the end-effector, there by obtaining the equations of motion.

In order to make sure that the robot follows a desired motion, a feedback controller is needed. The actual motion at the joints is measured by means of sensors and the measurement together with known desired motions are used as inputs to the controller. The outputs are torque and forces acting on the actuators for reducing the error of movement. Other micro-controllers can also be used for fast and better accuracy/precision of work.

Commonly used industrial robots have six basic elements as shown in Figure 16.3. These basic elements are: dynamic system, end-of-arm tool, computerised digital controller, actuators, feedback devices and sensor.

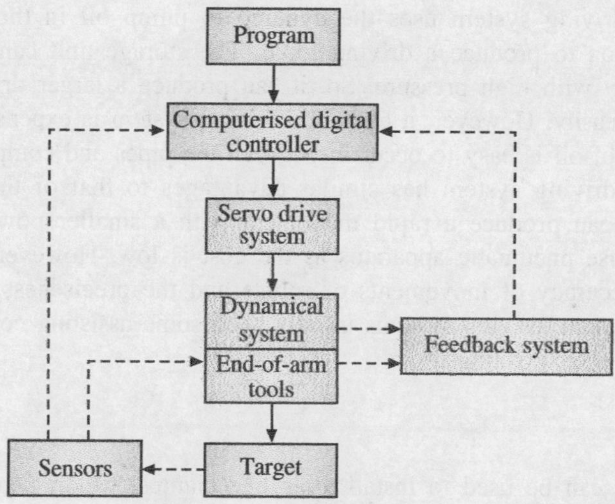

Figure 16.3 Basic elements.

Dynamic system

Dynamic system is needed for movement in the working area. Most important of which is the robotic arm. According to the designs of robots, the main axis can move linearly or rotationally. The number of axes represents the number of directions that a robot can move individually. It can be called 'the degree of freedom'. As shown in Figure 16.4, if the volumes of robots are the same, robot with 3 rotating axes or the highest degree of freedom has the larger working area. A robot usually has 2 to 10 axes, most of the robots have five to six degrees of freedom.

Usually, there are three types of drive systems, they are: electrical, hydraulic and pneumatic. Electrical system uses the dynamo to drive the robots. Simple electrical source is needed to provide energy, unlike hydraulic or pneumatic driving system that needs a lot of accessories. The advantages of an electrical system are simple, clean and silent, but it cannot lift or move heavy object. It is commonly used in middle sized or small sized robots.

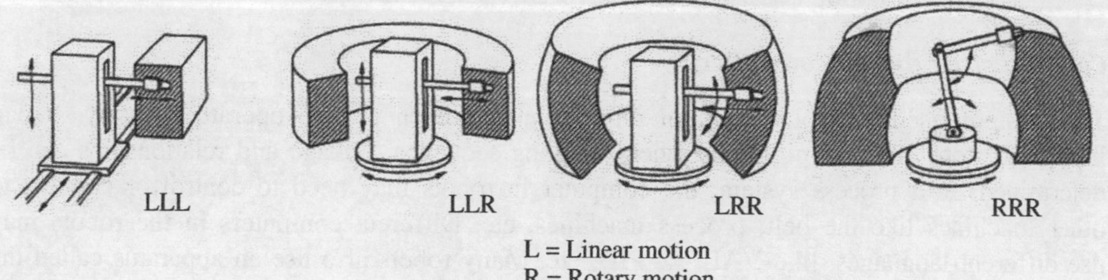

Figure 16.4 Triadic structure of industrial robot.

A hydraulic driving system uses the dynamo to pump oil in the storage unit, then compresses the piston to produce a driving force. The storage unit can produce an output driving force shortly with high pressure. So, it can produce a larger driving force and can be controlled more easily. However, a hydraulic driving system is expensive and complex in structures. Leakage of oil is easy to occur in between the pipes and components.

A pneumatic driving system has similar advantages to that of the hydraulic driving system. Pneumatic can produce a rapid movement with a smaller power. Most automatic chemical factories use pneumatic apparatus as the cost is low. However, since air is easier to compress, the accuracy of movements of robots and the preciseness of moment control are lower. So, pneumatic driving systems usually need some assisting components or special design to determine their positions.

End of arm tool

The arms of robots can be used to install tools like clamps, welding guns, electromagnets, suction pads, hooks or paws, etc., in order to operate any specific procedure. A clamp is a multifunctional robotic arm tool. There are different designs as shown in Figure 16.5. The clamp can be used to hold tight, keep and transmit components or tools, and make them to direct in a suitable position. For instance, a clamp with three axes can perform complex actions like tilting forward and backward, swirl and swinging from left to right.

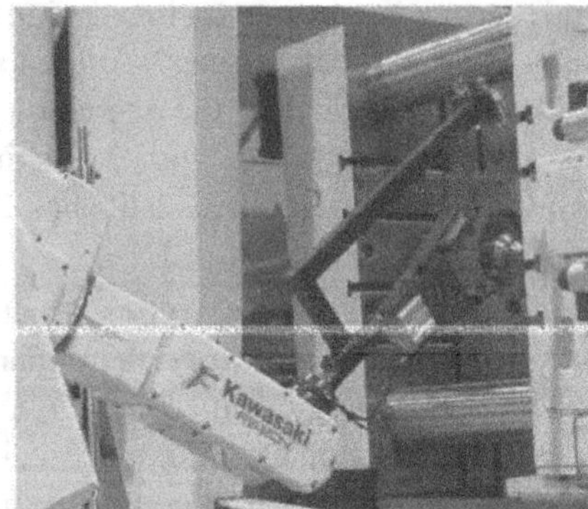

Figure 16.5 Robots with different clamps.

Computerised digital controller

The computerised control system in robots controls each part to operate properly. It can input and store different programs where working sequence, linkage and relationships can be determined. In a process system, the computer in robots may need to control or connect to other machines like the belt, process machines, etc. Different computers in the robots may use different languages, like VAL, RAPID, etc. Many robots also use an apparatus called the trainer which is connected to the controller. The trainer can control some of the motions of a robot, for instance, to direct the welding gun in the robotic arm to each welding tip. Robots also record information and store it in the computer, so that the information can be used repeatedly when the robots run. This can save time in programming.

Actuators

Actuators control each of the axes and maintain the direction of moving. Since robots need to handle weights from various components, it has a larger change in motion. Therefore, the motion of actuators is very important. Commands about the direction of movement are sent from the computer of robots and feedbacks are checked at the same time to ensure that the moving parts follow the correct path. This process requires a high-speed computer to send out commands, and reading the machine's motion at reasonable speed.

Feedback devices

Feedback devices of a robot measure the position of each axis, moving velocity and its acceleration. Each axis can set any point as zero as the reference point, then relative linear movement can be performed. However, the clamp usually uses absolute measurement values. It is because when the robot begins to operate, the position of clamp must be ensured.

Sensor

Robots use sensors to identify and evaluate the change in position, difference in patterns and the hinders or failure of external systems. Table 16.1 lists some commonly used sensors.

TABLE 16.1 Commonly used sensors

Sense	*Sensor*	*Detection*
Hearing	Microphones	Sound
Vision	Light dependent resistors	Light
Touch	Thermometer and contact tip	Heat, weight, pressure, shapes, position
Smell	Smell detectors	Smell

16.3 Operation

There is a lot of flexibility in operations of industrial robots, Infra Red (IR) can handle demands of different tasks. This is because IRs are controlled by computer programs. Therefore, by changing the program, operations can be changed accordingly. This is similar to numerically controlled machine tools. Both the computer and numerically controlled machine tools of robots need lot of memory, repeatable execution programs, program editing apparatus, accessory functions in input and output, and rapid calculating ability.

However, computers in IR require some special functions like input programs for trainers, remain horizontal movement in swinging arms (Figure 16.6), and automatic force compensation when holding heavy objects. The operation mode of IR can be classified as: point-to-point and continuous path.

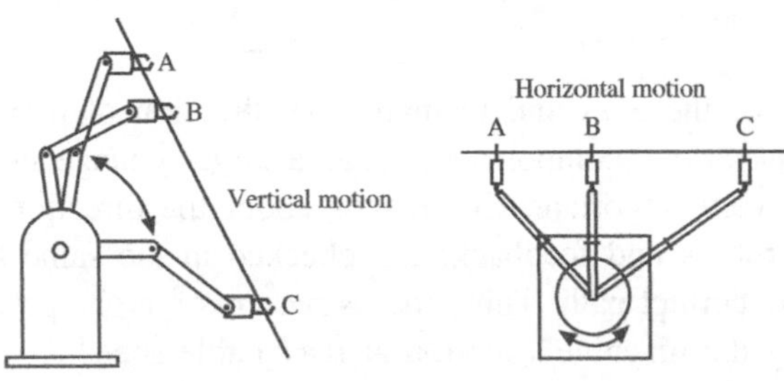

Figure 16.6 Clamp maintenance.

Point-to-point

When a robot moves in various individual and specific positions according to the instructions in a program, the operation mode is called point-to-point. This type of robot has a feedback and servomechanism in each of its axes, but the motion and moving path from one point to another is independent of other points. When using this operation mode, programs can be edited by trainer method or manual input. The control system for point-to-point is relatively simple, so it is more suitable to control components in the numerically controlled machineries. Figure 16.7 gives point-to-point operation of a robot movement.

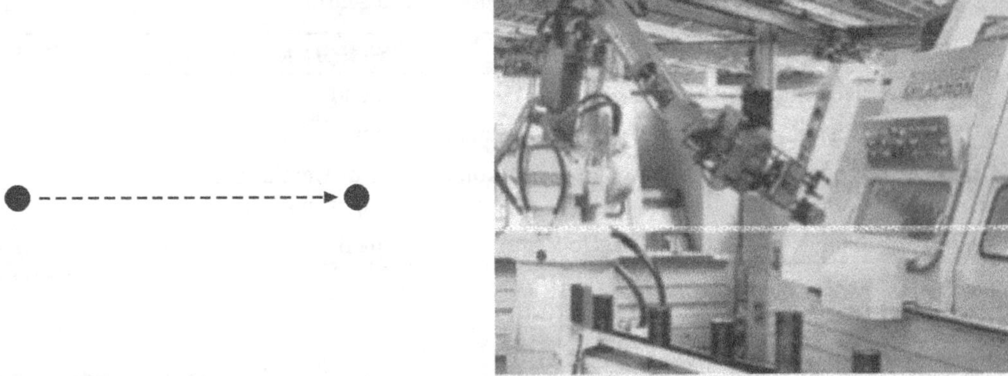

Figure 16.7 Point to point operation.

Continuous path

If a robot has to follow the path of program totally, this operation mode is called continuous path. This type of robot has a feedback and servomechanism in each of its axes. It can also move to any points between the polar positions. Robots with continuous path as operation mode can handle more complicated tasks, like welding of metal in a particular path (Figure 16.8), spraying paints on the surface of a car, etc. When using this type of operation mode, it is better to use method of trainer to write the program.

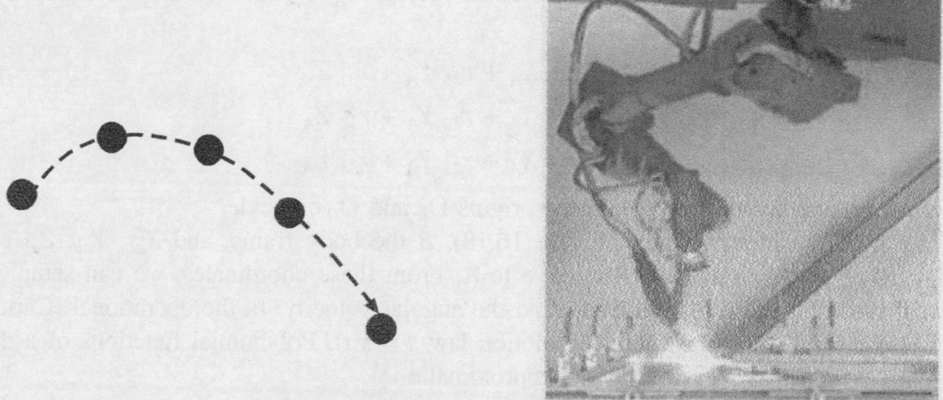

Figure 16.8 Continuous path operation.

16.4 Mathematics of Robots

Robots are designed to do specific tasks. These task are taught to them using the basic mathematics. Whenever or wherever we say about robotics, the backbone of each and every action is the embedded mathematics. A rigid body in 3D space has six degree of freedom and is described by its position and orientation. A position in rigid body is described by a vector drawn from a reference point, located in another rigid body to the point under consideration. Figure 16.9(a) shows fixing position of the rigid body.

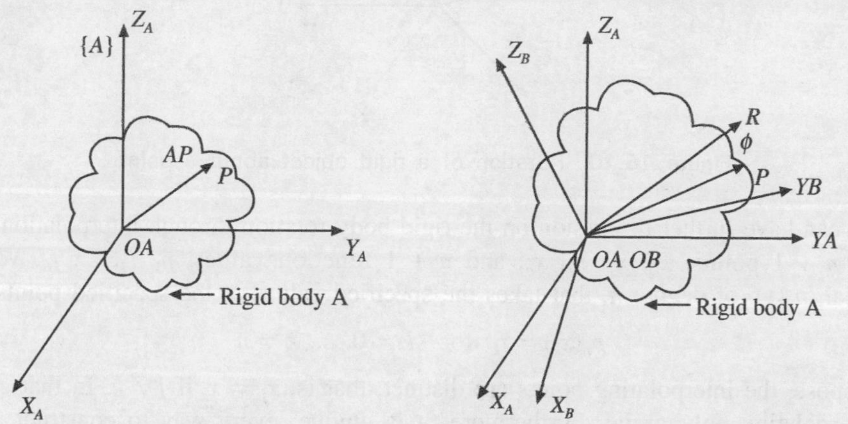

Figure 16.9 (a) Position of rigid body (b) Orientation of rigid body.

In the figure, origin of A is denoted by O_A and the coordinate axes are denoted by X_A, Y_A and Z_A. The position vector of the point P, in the rigid body A, is denoted by A_p. Obtaining the position of any one point is clearly not enough to describe the rigid body. We should know the orientation also. Orientation is obtained by attaching the coordinate system with respect to the rigid body. Orientation is shown in Figure 16.9(b).

In terms of components, each of the unit vectors X_A, Y_A and Z_A attached to rigid body B, can be described as:

$$X_B = r_{11}\,X_A + r_{21}\,Y_A + r_{31}\,Z_A$$
$$Y_B = r_{12}\,X_A + r_{22}\,Y_A + r_{32}\,Z_A$$
$$Z_B = r_{13}\,X_A + r_{23}\,Y_A + r_{33}\,Z_A$$

This relation is true only when the origins O_A and O_B coincide.

Let A be the inertial frame (Figure 16.10), B the body frame, and X_{ab}, Y_{ab}, Z_{ab} be the coordinates of principal axes of B relative to A. From these coordinates, we can setup 3×3 rotational matrix which can be used to find the angular velocity. In the operational (Cartesian) space, $p(s) = (x(s), y(s), z(s))$ with the motion law $s = s(t)$. Polynomial functions of a degree n are employed usually for way-points approximation.

$$s(t) = a_0 + a_1t + a_2t_2 + \ldots + a_nt_n$$

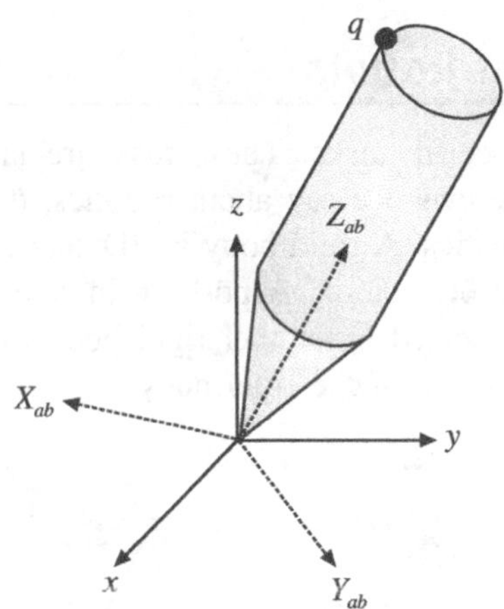

Figure 16.10 Rotation of a rigid object about a point.

We can have further discussion on the rigid body rotation through interpolation. Suppose we have $n + 1$ points, x_0, x_1, ..., x_n, and $n + 1$ function values, f_0, f_1, ..., f_n. We want a polynomial, $p_n(x)$, of degree n, that takes the specified values at the specified points:

$$p_n(x_k) = f_k \quad \text{for} \quad k = 0, \ldots, k = n \tag{16.1}$$

Suppose, the interpolating points are distinct, that is, $x_j \neq x_k$ if $j \neq k$. In that case, there is an interpolating polynomial. Furthermore, it is unique; every way to construct a solution leads to the same polynomial.

The construction of interpolating polynomial mostly used in scientific computing is the Newton construction. This construction builds a sequence of polynomials, each interpolating one extra point. The first one is $p_0(x)$, which has degree zero—that is, a constant. We choose this constant so that p_0 satisfies the first interpolating condition for $k = 0$:

$$p_0 = f_0$$

Next, we find a polynomial $p_1(x)$ of degree 1 that interpolates both f_0 and f_1. We choose $p_1(x)$ of the form

$$p_1(x) = p_0(x) + (x - x_0) \cdot c_1$$

so that the previous interpolation condition at x_0 is not disturbed, $p_1(x_0) = p_0(x_0) = f_0$. Then we use the remaining interpolation condition, $p_1(x_1) = f_1$, to solve for c_1:

$$f_1 = f_0 + (x_1 - x_0) \cdot c_1,$$

$$c_1 = \frac{f_1 - f_0}{x_1 - x_0}, \tag{16.2}$$

$$p_1(x) = f_0 + \frac{f_1 - f_0}{x_1 - x_0}(x - x_0) \tag{16.3}$$

The formula shows that coefficient c_1 is a divided difference. The construction of p_2 will make the general pattern clear. First, choose a form for $p_2(x)$ that ensures that the interpolation conditions at x_0 and x_1 are preserved:

$$p_2(x) = p_1(x) + (x - x_1)(x - x_0)c_2$$

Then choose c_2 to respect the final interpolation condition $p_2(x_2) = f_2$:

$$c_2 = \frac{\dfrac{f_2 - f_0}{x_2 - x_0} - \dfrac{f_1 - f_0}{x_1 - x_0}}{x_2 - x_1}$$

The general construction proceeds by induction. If we have a polynomial of degree $j - 1$ that satisfies the interpolation conditions for $k < j$, then we look for $p_k(x)$ of the form

$$p_j(x) = p_{j-1}(x) + (x - x_{j-1}) \ldots (x - x_0)c_j, \tag{16.4}$$

and use the interpolation condition $p_j(x_j) = f_j$ to solve for c_j. If $p_{j-1}(x)$ has degree $j - 1$, then give p_j degree j because the second term has j factors involving x. If we continue in this way, eventually we construct $p_n(x)$ that satisfies all the conditions.

This is essentially Newton's construction of the interpolating polynomial. Another way to construct the interpolating polynomial is through the Lagrange interpolation formula. For this purpose we use specific polynomials as follows:

$$m_k(x) = \prod_{j \neq k, j=0}^{n} (x - x_j), \tag{16.5}$$

$$l_k(x) = m_k(x)/m_k(x_k) \tag{16.6}$$

The denominator in Eq. 16.6 is not zero because the points x_j are never equal to x_k if $j \neq k$. The polynomials $m_k(x)$ and $l_k(x)$ have degree n because the products have n terms. Moreover, $m_k(x_j) = l_k(x_j) = 0$ if $j \neq k$. This is the purpose of the formula. The normalisation makes $l_k(x_k) = 1$. We express the interpolating polynomial $p_n(x)$ in terms of the $l_k(x)$ simply as:

$$p_n(x) = \sum_{k=0}^{n} f_k l_k(x) \tag{16.7}$$

If x is one of the interpolating points, say $x = x_m$, then only the term on the right side with $k = m$ is different from zero. Thus, satisfies the interpolation conditions. The formulae (Eqs. 16.5, 16.6 and 16.7) together are called the Lagrange interpolation formula.

The third approach we have to discuss is the least elegant but the most general. It is a method of last resort when elegant solutions are not available. We simply pose the interpolation problem as a system of linear equations and use general theorems from linear algebra. The fact that interpolating polynomial is unique falls out immediately as a consequence of what we already know.

Write the polynomial $p_n(x)$ in terms of its coefficients:

$$p_n(x) = a_0 + a_1x + a_2x^2 + \cdots + a_nx^n.$$

The interpolation conditions (Eq. 16.1) take the form

$$\begin{aligned} a_0 + a_1x_0 + \cdots + a_nx_0^n &= f_0 \\ a_0 + a_1x_1 + \cdots + a_nx_1^n &= f_1 \\ &\vdots \\ a_0 + a_1x_n + \cdots + a_nx_n^n &= f_n \end{aligned}$$

This system may be written in matrix form as

$$Va = f \tag{16.8}$$

where

$$V = \begin{pmatrix} 1 & x_0 & \cdot & \cdot & \cdot & x_0^n \\ 1 & x_1 & \cdot & \cdot & \cdot & x_1^n \\ \cdot & \cdot & \cdot & \cdot & \cdot & \cdot \\ \cdot & \cdot & \cdot & \cdot & \cdot & \cdot \\ \cdot & \cdot & \cdot & \cdot & \cdot & \cdot \\ 1 & x_n & \cdot & \cdot & \cdot & x_n^n \end{pmatrix}$$

$$a = \begin{pmatrix} a_0 \\ a_1 \\ \cdot \\ \cdot \\ \cdot \\ a_n \end{pmatrix} f = \begin{pmatrix} f_0 \\ f_1 \\ \cdot \\ \cdot \\ \cdot \\ f_n \end{pmatrix}$$

The matrix V is vandermonde matrix. It is $(n + 1) \times (n + 1)$. This is just a reformulation of the polynomial interpolation problem. Interpolating polynomials exist and are unique if the solution of system of linear equations exists and is unique.

16.5 Line Following Robot

Line follower Robot is one of the first robots that beginners and students would get their first robotic experience with. The major components required are an Arduino board, motor driver, geared motors, robot chassis, IR sensor, wires, power supply and battery connectors. The line following robot can be divided into four units—like sensors for line detection, decision making unit, motor driver and movement module. A basic circuit for line following robot is given in Figure 16.11.

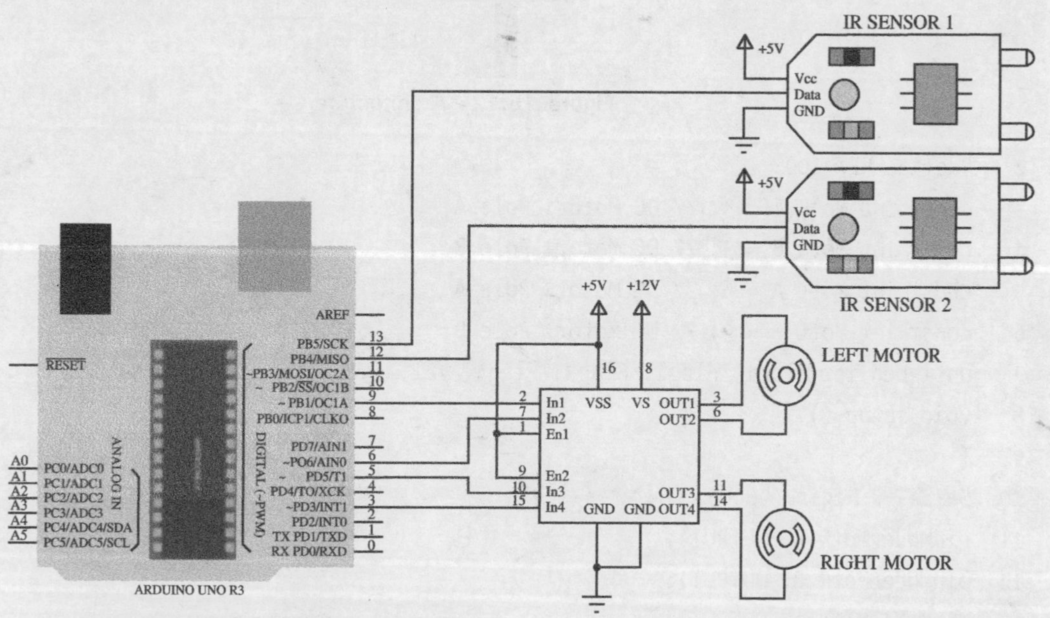

Figure 16.11 Circuit diagram.

The IR sensor is a major component of line follower. It detects the line drawn on the floor and gives input to the movement module. From checking whether the images are dark or bright it decides to go straight, go left, right or stop. Arduino continuously monitors the data from both the sensors and turns the robot as per the line detected by them. The basic architecture of the robot movement is shown in Figure 16.12.

A sample arduino code for line follower robot is mentioned below. This is a sample code which just does the simple robo movement activities.

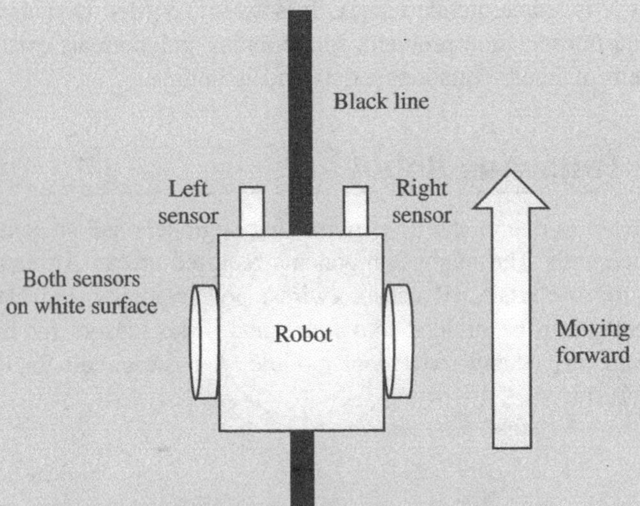

Figure 16.12 Architecture.

```
#define REF 200
const int MotR_A = 3; // DC Motor1 Pole_A
const int MotR_B = 5; // DC Motor1 Pole_B
const int MotL_A = 6; // DC Motor2 Pole_A
const int MotL_B = 9; // DC Motor2 Pole_B
unsigned int Flag1, Flag2, Flag3, Flag4, Val;
void setup ()
{
Serial . begin (9600);
pinMode(MotR_A, OUTPUT);
pinMode(MotR_B, OUTPUT);
pinMode(MotL_A, OUTPUT);
pinMode(MotL_B, OUTPUT);
}
void loop ()
{
int sen1 = analogRead(A0);
int sen2 = analogRead(A1);
int sen3 = analogRead(A2);
int sen4 = analogRead(A3);
if (sen1>REF)
```

```
{ Flag1 = 1; }
else
{ Flag1 = 0; }
if (sen2>REF)
{ Flag2 = 1; }
else
{ Flag2 = 0; }
if (sen3>REF)
{ Flag3 = 1; }
else
{ Flag3 = 0; }
if (sen4>REF)
{ Flag4 = 1; }
else
{ Flag4 = 0; }
Val = (Flag1 | (Flag2<<1) | (Flag3<<2) | (Flag4<<3);
Serial . print (Val);
Serial . println ();
switch (Val)
{
case 0:
Robo_Right ();
break;
case 1:
Robo_Left ();
break;
case 3:
Robo_Left ();
break;
case 6:
Robo_Front ();
break;
case 7:
Robo_Left ();
break;
```

```
case 8:
Robo_Right ();
break;
case 12:
Robo_Right ();
break;
case 14:
Robo_Right ();
break;
case 15:
Robo_Front ();
break;
default:
break;
}
}
void Robo_Front ()
{
digitalWrite (MotR_A, LOW);
digitalWrite (MotR_B, HIGH);
digitalWrite (MotL_B, LOW);
digitalWrite (MotL_A, HIGH);
}
void Robo_Back ()
{
digitalWrite (MotR_A, HIGH);
digitalWrite (MotR_B, LOW);
digitalWrite (MotL_B, HIGH);
digitalWrite (MotL_A, LOW);
}
void Robo_Right ()
{
digitalWrite (MotR_A, HIGH);
digitalWrite (MotR_B, LOW);
digitalWrite (MotL_B, LOW);
```

```
digitalWrite (MotL_A, HIGH);
}
void Robo_Left ()
{
digitalWrite (MotR_A, LOW);
digitalWrite (MotR_B, HIGH);
digitalWrite (MotL_B, HIGH);
digitalWrite (MotL_A, LOW);
}
void Robo_Stop ()
{
digitalWrite (MotR_A, LOW);
digitalWrite (MotR_B, LOW);
digitalWrite (MotL_B, LOW);
digitalWrite (MotL_A, LOW);
}
```

16.6 Applications and Risks

Robots as mentioned are used in various applications starting from agriculture to automotive to power plants. In this section, we are just briefing the major areas, which have not been touched so far. The applications of robot change from day to day. As you read this book the applications and uses mentioned in this section may be outdated due to the amount of innovation and changes happening in this area.

Roomba is a robot vacuum cleaner which is available commercially for some years. It is an autonomous robot and moves randomly through the area and cleans the surface. It also finds the adapter space and does charging whenever it is out of charge. This is one of the major cost effective robots that is available around. There are robots like ASIMO, Nesbot Monty, etc. which are humanoid robots which actually display human like behaviour. Behaviour starts with climbing stairs to dish washing, reading books, etc. These robots actually replace the need of an assistance at home.

Many robots have been developed to help human during natural calamities. Such robots are installed by sensors which can detect humans or animals trapped somewhere and send out signals to the emergency rescue team and report the live locations. Such sort of robots are also used in bomb diffusion and other skilled and risky tasks. HULC is a robot which carries weight for humans. This can be used in work spaces and in construction sites. The amount of weight that can be lifted is programmed and the accuracy of delivery is found to be cent percent.

For military purpose, smart dust is being used for surveillance and for detecting motion. They are suspended in the air and can take photographs and transmit to a remote place. This can also be seen as a treat when it is used for unethical things. Robotic arm designing is another emerging area. It is used in assembly line and for pick-and-drop items. In medical field, physicians use robots to help them examine and treat patients in rural or remote locations, giving them a 'telepresence' in the room. Specialists can be on call, via the robot, to answer questions and guide therapy from remote locations.

A pair of lightweight, robotic planes have made first unmanned flights over Antarctica's icy expanses. Driven by propeller, the machines made 20 low-altitude sorties, including four over the Weddell sea. Unmanned Aerial Vehicles (UAVs) were launched by catapult but flew autonomously until landing. During some of the test flights, machines were fitted with miniaturised instruments to collect data for using in predictive climate models.

The whole data storage and retrieval is going to be replaced by autonomous robots. The entire network is connected with robots from various streams and data is pushed to the cloud servers. The database and query processors need to be so smart that the response time for each and every action is transferred to the robots with the speed of light. The major challenge next database management system is going to face is the speed in query analysing. As the data is pushed by the robots, sensor data is in high speed. A sample data flow diagram is shown in Figure 16.13.

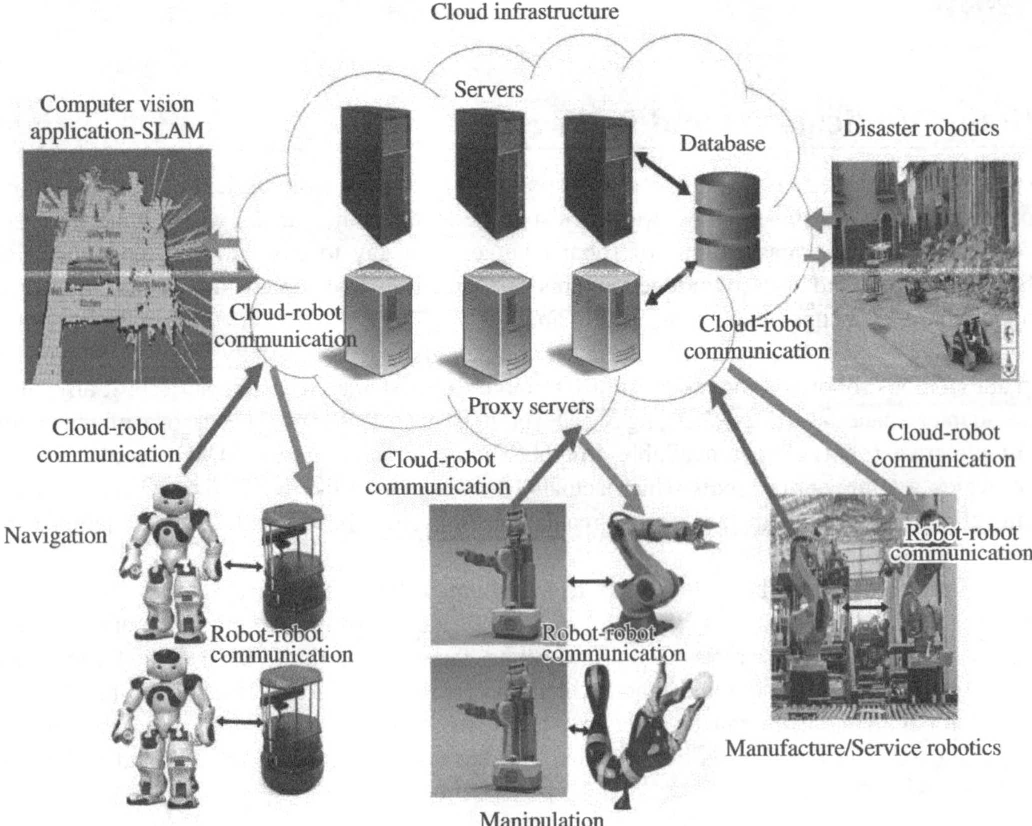

Figure 16.13 Application of robots.

The next generation of robots will not be metallic machines, but instead will be made of soft materials with a wide range of motions that react to applied electricity. They will be able to adapt to dynamic environments and be well-suited to work closely with humans. Such soft devices can perform a variety of tasks, such as grasping delicate objects and lifting heavy materials.

A central challenge in the field of 'soft robotics' is the lack of actuators, or artificial muscles, that can replicate the versatility and performance of the real thing. One design of soft actuator, known as a Peano-HASEL, is transparent and self-sensing, and highly controllable. It is strong and versatile but reportedly cheap to produce, using inexpensive commercially available materials. Remarkably, it is able to lift more than 200 times its weight.

The robotics industry, while in development for half a century, is still relatively infancy and faces a number of challenges in the years ahead. Besides the technological and cultural hurdles to overcome, questions remain unanswered regarding their economic and environmental impacts as well as the ethical issues of human and robot interaction. What is obvious is that robots, whatever form they take, will increasingly play a role in societies around the world and that the ecosystem of services and capabilities will offer increasing opportunities for designers in the years to come.

16.7 Case Studies

16.7.1 Stability Management of Multi Axle Vehicle

Vehicles normally have double axle. Some load bearing vehicles have more than two axle and are generally included in the class of multi-axle coach. Multi-axle vehicle also known as a twin-axle coach or tri-axle coach, or more rarely, quad-axle. Extra axles are usually added for legal weight restriction reasons, or to accommodate different vehicle designs such as articulation, or rarely, to implement trailer buses.

When load in the vehicle is more, the driver needs to manually select which all axle need to be deployed and in which order. This is normally a driver's task where an experienced driver does this very easily. But there are cases where the loads are not uniformly distributed and the inexperience of drivers causes stability issue to the vehicle. Figure 16.14(a) shows a general image of beam axle front suspension.

There are links connecting between the wheels and shocks that help in reducing the vehicle displacement when encountering any sudden change in vehicle's speed. Consider a truck carrying five ton of load. The driver has to make sure, where the loads are basically entered and which all axles need to be in place. If not done properly, then there are issues of vehicle breakdown and even roads may get damaged.

As the technology has progressed, one major advancement is automating the axle management. There are robots which actually calculate the actual mass the truck is holding and find the centroid of acting. This is calculated effectively once the truck is fully loaded. After this, signals are given to the dashboard indicating axles which need to be deployed for the current load and also inclinations that the axles must deploy while taking sharp turn or making a hill ascend or descend.

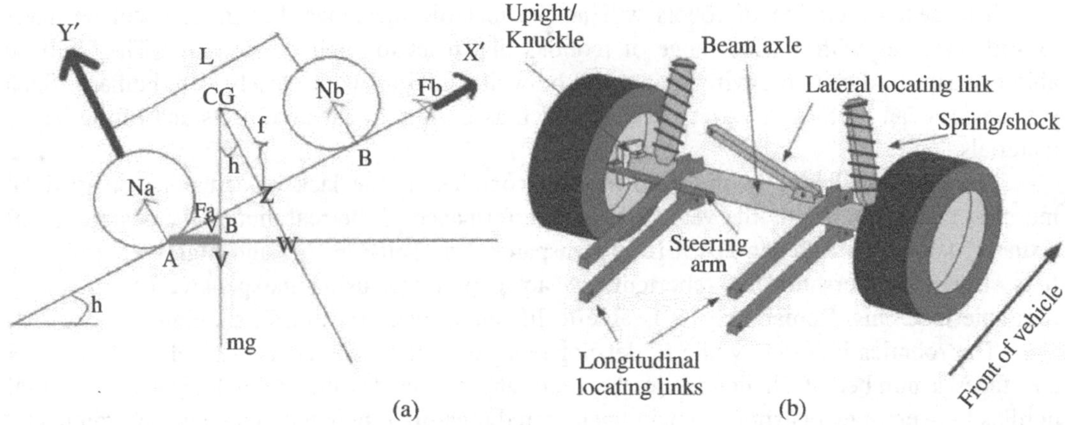

Figure 16.14 (a) Beam axle (b) Forces acting.

There is a camera monitoring the road topology, whenever there is a change in lane or whenever it detects a hill ascend or descend, it will send signal to the dashboard regarding the nature of road ahead. These signals are processed and an automated internal board regulates the spinning speed of individual wheels. During the hill climb, with such a heavy load what velocity should be the maximum that a tyre should acquire. Similarly during descend what is the strategy that needs to be taken. All these are programmed and a robotic arm basically overlooks the process.

Figure 16.14(b) shows a simple bike on an inclined surface, let *L* is length of the bike and distance between centre of gravity and centre wheel centres *A*, *B* is *L*/2. Normal forces perpendicular to the plane Na, Nb, friction forces Fa, Fb and the gravity force is acting at the centre of gravity point. The automated monitoring system helps in making sure that the forces acting are in equilibrium and the bike moves through the hill safely. Why is this stability management system very important? It creates an integrated active safety system to ensure highest levels of safety by coordinating Traction Control System (TCS), Electronic Stability Control (ESC), Brake Assist (BA), Anti-lock Braking System (ABS) and Electronic Brake-force Distribution (EBD). Traction control is basically to take care of this automated unit which decides the velocity and rotation speed of wheels.

16.7.2 Household Robots

The robot during its inception stage was basically used for doing household chores. It can start from just cleaning the dish to cooking dishes. Such robots were less programmed and were less intelligent. HERO Jr was the first affordable, personal robot with a dynamic personality. People could use it to play songs, games, wake people up in the morning, notify important events, and even guard the home. As a private robot, people do not need program skills to operate the robot and if they want to re-program the robot, people can do it simply with a home computer and optional RS 232 accessory and BASIC cartridge. Figure 16.15(a) shows the HERO Jr robot.

Figure 16.15 (a) HERO Jr robot (b) Moley robot.

HERO Jr. has a 6808 processor, but only 2 kB of RAM. It has onboard speech synthesis, a Polaroid sonar range sensor, a light sensor, a sound sensor, and an optional infra-red sensor. Other optional components include a pair of extra batteries to double the operational time between charges, from an estimated 4 hours to 8 hours. A remote control accessory allows users to drive the robot around. It includes a motion sensor that causes the robot to croak 'SOME-THING-MOVE' when it detects a source of motion. It was this feature which made robots useful for guarding the house.

Technologies have changed, from a simple household robots to intelligent robots which can schedule and perform jobs of their own. One of the major technological advancements is in the area of cooking. A London-based company has developed Moley—a kitchen robot designed for the home. Rather than cooking like a machine, the system works by first recording human actions in 3D and then converting these into highly precise movements. The prototype was trained by chef Tim Anderson, who spent time whipping up original dishes in a motion-capture studio. We can control it remotely using an app, which means after a tiring day at office, you could order your dish online and it is ready to serve when you get home. Figure 16.15(b) shows the Moley robot.

Yet another innovation is the beam system robot that navigates and moves on its own with a physical body attached to the video screen. Bosses can check on employees when they are not in the office, parents can watch if their kids do homework while at work, far away friends can hang out easily. Essentially, this thing will replace the need to leave your house ever again. All these are realistic robots that are available now. Figure 16.16(a) shows the beam system robot.

How about robots in Olympic? The 2020 Tokyo Olympics is doing its best to embrace Japan's heritage of technology. Toyota, an official Olympic partner, is leading the charge with what it calls the 'Tokyo 2020 Robot Project'? Robots are going to be front and centre at the Olympics, including special-use bots that will be helping out during the events. Known as Field Support Robots (FSRs), they will carry javelins, shot-puts, and other items that can be thrown to athletes in the Olympic stadiums. Toyota hopes that the robots will be able to reduce the time needed to retrieve items, as well as lighten the workload for official support teams.

Figure 16.16 (a) Beam system robot (b) Tokyo 2020 Olympics robots.

Robotic butlers will also be on display in the form of Human Support Robots (HSRs) and Delivery Support Robots (DSRs). They will be able to guide attendees to their seats and bring light refreshments and drinks. Most importantly, they will be able to help attendees with mobility problems in getting to their seats. Figure 16.16(b) shows the entire robots revealed so far by Toyota made for Tokyo 2020 olympics. Olympic officials are hoping that the robots will be the part of a grandeur that leaves attendees and athletes thrilled. We can be fully assured that the era of robots has just begun.

16.8 Exercises

1. Differentiate data, information, knowledge and wisdom.
2. Write a unit vector in matrix form that describes the direction of cross product of $p = 5i + 3k$ and $q = 3i + 4j + 5k$
3. How does a sensor work in line follower robot?
4. Mention major robotic applications in the medical field.
5. Derive a general equation for circular movement robotic arm.
6. What are the major challenges while designing robot in a cost-effective manner?
7. What are the major robotic components?
8. Differentiate between joint space and cartesian space.
9. Define robot manipulator and effector.
10. What is meant by robot degree of freedom?
11. What is inverse kinematics of robots?
12. What are actuators?
13. It is desired to have the first joint of a six axis robot go from initial angle of 75 degree in 5 second. Using third order polynomial, calculate the joint angles at 1, 2 and 3 seconds.
14. Given arm configuration, work volume and work envelope of different robotic structures. Define all the performance parameters like resolution, repeatability, payload and precision?

CHAPTER 17

Futuristic Artificial Intelligence

Though this article focuses on futuristic Artificial Intelligence, let us begin with a brief on past and present scenarios.

Any technology ever developed is to reduce human effort and as an increment to convenience. The fire to heat, cook or to distract and repel animals, the wheel which allowed easy manoeuvre of heavy objects and many more.

The old technology is under an update, wheels started to be powered by animals, later engines. Fire is now produced by flick of a finger. Financial era arrived and stock markets, taxes, insurance, documentations, etc., which were once managed by handwritten texts, were simplified by typewriters, which were again replaced by computers that write, read, analyse, manipulate and store data. Are the efforts over-simplified?

With just two symbols '0' and '1', many industries were conquered. Computers were then upgraded in terms of performance, scaling its processing power and storing capacity while remaining miniature enough. During the mid 20th century, for the first time term 'Artificial Intelligence', officially took off. Alan Turing's statement, during *with combinations of '1's and '0's could simulate any conceivable act of mathematical deduction*. As a field of study under mathematicians, artificial intelligence grew amongst philosophers and mathematicians. Artificial intelligence works itself, without human intervention, takes decisions based on experience, and however coded, it uses statistical analysis on input, based on the previous experiences for output.

Artificial intelligence as a field had its recession period called artificial intelligence winter due to the lack of funds invested on research and development. Later acquired by computer scientists, artificial intelligence regained its market value. But again fell back due to the lack of resources, currently available as GPU or TPU, for highly scaled computing power.

17.1 Robots as a Fantasy

Humans have always dreamed big and have transformed many of those dreams into reality. When did we think of robots? It was not in 2000 or not in 1980s. During 1963 a Japanese television drama based on a comic strip series started telecasting. This was based on the adventures of a young boy named Shotaro Kaneda, who controls a giant robot named Tetsujin

28, built by his late father. The series is credited with featuring the first humanoid giant robot controlled externally via remote control by an operator. A comics strip on late 1950's featured robots—which shows the kind of vision we had.

Johnny Sokko and his Flying Robot (Figure 17.1) is a series created by Mitsuteru Yokoyama and is similar to Yokoyama's Tetsujin 28. The plot of the series is as follows, 'Earth is invaded by an interstellar terrorist group and a young boy with the help of a voice controlled robot saves the Earth'.

The vision on creating a robot with a wireless voice controlled module in early 1960's has improved like anything when we are at 2020. Artificial intelligence, Machine learning, NLP, Robotics, etc. have contributed much in developing enhanced machineries.

Figure 17.1 Characters of the series.

17.2 Tip of the Iceburg—Current Era

Any reference to artificial intelligence includes its subsets Machine learning and Deep learning. As it is about the machine to take decisions itself, there is a scope for self learning. After its second fall, technology grew exponentially. Emergence of Graphical processing units capable of handling multiple processes using its CUDA cores in NVIDIA GPUs, and Stream processors of AMD. This growth allowed artificial intelligence to regain traction.

Though digitalisation is not completely saturated, artificial intelligence and its subsets boom across the market. The demand meets supply, for the users, either naive or professional, the term artificial intelligence is a buzzword in any area of interest, say from farming to coding. Artificial intelligence is currently used to solve many unsolved problems or optimise the constraints.

Systems that are integrated with intelligence usually suggest the user, rather than taking the whole decision. Technologically, artificial intelligence is embedded into many applications and services. Let us see some of the major applications and areas where artificial intelligence is used in abundance.

Voice enabled computing

Alexa: it is a smart voice assistant developed by Amazon. It is embedded into a speaker and is optimum for a room. Alexa fetches music, news, stories, etc. Alexa also offers users a platform to integrate programs, applications as well as compatible smart home devices. It holds its popularity under customer services.

Siri: Apple's computing systems are embedded with an intelligent voice assistant, converting our voice command to the stream of text to fetch the desired or most relevant results. It uses natural language user interface to answer, make recommendations and perform actions by delegating relevant services. It holds its popularity under customer services.

Google Assistant: Similar to Alexa, Google developed its own device, also embedded the artificial intelligence based voice assistant, into its android devices. Google operates on and consumes the meta-data from user interaction. The assistant is developed based on such meta-data. Any voice command is directly mapped to the Google search engine. It is developed to a stage of making human like calls. Figure 17.2 shows the various voice assistants.

Figure 17.2 Alexa, Siri and Google assistant.

Cogito

Cogito uses real-time emotional intelligence software that incorporates artificial intelligence and machine learning to analyse voice calls and assist call centres to better engage with their customers. It has a market under customer services.

Transportation

Transportation has seen major change. Let it be cars, bus or trains. The change in design, efficiency and reliability has had an exponential growth. Figure 17.3 shows the transition of train from past to the present and future. It started off with the train running on coal for very limited travellers. Later, it became electrical with the speed of travels getting improved. Now, we are in the era where we have super fast trains which travel around 450 kilometers per hour. The future may be train pods, which the travellers might be able to control with no designated tracks. Speed of such trains may be beyond imagination.

Tesla is the most known innovation in transportation sector. Autonomous driving platform for cars, produced and marketed by Tesla. An Electric car, with the ability to self-steer over street using sonic signals and signal recognition engines based on artificial intelligence. Autopilot is Tesla's advanced suite of active driver assistance which is capable of controlling the throttle, brakes and steering in certain situations. It can even make lane changes if the driver activates vehicle's turn signal. The car is internet-connected to provide software updates, Tesla was also able to bundle in audio-streaming services such as Slacker and TuneIn. Slacker allows you to create radio stations and personalise them by thumbing songs up or down.

Figure 17.3 Transition of train.

Financial—Bitcoin

Bitcoin is a new digital currency across the globe, picking up very fast through its journey. Many countries have made it legal to use bitcoins for payments. Bitcoin is a digital currency that works on a peer-to-peer basis. It was invented somewhere in the year 2008 by an unknown programmer, or a group of programmers, under the name 'Satoshi Nakamoto'. It is designed for secure financial transactions taking place between users directly, without an intermediary which means no central authority, no banks, and no government regulators. People who bought bitcoins in the early stages have now become millionaires and authorities seized millions of dollars worth of bitcoins from the Web's notorious black market, i.e. the Silk Road. There have been huge efforts made to uncover the identity of bitcoin creators. Figure 17.4 shows the bitcoin.

Figure 17.4 Bitcoin.

It is really important to understand how the bitcoins actually work. You can send bitcoins digitally to anyone who has a bitcoin address anywhere across the globe. One person could have multiple addresses for different purposes—personal, business and the like. To put it in simple words, bitcoin is the internet's version of money.

Bitcoin at its core, is a cryptographic protocol, which is why it is also referred to as a 'crypto-currency.' The protocol creates unique pieces of digital property that can be transferred from one person to another. The protocol also makes it impossible to double-spend a bitcoin, meaning you can not spend the same bitcoin twice. Each bitcoin is based on a cryptographic logic having a public address and a private key, which are long strings of numbers and letters that give each bitcoin a specific identity.

In addition to having a unique digital footprint, Bitcoins also have all the transactions recorded in a public ledger of all bitcoin transactions known as the blockchain. Buying or spending a bitcoin records your purchase publicly and permanently. The blockchain is maintained by a distributed network of computers around the world. This decentralisation means no one entity, such as a government, controls it. Transactions happen digitally from person to person, without middlemen such as banks or clearing houses.

Video processing—DeepFake

DeepFake has trained its model to use an image to generate a fake video out of it. It uses facial features of a human to analyse. Input the statements and an image of a human face, it controls facial features of the image to generate videos based on the context given. This is a perfect example of misusing the technology. For example, if we have a video of Sachin Tendulkar playing cricket, his mannerisms can be cloned to produce a fake video with someone else's face.

GauGAN

A gaming or video industry product developed by Nvidia, where a simple naive input of a landscape drawing is converted to photo-realistic image. It is a software, that uses GAN networks to read the input and generate a realistic landscape. It uses real-time ray tracing and shadow rendering to create foolproof images. It holds its popularity in the manufacturing sector.

Boston dynamics

Robotics field of study is completely exploited by this start-up. It focusses on developing bots like humans or animals, mimicking their motions and actions. It uses artificial intelligence engines to analyse a real human actions and reproduce through the bot. It holds its popularity in Research and Development sector. Figure 17.5 shows a robot which mimics the pet dog.

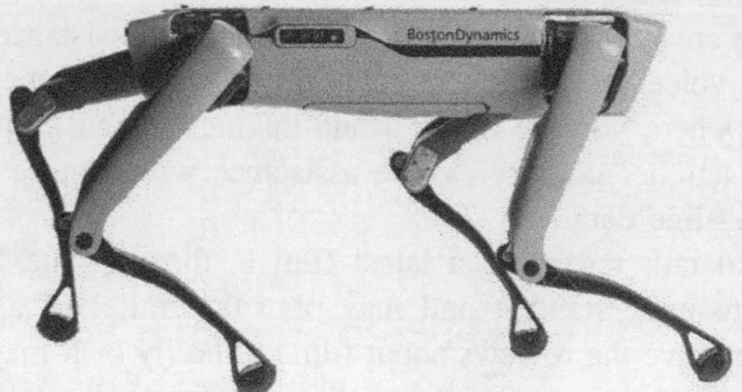

Figure 17.5 Robot mimicking the behaviour of dog.

Technologies listed above are the most popular among users. Yet there is only minute differences that exist with other unlisted applications, as in business logic, area of interest or data and environment it works on.

Artificial intelligence in current status will not be developed to do generalised tasks, like humans. Not adaptive nor creative enough, there exists statistical creativity, but not human like. Generalisation of artificial intelligence can be achieved with more complex algorithms and quantum computing change the whole perspective of artificial intelligence. Quantum computers work on probability or superposition of classical states which indeed will support artificial intelligence by dramatically accelerating computational tasks.

17.3 To the Base of Iceburg—Futuristic Thoughts

Let us start by sailing to 10 years ahead. You are just waking up in the spring of 2030. Your Internet of Things bedroom opens solar powered e-windows and plays gentle music while your smart lighting displays a montage of beach front sunrises from your recent vacation. Your shower uses very little water or soap. It recycles your grey water and puts the excess heat back into your home's integrated operating system. While you dress, your artificial intelligence assistant shares your schedule for the day and plays your favourite tunes. You still start your day with caffeine but it comes from your IoT refrigerator which is capable of providing a coffee house experience in your home. A hot breakfast tailored to your specific nutritional needs (based on chemical analysis from your trips to the 'smart toilet') is waiting for you in the kitchen. When it is the time to leave, an on-demand transport system has three cars waiting for you, your spouse and your kids. On the road, driver-less cars and trucks move with mathematical precision, without traffic jams, routine maintenance or road rage. Accident rates are near zero. Is this the future? Time will prove.

We have realised many sectors where artificial intelligence can be implemented, but there is more to it. More scaled, fast and accessible systems. We have not taken much steps into security and surveillance like cryptography, cryptanalysis, etc. Artificial intelligence can also be used as a modular tool, embedded into Augmented reality, Virtual reality or Robotic systems. In this section, some industry and applications which will see tremendous change in the next few years are mentioned. Many of the prototypes are getting ready and will be very common in the coming years.

Voice enabled computing

We have discussed the emergence of Siri, Alexa and Google assistance. What can be its near future? As such these voice recognition systems do things which they are instructed to do. Consider the situation where you are in a car and talking to your co-passenger regarding the weather conditions in another country. Voice assistance will capture such talks and readily make available the real-time data.

Consider a casual talk regarding a latest film in theatre. The future voice assistance devices will be able to understand it and may play the trailer of a movie in a very near connected device or can give the reviews about film on the fly or it may even open up options for you to book ticket for the same. The scope of expansion is beyond the present.

Transportation

Although it could take a decade or more to perfect them, autonomous cars will one day ferry us from place to place. Tesla and many others, already into this industry, are trying to globalise it, for any type of road, climate and speed. Transportation industry is all set to take a fly. From high performance amphibian vehicles to invisible transportations.

Invisible car: Car turning invisible? Well, almost. It is quite an example of science and marketing coming together to catch your eye by, ironically, not letting you see the product. The researchers at Mercedes Benz created the illusion that their new zero emissions F-cell Mercedes Benz is not even there at all. In this clever publicity stunt, Mercedes wanted to emphasise that its fuel cell vehicle has no omissions, making it virtually invisible to the environment.

This was achieved by placing a mat of LEDs across one side of the vehicle, which received a live signal from a video-shooting digital SLR camera on the other side of the car. The result was stunning to say the least. This can be the start of another revolution. Figure 17.6 shows an invisible car with the above techniques.

Figure 17.6 Invisible car.

Talking Car: One of the major upcoming revolutionary idea is cars talking to one another. While moving through roads, car create a adhoc network among them to initiate B2B or B2C communications. Figure 17.7 shows a network building process. Each node will be able to transfer signals. It may be related to the traffic, or some emergency help or reporting accidents, etc. Such a platform may also be used as a replacement for bill boards. Technologies are elevating in a speed where any distance can be covered in the required time. Who knows, time travel will be a reality in coming years!

Figure 17.7 Talking car—Adhoc network creation.

Hyperloop: A hyperloop (Figure 17.8) is a proposed mode of passenger and/or freight transportation, first used to describe an open-source vactrain design released by a joint team from Tesla and SpaceX. Drawing heavily from Robert Goddard's vactrain, a hyperloop is a sealed tube or system of tubes through which a pod may travel free of air resistance or friction conveying people or objects at high speed while being very efficient, thereby drastically reducing travel times over medium-range distances. The hyperloop concept operates by sending specially designed 'capsules' or 'pods' through a steel tube maintained at a partial vacuum. Linear induction motors located along the tube would accelerate and decelerate the capsule to appropriate speed for each section of the tube route. With rolling resistance eliminated and air resistance greatly reduced, the capsules can glide for the bulk of journey.

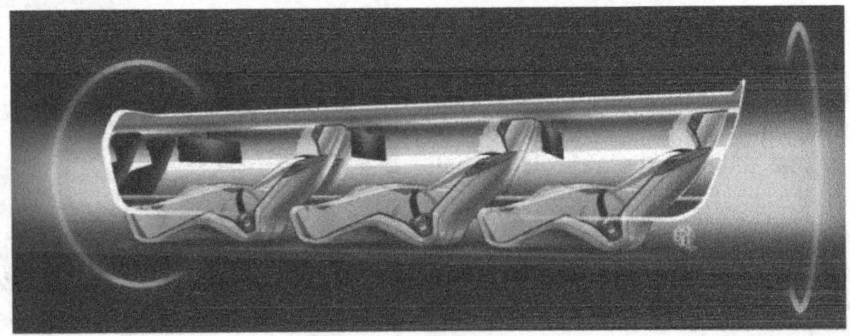

Figure 17.8 Hyperloop model.

Manufacturing

Artificial intelligence powered robots work alongside humans to perform a limited range of tasks like assembly, stacking, predictive analysis and for sensors keeping smooth equipment running. Boston dynamics, UIPath, are few known brands that incorporate artificial intelligence into robotics and for robotic process automation. In many manufacturing industries, they have started using technologies of next generation. to keep the industry profitable. As a result, six key themes emerged in manufacturing technology as follows:

- **Robotics:** auto-guided vehicles, sorting arms
- **Software as a service (SaaS):** vision software, predictive maintenance
- **Digitisation/data software:** marketplaces, data analytics solutions
- **Augmented/virtual reality (AR/VR):** wearable headsets, vision picking software
- **Material science:** additive manufacturing, innovative materials
- **Batteries:** contactless charging systems, ultrasound transmitters

The first wave of disruption in manufacturing technology came with sensors and big data analytics, but this is just the beginning. Manufacturing tech startups is now creating cutting-edge, disruptive technology to add tremendous value to clients and generate lucrative returns to investors.

Healthcare

In comparative artificial intelligence, a nascent field of healthcare, diseases are more quickly and accurately diagnosed. Drug discovery is sped up and streamlined, virtual nursing assistants monitor patients and big data analysis helps to create a more personalised patient experience. Nvidia itself has huge share in this industry, they develop devices which are cheap and provide real-time results.

With Big data and well trained artificial intelligence system; patient diagnosis, precision medicine, patient data security and clinical decision support accuracy increases. When it comes to artificial intelligence, we always talk about the rise of intelligent robots, taking our jobs. Yes, it surely does. Figure 17.9 represents a future healthcare centre. Artificial intelligence systems are capable of precise diagnosis with little inspection of organs and symptoms. Some recent researches in this area are, Ultromics—an artificial intelligence diagnostics system which is more accurate than the cardiologist at diagnosing heart diseases. It saves time, money and works with high precision, and this is what we expect from an artificial intelligence system. Commercialised artificial intelligence system developed by Optellum, diagnoses lung cancer in as many as 4,000 lung cancer patients per year than as compared to doctors.

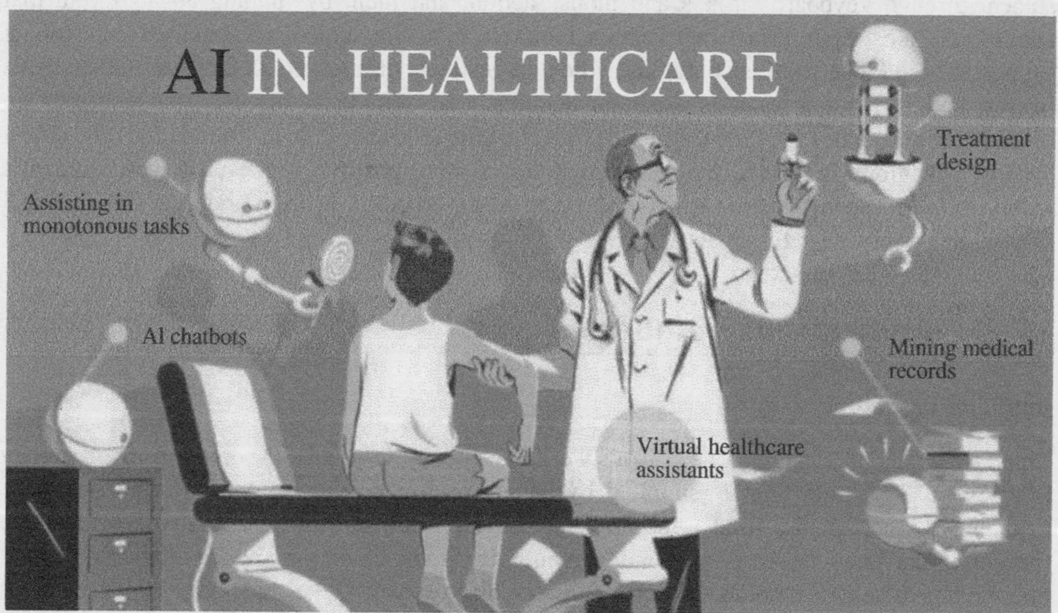

Figure 17.9 Future of artificial intelligence in healthcare.

Precision medicine is an approach for the treatment of diseases and prevention that takes data of individual variability in genes, lifestyle, and the environment they live in for each person. It further helps doctors and researchers to correctly predict the treatment and prevention strategies for a particular disease. Algorithms that can learn by themselves require significant computing power since it has to train the model with huge data. These algorithms can learn by themselves at an unprecedented rate and generally an approach that uses the cognitive capabilities of physicians on a new scale. Robot-assisted surgery, Virtual nursing assistants, Administrative workflow assistance, Fraud detection, Dosage error reduction, Connected machines, Clinical trial, Participant identifier, Preliminary diagnosis, Automated image diagnosis are going to be the highlights of artificial intelligence in this industry.

Artificial narrow intelligence is an area which will see the usage of artificial intelligence in streamlined applications. Healthcare will be the area which will see a lot of artificial narrow intelligence for specific tasks.

Education

Textbooks are digitised with the help of artificial intelligence. Early-stage virtual tutors assist human instructors and facial analysis gauges the emotions of students to help determine who is struggling or bored and better tailor the experience to their individual needs. The future of education is artificial intelligence-assisted and not artificial intelligence-led.

A child's emotional state affects how well or poorly he is able to focus, engage and stay motivated to learn. With this in mind, a robot called ARTIE (Affective Robot Tutor Integrated Environment) is being developed. ARTIE's chief role is to identify the emotional state of a student through keyboard strokes and mouse action, and then, by running an algorithm that chooses the most appropriate intervention required, give the student personalised educational support. These range from encouraging words, to gestures, or attempts to increase the student's interest and motivation towards a certain learning goal. ARTIE is designed by focussing on three cognitive states: concentrating, distracted and inactive. It was found that the social support behaviours provided by these robot tutors have positively impacted these student's ability to learn. Such assistant robots are going to be common in each and every educational institution in very near future itself.

17.4 Applications Tomorrow

Artificial money

This is the era of virtual money. Now we carry money as cards. Another form of money is the virtual currency that is supplied by some specific providers to those who subscribe to their services. Money used tomorrow may be different. It will be a virtual world where the entire currency is stored in some manner in block chain or similar platform. Every one will have separate nodes, which calculates his balance. Time and money may be used interchangeably. Similar to bitcoins, there will be several individual owned currencies, which may be standardised.

Neurohacking

There are psychiatrists who can read the minds after hypnotising. Even then the amount of things he gets is limited. How about a tomorrow, where anyone can read through your mind? Neuroscientists are finding ways to read people's minds with machines, and although this has been in the works for decades, real progress is being made by researchers at the University of California, Berkeley, and elsewhere. Translating electrical activity from the brain by means of decoding brainwaves is one way to help sufferers of dementia, for example, who have complications with neurotransmitters relaying thoughts into comprehensible speech or holding thoughts long enough to get them out verbally before they are forgotten.

On the other hand, it is more than little frightening to know that science and machines could soon have access to our innermost thoughts. Implications of neurohacking people's thoughts have also been studied in relation to neuromarketing, which targets people's brains by manipulating their wants and desires through marketing and advertising. Our thoughts and actions could actually be hijacked by a form of media that makes us think we are getting what we want, when really, we are going for something that our brains may only think is supposed to be good.

Universal translators

All roads lead to rome and everyone talks the same language. Universal translators reunite the nations and put an end to global language confusion. The day is near when two people speaking different languages can communicate with one another in their own voices but in languages they do not know or understand. Just a pod and a speaker with a highly trained NLP engine can accomplish this task.

High raised farm and laboratory meats

As the population is getting increased day by day, space is not just an issue for humans. The plants, animals all are fighting for their living space. So, the world is not that far away from witnessing high raised farms—farms on the top of skyscrapers and bigger buildings right in the middle of the city. Future will see technologies developed for farmland in unusual places.

Similarly the use of meat products is so much increasing that one day there might not be enough supplies. Labs will take up this challenge and meat of any animal may be produced. If scientists can develop a cost-effective way to grow edible meat in the lab it would change the way we live and eat forever. This change not only cuts down the ecological damage meat farming does to the world, but also makes for an ethical alternative to meat eating that many people could enjoy.

Gesture computing and convenient projections

The concept of gesture-based computing has been around for a while. It is also already appearing in multiple different technologies and is almost second-nature for most anyway. How many gestures do we already use on a day-to-day basis on our smartphones, tablets and other smart devices? It is reasonable to see a future where we are no longer restricted by mouse and keyboard and instead rely on voice and action to manipulate data in a virtual space. Figure 17.10 shows an idea of future computers. The time to switch-on the laptop and wait to get the screen up is over. We can project the screen to anywhere we want with zero size technology. Super computer chips embedded with projectors help in making any place our workbench. The integrated display system and high efficient processing will make this type of computing the most reliable one.

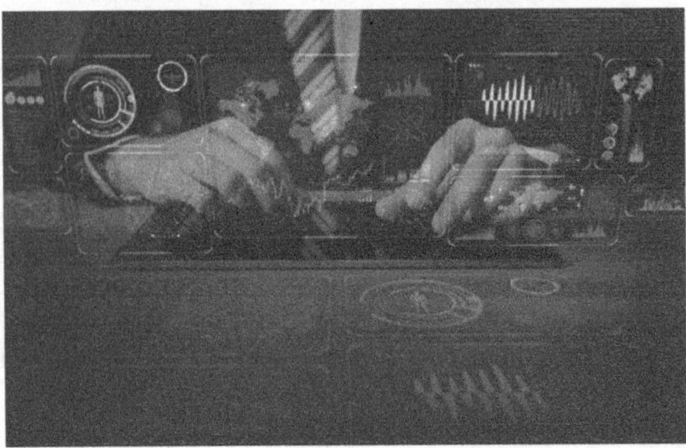

Figure 17.10 Future computers.

Healthcare robots

The scope of artificial intelligence in healthcare is unlimited. We can have various gadgets which will be smart enough to get data from our body and giving suggestions for our lifestyle. You can walk into the bathroom, whose mirror will check your pulse and blood oxidation. An ultrasonic teeth cleaner samples and analyses your saliva, while bathroom scales, built into the floor, check your weight and skin salinity. Your toilet analyses all body matter issued, and all data is checked against your wearable to consider the recorded activities of the previous day, including food and drink intake.

By the time you are dressed and enter the kitchen, your general health and bodily needs have been assessed and a suitable smoothie and coffee will have been prepared, along with a suggested breakfast menu of food, which is optimised to match the day's activities, pulled from your diary and messaging systems. If the supplies in fridge go below the specified quantity it will automatically order the things. Even if the fridge gets empty, you can have 3D food printing machine which can print your morning breakfast. Is a cake or pancake enough for the day?

Homes may also have robot which cooks you the meal by checking your health. Everything will be tailored to your own needs, and by the time you leave the house, you will be completely refreshed and energised for the day ahead.

Space tours and colonisation in other planets

Can we have a 3 week trip to our nearest star. Space tourism has long been seen as the last frontier for tourism, and with advancements in technology, non-astronauts out there can soon experience a space mission. The one thing that has held back the development of space tourism is developing a low-cost passenger launch vehicle. Currently, a rocket launch is a very expensive ordeal, with a very high cost of fuel and resources. Right now, space tourism is in pioneering phase in which customers are very few, and the cost is fairly high. These trips will be more adventure travel rather than a luxury filled experience. These trips will be focussed more on the safety of passengers as opposed to the ones more comfort focussed.

If legendary astronaut Buzz Aldrin is believed, humanity will have established a Martian colony by 2040, with the first visitors to the Red Planet arriving with no intention of coming back. However, he does not expect the first off-Earth colony to be self-sufficient within this time scale. Instead, early colonists would need to survive on regular shipments from the blue marble until farming and other similar facilities could be established.

A third eye

Sometimes the world around us is so beautiful that we may have to have a third eye. Google glasses available today show about the navigation and other limited features. But with a bionic eye, we can experience the AR-VR at its best. Each and every thing that we see is transmitted as data and is stored for future use. These bionic eye may be visualised as a camera which records everything we see. The features that can be added to bionic eye can vary from time to time.

Dream linking

Using pillows with conducting fibres in the fabric, it will be possible to see and monitor electrical activity of the brain. This will not only show when someone is dreaming, but recent developments indicate that we will also be able to tell what they are dreaming about. It is also possible (with prior agreement presumably, and when both people are in a dream state at the same time) for two people to share dreams. One could try to steer a friend's dream in the same direction, so that they could effectively share a dream, and may even be able to interact with it.

Internet revolution

Early research on the sixth generation (6G) had started during the late 2010s when China, the USA and other countries investigated the potential for working at higher frequencies. Whereas the first four mobile generations tended to operate at between several hundred or several thousand megahertz, 5G had expanded this range into the tens of thousands and hundreds of thousands. A revolutionary technology at the time, allowed vastly improved bandwidth and lower latency. However, it was not without its problems, as exponentially growing demand for wireless data transfer put ever-increasing pressure on service providers, while even shorter latencies were required for certain specialist and emerging applications.

This led to development of 6G, based on frequencies ranging from 100 GHz to 1 THz and beyond. Figure 17.11 shows the cellular frequency from 1G to 6G. A ten-fold boost in data transfer rates would mean users enjoying terabits per second (Tbit/s). Furthermore, improved network stability and latency achieved with artificial intelligence and machine learning algorithms could be combined with even greater geographical coverage. The Internet of Thing, already well-established during the 2020s, now had the potential to grow by further orders of magnitude and connect not billions, but trillions of objects.

Following a decade of research and testing, widespread adoption of 6G will occur in the 2030s. However, wireless telecommunications are now reaching a plateau in terms of progress, as it becomes extremely difficult to extend beyond the terahertz range. These limits are eventually overcome, but wholly require new approaches and fundamental breakthroughs in physics. The idea of a seventh standard (7G) is also placed in doubt by several emerging technologies that support the existing wireless communications, making future advances iterative, rather than generational.

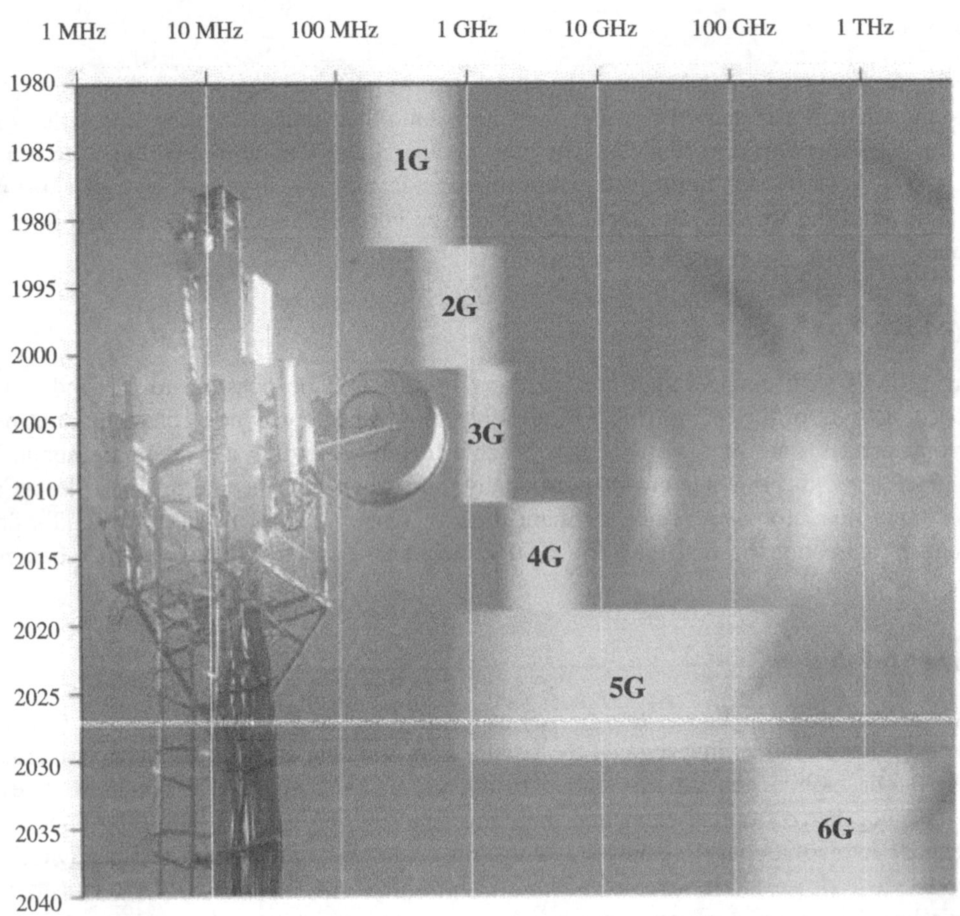

Figure 17.11 Cellular frequency.

Drones—Everywhere

We order many things online. Starting from books, cloths, electrical appliances to medicines and even food. We can witness delivery drones at our house doors, which delivers the products. Drones can also be seen in military operations. Face recognition drones may be used for personalised message delivery. After training the drones could be let free. It will search around, find the similar face and do the needful. In war ground also, drones will be very common. It may also be seen as a possibility of drone transportation.

Communication

Our smart phones will have more or less disappeared, replaced by control centres which we will wear in a series of devices around our body. For example, we will wear smart contact lenses, with texts floating in front of our eyes, and earrings that send messages from a virtual assistant into our ears. We will not look as if we are wearing anything extra but it will be as if we are looking through a smart phone at the real-world, albeit one more powerful than anything we know today. Our social networks will also become integral to the real-world. We may see a stranger in the street and, using facial recognition software linked to our control centres, will instantly know their name and be able to access their profile.

3D printing

3D printing has already started to make its way into the medical industry, but it is limited to implantable devices and things like replacement hips and knees. Between now and 2040, 3D printing could allow surgeons to create transplant organs on demand with the patient's own cells. By utilising the patient's own cells, there is no chance of organ rejection, eliminating the need for immune suppressing rejection drugs.

Currently, in 2019, researchers are working toward creating 3D printed blood vessels. By 2040, this technology could make up for the lack of organ donors, potentially saving up to 21 people every day. Figure 17.12 shows a possible prototype of 3D heart printing.

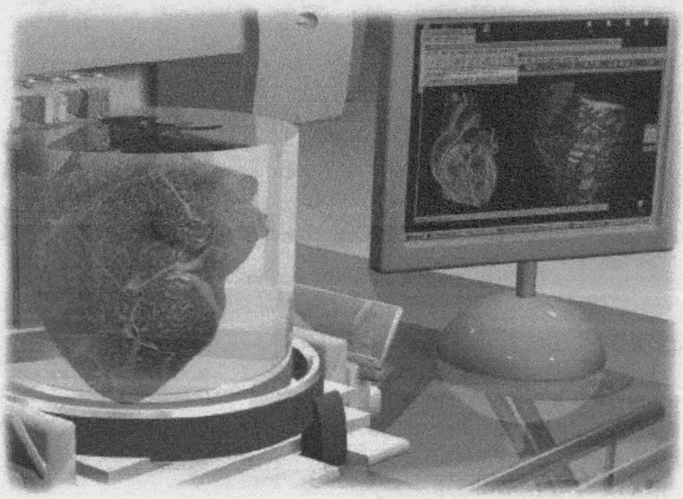

Figure 17.12 3D heart printing.

Floating cities

According to a 2007 report by the Paris-based Organization for Economic Cooperation and Development, by 2070, rising sea levels due to climate change could have a devastating effect on coastal cities around the globe. As many as 150 million people would be at risk of having to flee flooded homes, and as much as $35 trillion in property would be at risk of ending up underwater. We still might be able to stave off such a catastrophic scenario by dramatically reducing greenhouse gas emissions, but time is running out. That means low-lying communities may have no choice but to build higher and higher seawalls, or else relocate their populations.

What if, instead of fleeing the rising seas, we simply build new cities that float on them? In 2008, Callebaut unveiled on the web his design for Lilypad, a 50,000-inhabitant floating city modelled in shape after the giant water lily native to the Amazonian basin. As a haven for climate change refugees, Lilypad would be a totally self-sustaining community, with aquatic gardens for growing food, a desalination plant to produce drinking water, and energy generation through solar, wind and wave power. Better yet, Lilypad would be outfitted with a titanium dioxide skin, capable of absorbing carbon dioxide from the atmosphere and making atleast a small dent in global warming. Figure 17.13 shows a floating city prototype.

Figure 17.13 Floating city prototype.

Unseen Computers

As noted, 'ubiquitous computing' also known as 'pervasive computing' and 'everywhere' is coming. Today we already have computers in our cars, our phones, our toys, and even our fridges. But they are still very obvious. We often have to hold them or use keyboards to input information into them.

These devices, however, are getting steadily smaller owing to the miniaturisation revolution that is in full swing (For example, the shift towards micro electromechanical systems, or MEMS). In short order, we will be living in a Rainbow's End world, where information processing devices will be virtually everywhere, but completely invisible—absorbed into our surroundings. These computers will be in our clothes, our fashion accessories, and even in our contact lenses. And to use them we will use natural language and haptic technologies (i.e., tactile feedback). Or better yet, these devices will be endowed with a certain level of 'ambient intelligence' to help them perform autonomously under specific conditions. So, in next 5–10 years, we will be completely surrounded by computers, but utterly unaware of their presence.

17.5 Conclusion

Artificial intelligence has the capability to automate tasks redundant and tiresome to humans. Suggest new pathways to break down problems, optimise the solution, simplify the inference of complex entities, enable users focus on tasks that are human centric. It helps in upscaling the complexity of job in development and maintenance. We are not that far away from the time when we see each and every application being taken over by intelligent machines.

Let us close this discussion by mentioning some of the possible job positions that may come up in next 20 years. Alternative vehicle developer, Avatar manager, Body part maker, Climate change reversal specialist, Memory augmentation surgeon, Nano medic, Narrowcaster, 'New Science' ethicist, Old age wellness manager, Quarantine enforcer, Social 'networking' officer, Space pilot, Orbital tour guide, Vertical farmer, Virtual clutter organiser, Virtual lawyer, Virtual teacher, Waste data handler, Orbit traffic controller, etc. It is a hard fact that more than 50% of the jobs in next 20 years will be those which we have no idea about as of now.

Index